SCIENCE,
TECHNOLOGY, AND
ECONOMIC DEVELOPMENT

edited by
William Beranek, Jr.
Gustav Ranis

Science, Technology, and Economic Development

A Historical and Comparative Study

PRAEGER PUBLISHERS
Praeger Special Studies

New York • London • Sydney • Toronto

Library of Congress Cataloging in Publication Data

Main entry under title:

Science, technology, and economic development.

   1. Science. 2. Technology. 3. Economic
development. I. Beranek, William, 1946-
II. Ranis, Gustav.
Q172.S33   500   78-5660

500
S4161

PRAEGER PUBLISHERS
PRAEGER SPECIAL STUDIES
383 Madison Avenue, New York, N.Y. 10017, U.S.A.

Published in the United States of America in 1978
by Praeger Publishers,
A Division of Holt, Rinehart and Winston, CBS, Inc.

1234567890 038 098765432

81- 4037

Printed in the United States of America

FOREWORD

In view of the fact that the United States observed its bicentennial in 1976, the International Council of Scientific Unions (ICSU) accepted an invitation from the United States National Academy of Sciences (NAS) to hold its General Assembly in Washington, D. C., during October 10-16. This brought to the national capital prominent representatives from many of the more than 50 academies, royal societies, research councils, and other scientific bodies constituting the national members of the ICSU, the principal officers of the 17 scientific unions that make up the scientific members, and representatives of the scientific committees, special committees, commissions, and permanent services that have been established by ICSU to deal with matters that transcend a single discipline.

In addition to the regular business of the General Assembly and the social and cultural events that traditionally contribute notably to international understanding at an ICSU General Assembly, the NAS presented, as a part of the nation's bicentennial observance, a three-part symposium of special presentations aimed at illuminating the interaction of science and societal affairs.

Underlying the title of the symposium, "Science: A Resource for Humankind," is the conviction that major developments in science and technology over the past few hundred years have brought society to a crossroads.

The symposium was developed in three topics. The first examined, retrospectively, the role of science and technology in the social and economic development of seven selected countries, with the objective of providing new insights into this complex interaction. The second topic addressed the contemporary problem of successfully managing the quality of the human environment and the global supply of natural resources. It focused on the adequacies and inadequacies of our knowledge base for this task. The third topic dealt with innovations in science and technology, desirable or otherwise, that over the balance of this century may affect such interrelated problems as world food supply; the nutrition, health, and size of human populations; utilization of natural resources; and adequacy of energy supplies. The purpose was to realize the attractive opportunities and avoid the hazards that appear to lie ahead.

Several audiences were addressed: the formal delegates representing major scientific bodies and institutions at the General Assembly of ICSU, especially invited decision makers from the

v

public and private sectors, selected promising young individuals from developing countries, and interested general citizenry reached through the public media.

Dr. William Beranek, Jr., of the Holcomb Research Institute, Butler University, Indianapolis, provided unifying guidance for the studies prepared for the symposium, helping to select, and then working closely with, the directors of each topic.

The year-and-a-half-long study "The Retrospective Look at Science, Technology, and Economic Development" was headed by Dr. Gustav Ranis, professor of economics at the Economic Growth Center of Yale University, who has had extensive practical and scholarly experience in the field of developmental economics. He drew around him a distinguished advisory council made up of Julian Engel, Board of Science and Technology for International Development (presently at the Organization for Economic Cooperation and Development); Dr. Melvin Kranzberg, Georgia Institute of Technology; Dr. Richard Nelson, Yale University; Dr. Derek de Solla Price, Yale University; Dr. Eugene Skolnikoff, Massachusetts Institute of Technology; and Dr. E. Nancy Stephan, University of Massachusetts.

A two-day meeting of the authors and advisers was held in December 1975, in Washington, D.C., to focus the questions and to coordinate the research. A synthesis paper, based on the results of these seven country studies, was prepared by Dr. Ranis. Dr. Simon Kuznets, professor emeritus of economics at Harvard University, then reviewed the entire project and contributed his response. The nine papers were presented on October 11, 1976, at the NAS Bicentennial Symposium.

Unfailing encouragement and support for the project were provided by NAS Foreign Secretary Dr. George S. Hammond, ICSU President Dr. Harrison Brown, NAS President Dr. Philip Handler, the Bicentennial Symposium Steering Committee, and the NAS Board of International Organizations and Programs. The staff of the Board of Science and Technology for International Development, especially Dr. Victor Rabinowitch and Wesley Copeland, helped to develop the initial idea of the project and contributed to the effort throughout. Murray Todd, Edmund Rowan, Kathleen Behr, Samuel B. McKee, and all members of the staff of the Office of the Foreign Secretary provided invaluable assistance for the entire symposium. Technical assistance was provided by Cheryl Hunt of Yale University and Barbara Whitcraft of the Holcomb Research Institute in such a cooperative and competent manner as to lighten the burden of manuscript preparation enormously.

Thomas F. Malone, Chairman
United States National Academy of Sciences
Bicentennial Symposium Steering Committee
Holcomb Research Institute, Butler University

ACKNOWLEDGMENTS

Many organizations and institutions have provided direct financial support and other services for the National Academy of Sciences (NAS) during the preparation of the Bicentennial Symposium. The opinions, findings, and conclusions or recommendations expressed in this publication are those of the authors and do not necessarily reflect the views of the NAS or any of the symposium sponsors.

The NAS gratefully acknowledges the generous financial or professional support for the symposium provided by many individuals and institutions, including the following:

Agency for International Development
Agricultural Research Service
AMAX, Inc.
Bechtel Foundation
Butler University
The Coca-Cola Company
Council on Environmental Quality
Deere & Company
Walt Disney Productions
Exxon Corporation
General Electric Company
General Mills Foundation
General Motors Corporation
Gulf Oil Foundation
Hoffman-LaRoche, Inc.
International Business Machines Corporation
International Telephone and Telegraph Corporation
Johnson & Johnson Associated Industries Fund
Lever Brothers Company, Inc.
The Arthur D. Little Foundation
Lockheed Aircraft Corporation
Mobil Oil Corporation
Monsanto Company
National Aeronautics & Space Administration
National Institutes of Health
National Oceanic & Atmospheric Administration
National Science Foundation
The Rockefeller Foundation
Rockwell International Corporation

G. D. Searle & Company
Sperry Rand Corporation
Stauffer Chemical Company
Texas Instruments Foundation
The Times Mirror Company
T. R. W. Foundation
United States Geological Survey
United States Steel Foundation

CONTENTS

Page

FOREWORD
Thomas F. Malone   v

LIST OF TABLES AND FIGURES   xi

LIST OF ABBREVIATIONS   xiii

INTRODUCTION
William Beranek, Jr. , and Gustav Ranis   xiv

Chapter

1 SCIENCE, TECHNOLOGY, AND DEVELOPMENT:
   A RETROSPECTIVE VIEW
   Gustav Ranis   1

2 SCIENCE, TECHNOLOGY, AND ECONOMIC
   DEVELOPMENT: THE BRITISH EXPERIENCE
   D. S. L. Cardwell   31

3 THE ROLE OF SCIENCE AND TECHNOLOGY IN THE
   ECONOMIC DEVELOPMENT OF MODERN GERMANY
   Wolfram Fischer   71

4 THE ROLE OF SCIENCE AND TECHNOLOGY IN THE
   NATIONAL DEVELOPMENT OF THE UNITED STATES
   Nathan Rosenberg   114

5 THE ROLE OF SCIENCE AND TECHNOLOGY IN
   HUNGARY'S ECONOMIC DEVELOPMENT
   María Csöndes, Lajos Szántó, and Péter Vas-Zoltán   169

6 SCIENCE AND TECHNOLOGY IN MODERN JAPANESE
   DEVELOPMENT
   Shigeru Nakayama   202

Chapter   Page

7   SCIENCE AND TECHNOLOGY IN BRAZILIAN
     DEVELOPMENT
        José Pastore   233

8   THE ROLE OF SCIENCE AND TECHNOLOGY IN THE
     ECONOMIC DEVELOPMENT OF GHANA
        Edward S. Ayensu   288

9   GAPS IN THE SCIENCE-TECHNOLOGY-DEVELOPMENT
     SEQUENCE: A COMMENT
        Simon Kuznets   341

ABOUT THE EDITORS AND CONTRIBUTORS   348

LIST OF TABLES AND FIGURES

Table		Page
4.1	Share of Finished Manufactures in U. S. Exports and Imports, 1770 to 1964-68	123
4.2	Lumber Consumption in the United States and United Kingdom, 1799-1869	128
4.3	Commercial Fertilizers: Quantities Used and Average Primary Plant Nutrient Content, 1940-66	135
6.1	Foreign Holdings of Japanese Patents, by Country, March 1910	214
6.2	Degree of Native Japanese Patent Holdings, 1899-1905	215
6.3	Economic and Functional Classification of Japanese Public Expenditure, 1900 and 1960	226
8.1	Area and Production of Major Staples, Ghana, 1968-72	305
8.2	Output and Export of Logs, Timber, Veneer, and Plywood, Ghana, 1968-72	309
8.3	Ghana Domestic Fish Catch and Consumption, 1968-72	312
8.4	Recorded Unemployment in Ghana, 1968-73	330
8.5	Functional Classification of Ghana Central Government Capital Expenditures, 1968/69-1972/73	332

FIGURE

5.1	Organization of Science in Hungary	193
6.1	Japanese Research Expenditure, by Type of Institution, 1952-62	225
7.1	Coffee Productivity	246

Figure		Page
7.2	Sugarcane Productivity	249
7.3	Cotton Productivity	252
7.4	Rice Productivity	254
7.5	Bean Productivity	255
7.6	Corn Productivity	257

LIST OF ABBREVIATIONS

BASF	Badische Anilin- und Sodafabrik
CSIR	Council of Scientific and Industrial Research (Ghana)
DSIR	Department of Scientific and Industrial Research (England)
GDP	gross domestic product
GNP	gross national product
ICSU	International Council of Scientific Unions
NAS	National Academy of Sciences (U. S.)
NNP	net national product
NRC	National Redemption Council (Ghana)
NRDC	National Research Development Corporation (England)
OECD	Organization for Economic Cooperation and Development
R&D	research and development
UN	United Nations

INTRODUCTION

William Beranek, Jr.
Gustav Ranis

Routinized technological change associated with continuous progress in science clearly represents one of the key characteristics of the advanced industrial society. Europe's successful transition from agrarianism to economic maturity, achieved in the course of what is usually called the Industrial Revolution of the eighteenth century, amounted to a fundamental change in the rules of societal behavior. Regardless of the assumed strength or even the order of causation among science, technology, and development, it is this new and enhanced role of science and technology that has become a symbol of that change.

The two-thirds of humanity that is today trying to effect a similar kind of transition is asking itself how to harness these forces most effectively. To what extent, and how, can science and technology be transferred from the rich to the poor? To what extent, and how, must progress be indigenous? With all the talk about the appropriateness of technology, is there such a thing as appropriate science as well? These are the questions being asked in the finance ministries and planning commissions of the Third World, among aid agency officials in the rich countries, even in the board rooms of the multinationals. It is the principal subject to which the major United Nations Conference on Science, Technology and Development in 1979 will be devoted.

This symposium, "The Role of Science and Technology in Economic Development," was undertaken as part of the U. S. bicentennial celebration, to help provide a more factual historical underpinning for the emerging debate on these vital issues. In the course of this effort the growth process in seven rather different historical country situations was analyzed. The country sample included Great Britain, the acknowledged historical leader in the transition to modern growth; Germany and the United States, two early followers; Japan and Hungary, two relatively late followers; and Brazil and Ghana, two contemporary developing economies. The focus in all cases was on the role of science and technology in the early transition effort; the basic purpose was to try to achieve some capacity for generalization that might prove useful for today's developing world.

Economic conditions, the international environment, and the social, political, and scientific institutions within each country of course differ markedly over time and across societies. Caution must therefore be exercised in extrapolating trends or attempting to extract ready lessons. Nevertheless, we believe that explorations such as this into the hitherto relatively neglected historical laboratory can prove exceedingly useful as an important input into today's debates and decision-making processes.

1

SCIENCE, TECHNOLOGY, AND DEVELOPMENT: A RETROSPECTIVE VIEW

Gustav Ranis

The difficult task I have set for myself in this paper is to attempt to illuminate the relationship among science, technology, and development, using as building blocks the experience of seven countries having widely different characteristics and operating at different stages of development.* The issues to be addressed are inherently complicated. Relevant theory is, by consensus, still in its infancy and modest, step-by-step empirical approaches consequently cannot help very much. This is also an area of major discontinuities in behavior, always more difficult for scientists to handle, and one in which the temptation to perceive all dimensions of human progress as relevant is both natural and bound to lead in too many directions at once.

Nevertheless, the effort should be made—not only to attempt to improve our basic understanding of how we got here, and why, but also because the distillation of such an improved understanding may hold some lessons for the future, especially with respect to the achievement of modern growth by contemporary developing countries. The problems and aspirations of that two-thirds of humanity that does not yet enjoy modern growth, but is anxious to achieve it—while

*While in this sense this paper attempts, among other things, to synthesize the work of others, they are not to be held responsible for the interpretations made, conclusions drawn, and errors committed here.

The author wishes to acknowledge the assistance of H. T. C. Hu and the comments of Bill Beranek.

1

deeply puzzled as to the proper role of science and technology—represent our main concern.

The very nature of this effort, necessarily eclectic and general in character, means that we should not expect to find a new, definitive set of answers to these old and perplexing problems. It also means that the temptation to force observations into consistency with some premature unifying theme or preconceived theoretical mold must be resisted. All I can do, and have tried to do, is selectively organize the information and insights gathered by all the participants, in order to improve somewhat our understanding of these complicated interrelationships and to point the direction that further, more original and more basic, analysis might take. If, in spite of myself, I seem at times to have reached for some broader explanatory framework, it is with the understanding that the improvement, or even the reasoned rejection of such a framework by others, will serve to advance the common cause.

The approach of this paper is frankly historical, focusing mainly on the eighteenth- and nineteenth-century experience of the now advanced countries and the more recent experience of the currently developing countries. This is because I believe that history represents the most important, and as yet most underutilized, laboratory for the exploration of these issues.

Moreover, I recognize, at this most general level, that while all societies, both historical and contemporary, share the tyranny of their initial endowments, they face substantial alternatives with respect to their objectives and the way in which they decide to organize themselves. To keep the scope of this paper from becoming entirely unmanageable, I shall not concern myself very much with a comparative evaluation of organizational systems. I shall assume, especially when dealing with contemporary developing countries, that they are of the mixed, nonsocialist variety. Moreover, and not unrelated, I shall not be concerned with possible intercountry differences in social objectives. Instead, I shall assume that all societies may be located on some more or less continuous spectrum of institutional choice and that the "old-fashioned" development objective—sustained increases in per capita income—either is shared by all or, more satisfactorily, is not in necessary conflict with such other valid nontraditional concerns as distribution and employment.

In order to avoid a veritable parade of definitions, I shall equate "social and economic development" with per capita income growth; by "science" I shall mean the accumulation of basic systematic knowledge about the natural universe around us; and by "technology," the application of such knowledge to the construction of a pool of ideas useful in the production of goods and services. With respect neither to science nor to technology will I entertain the plausible notion that such activities may represent some sort of valid art form carrying its own cultural, aesthetic, or consumption values. While I recognize that the relationships among "science," "technology," and "development" as defined here may be indirect, lagged, multidimensional, uncertain, and, above all, complicated, it is with these relationships that I shall be mainly concerned.

The country sample includes Great Britain, the acknowledged historical leader in the transition to modern growth; Germany and the United States, two early followers; Japan and Hungary, two relatively late followers; and Brazil and Ghana, two contemporary developing economies.

I will first try to define more precisely the issues on which this volume is attempting to shed some light. I will then review the relevant evidence from the historical experience of the now mature early developers, of the late followers, and of the currently developing economies. The findings and conclusions are then summarized.

SOME OF THE ISSUES

Most scientists, whether natural or social, and most officials, whether from developed or developing countries, share the general conviction that strong relationships exist among science, technology, and development. There is, however, considerably less understanding—and, hence, agreement—on the precise nature of these relationships or even on the direction of the causal order. Consequently, with underlying behavior not well understood, it is natural that there should exist a good deal of uncertainty with respect to what constitutes appropriate government policy in support of a society's basic developmental objective.

The relatively "easier" part of the puzzle is undoubtedly that which focuses on the relationship between technology and development. A substantial amount of work, both theoretical and empirical, has been done in this area, mainly by economists. This work has permitted me to conclude fairly unambiguously that the association between technology and growth is indeed strong, that is, that changes in the quality of a society's processes and goods are highly associated with economic growth. The precise character of the technological change associated with growth remains a "measure of our ignorance"; we do not know whether it is manna from heaven (what the economists call "disembodied" and "exogenous") or whether it results from research and development (R&D) embodied in people or machines. The increased physical availability and application of homogeneous factors (that is, "more of the same"), in the absence of technology change, probably accounts for only a small portion, perhaps as little as 20 percent, of total growth in most of the advanced nonsocialist countries. As Simon Kuznets puts it, even when we acknowledge that new technology may have negative as well as positive impacts on society—including additional social costs and discomforts—"the nettest definition . . . would still show a rapid increase [of income] per head, against fewer working hours."[1]

Most aggregative studies—such as that by Robert Solow, John Kendrick, and Edward Denison—assign the label "technology change" to everything that cannot be explained through an augmentation of enumerated physical inputs. However, the strength of particular microassociations and the all-important "richer" issue of what causes technology change endogenously brings us, even here, onto shakier ground. Cross-country studies trying to relate expenditures on R&D (as a percent of gross national product [GNP]) to growth in productivity, for example, by the Organization for Economic Cooperation and Development (OECD), have failed to do so.[2] It is just as likely that the major causal chain is reversed: More growth, and hence affluence, may simply permit more R&D to be carried out. In other words, while we are fairly sure about the causal importance of technology change for growth, we do not yet understand the anatomy of technology change well enough to know how to affect its strength or character with any degree of precision.

A large part of this problem is that our understanding of the relationship between science and the other two members of our basic triplet is even more precarious, especially as far as the developing countries are concerned. What we do have is an act of faith on the part of some that science must precede technology, which causes growth. Others see relatively little evidence of a necessary causal relationship between science and technology, at least for any given country, and, rather, view science as something only rich countries should be able to afford, while poor ones borrow. Such extreme points of view naturally lead to equally extreme positions on policy. On one side are those who would advise developing countries, for example, to acquire and maintain frontier capacity in every major field of science, in order to be able to participate fully in the benefits of technology change. On the other are those who would counsel developing countries to let the developed nations spend their relatively ample resources on basic science, and then pick and choose only "appropriate" areas of science (if necessary), along with only "appropriate" types of technology (if possible), from the "free" international shelf of human knowledge. Undoubtedly, the truth lies somewhere between.

Third, the notion of "appropriateness" itself, whether with respect to science or to technology, must, if it has validity, be relatable to national endowments or capacities, even if these are interpreted from a dynamic and long-term perspective. Just as the product cycle, for example, seems to hold certain useful notions about the path of product and technology mixes across countries at different levels of development, is there a valid analogy in science? Do different resource endowments really induce different types or directions, as opposed to simply different quantities, of technology change? Again, is anything analogously valid in the field of science?

Fourth, and closely related to what has gone before, is the issue of the potential role of government in strengthening the links between technology and growth and between science and technology. This, of course, depends in large part on the illumination of the basic behavioral relationships that, one hopes, will result from our rummaging through the various available historical laboratories, particularly with respect to such issues as the relative appropriability or inappropriability of new scientific and/or technological know-how and its relation to the perfection or imperfection of goods and information markets.

Finally, there is the question of whether, whatever relationships existed between science, technology, and development in the eighteenth and nineteenth centuries, these relationships have become fundamentally altered in character in the twentieth. Here there are at least two major viewpoints in the literature: one, that technology used to be empirically based but is now science-based; the other, that technology always has been, and continues to be, science-based, except that science is now "bigger" and the gap between it and technology smaller.

THE EARLY DEVELOPERS: GREAT BRITAIN, GERMANY, AND THE UNITED STATES

It is generally accepted that Great Britain was the world's leader in both technology and growth in the eighteenth and early nineteenth centuries, followed by France and Germany on the Continent, and then by the United States in the "overseas territories." There appears to be substantial agreement on the factors to which Great Britain owed its original position of preeminence, but less on why it failed to persist, and least on the relationship (if any) that all of this had to the role of science.

The Industrial Revolution, associated with the substitution of machines and inanimate power for labor power, came first to Great Britain. This is frequently attributed[3] to that nation's relatively higher level and better distribution of income (and, hence, broader base of purchasing power); its favorable geographic position, associated with both greater immunity from war and greater access to less troublesome trading partners; relatively better endowment in natural resources, especially coal; and, most of all, its relatively greater progress in throwing off internal feudalistic and mercantilist interferences. Thus, while all of Western Europe was undergoing significant long-term change associated with urbanization, nationalism, and the generally enhanced application of reason to help man better manipulate his environment, Great Britain emerged with a clear lead in textiles, as well as in the machinery industry to which improvements in textile manufacture gave rise, until the middle of the nineteenth century.

Yet, in spite of all efforts to keep advances in technology "bottled up" in the British Isles, through prohibiting the export of workers before 1825 and of machinery before 1842, by the time of the Crystal Palace Exhibition in 1851 there were clear indications that the Continent, especially Germany, was taking the lead in the important chemical, pharmaceutical, and electrical engineering industries, with the United States forging ahead in mechanical engineering. By the time of World War I, as D. S. L. Cardwell points out in chapter 2, Britain had become an importer of skilled labor and technology. Explaining the more controversial "why" of this change in leadership in technology and growth is interesting not only for its own sake but also because it may help us to understand better their mutual relationship to science.

Cardwell attributes the decline of British leadership largely to the fact that British technology was substantially "empirically-based" rather than "science-based." Such a distinction between technology that arises from trial-and-error manipulations of the environment rather than from changes in our basic comprehension of the laws governing that environment is also made by Nathan Rosenberg (see chapter 4) and others. Britain's early lead, according to this view, was based on such industries as textiles, metals, and brewing, which developed on the basis of "tinkering" rather than new scientific insights. Even the smelting of iron ore presumably was done without knowledge of the chemistry of oxidation or reduction.

According to this view, one of the principal reasons for Britain's later relative decline is the fact that empirically based technology change—even if sustained for a time by sequential or "neighboring" problem-solving innovations[4]—ultimately is not sustainable if it is not replenished by basic scientific advances. Cardwell sees British science in a long post-Newtonian decline, with pronounced neglect of scientific education, at the very time the country is leading in steam power, textiles, and metallurgy. At the same time the British Empire is siphoning off energies and capital, and British entrepreneurs are becoming "fat" and more interested in the gentlemanly life than in the improvement of their mills and factories. When this situation is placed in the context of an increasingly strong Continental nationalistic response to the "British challenge," in the form of greater emphasis on science and on science-based education, the loss of leadership in "science-based" industries to France and, especially, to Germany can be explained. The French École Polytechnique had no equivalent in Britain; and the somewhat more pragmatically oriented post-Liebig research labs and engineering schools helped Germany to outdistance everyone in the chemical/pharmaceutical, iron and steel, and electrical machinery industries by the end of the nineteenth century.

An explanation of Britain's relative decline based heavily on its failure to perceive the existence of a direct causal link between basic science and technology does not, however, serve us very well when we examine the relative success of the U.S. experience beginning in the last half of the nineteenth century. The base for U.S. technology change and its associated growth pattern was clearly "empirical," in the sense that the U.S. exploitation of the idea of mass production with interchangeable parts, which gave it a commanding lead in the mechanical engineering industries, can also be said to have emerged from trial and error applied to largely imported technology.

Rosenberg sees the Americans borrowing "freely and extensively from Europe," with very little "genuinely inventive activity" in evidence during the colonial period. There was little government support of science. Beginning around 1850, however, the United States began meaningful innovations in production engineering and the application of improved mechanical skills; the McCormick reaper, the Colt .45, the cotton gin, and the typewriter were among the products that fundamentally revolutionized factory production methods in general. While "putting out" (subcontract cottage industry) and handicraft production persisted in Europe, the United States, as Cardwell and Rosenberg agree, quickly became the undisputed leader in industries that lent themselves to the introduction of labor-saving machinery for the mass production of a standardized product. Before World War I, Singer Sewing established a subsidiary outside Glasgow said to be the "most advanced" in Britain, if not in Europe. None of the industries in which the United States began to set the innovative pace can be said to be "science-based," certainly not in contrast with the industries in which Germany assumed the lead.

A somewhat different explanation of why Britain's leadership role was gradually eclipsed may simultaneously provide an approach to a better understanding of the relationship among science, technology, and growth. This explanation would essentially start by rejecting the notion that any sustained technology change can really be "empirically-based," as opposed to "science-based." There clearly are marked, and important, differences in the directness of the link, from either the physical or the temporal point of view. But we find it difficult to accept the notion that British technological advances in textiles and metallurgy were not firmly based on steam power or that the steam engine was not based on prior basic advances in man's understanding of physics.[5] Even if Watt's steam engine can be relegated (as Cardwell does) to the realm of an "isolated exception"—which we doubt—it is a most important one. And there were others. For technology change to occur, something has to be "in the air" in the form of recent or past improvements in our basic understanding of the universe—even if the innovator himself is not a scientist working in a laboratory. Richard Arkwright's water frame (1769) is viewed as "wholly barren of science" by Cardwell, yet the fact that he was previously an apprentice to a barber and a wigmaker does not mean the invention was not based on previously acquired science. As Cardwell puts it, "if [such innovations] have scientific content . . . it is so well known that it can be taken as common knowledge." But that is just the point. The fact that the idea utilized has not recently sprung from the garret or the research laboratory of a scientist does not make it any less science-based.

This is, of course, not to say that we cannot, or indeed should not, distinguish between relatively major or epochal types of technology change and relatively minor, successor (or adaptive) types.* The former may be more obviously and directly science-related, such as a new hybrid seed and Mendelian laws, or plastics and molecular chemistry; but the new combinations of fertilizer and water required to render the new seeds most effective, and the new industrial applications of plastic materials, surely are just as much related to science as the initial major technical advance.

Closely related to this question, and thus perhaps shedding additional light on it, is the possibility of a "reverse" causal ordering, running from technology change to scientific progress. As Kuznets has pointed out, science is likely to be stimulated by new data, new tools, and new "puzzles" that emerge during application and modification of technology. Thus, the original smelting of iron may have proceeded without full understanding of the chemistry of oxidation or reduction. Yet the fact that the Bessemer process initially worked in England but not on the Continent (where the iron ore had a higher phosphorus content) led to new scientific inquiries into basic metallurgy and, in turn, to the improved Thomas-Gilchrist steelmaking process. Similarly, the difficulties encountered in the planting of

---

*Cardwell himself contrasts "revolutionary inventions and evolutionary improvements."

improved seeds from one country in another country have led to substantial new breakthroughs in agricultural chemistry.

It may therefore be useful to think of science and technology as more of a closed, mutually reinforcing and mutually dependent circle, and for both scientific and technological advances to be viewed as moving points on a spectrum, some indicating major or epochal "jumps," others less spectacular advances in understanding and accomplishment. Does such a notion tend to obliterate the difference between the concepts of science and technology? We do not think so; the definitions previously adopted stand up rather well. What it does do is cast doubt on the usefulness of the distinction between "science-based" and "empirically-based" technology change. It might perhaps be more useful to speak in terms of "science-intensive" versus "engineering-intensive" technology change along that spectrum. This might help us to distinguish between what was happening during the last half of the nineteenth century in the chemical/pharmaceutical industries to the United States? It is perhaps more useful to seek machine technology-dominated industries in the United States.

But where does this leave us with respect to our search for an "explanation" of why Britain lost its lead in the "science-intensive" industries to Germany and its lead in the "engineering-intensive" industries to the United States? It is perhpas more useful to seek such an explanation in the realm of the changing impact over time of differences in the endowment and in public policy.

Britain's early leadership position was closely tied to its relatively abundant natural resources, in particular coal and iron ore, as well as to the relatively more pronounced laissez-faire position of its government—guaranteeing not only nonintervention at home but market access abroad. It is plausible to argue that some of these advantages became disadvantages.

Let us begin with natural resources. There is little doubt that Britain's advantage in coal, iron, and geography heavily contributed to the smugness and loss of entrepreneurial energy previously noted. But it also meant that the Continent, especially Germany, felt under great pressure to catch up. Even after the exploitation of the Ruhr's coal deposits began in earnest, fuel costs remained higher on the Continent. The same was true for iron ore. Consequently, "continental ironmasters were making more of their resources than their competitors across the Channel; and since fuel economy was the key to efficiency in almost every stage of manufacture, the tentative advances of the 1830s and 1840s were the starting point of a scientific metallurgy that was to pay off in major improvements a generation later."[6] There can be little doubt that Germany's spectacular success in the science-intensive chemical industry was very much related to a strong nationalistic drive aimed at finding substitutes for its deficiencies in natural resources at home and in its colonies.

The United States could increasingly take advantage of its relatively much more abundant wood supply to manufacture lighter textile and other machinery. Such machinery was first considered an amusing oddity, but later generally recognized as technically superior. The relative abundance of the natural resources base also gave the United States the continuing advantage of a cheaper supply of fuel, first based on steam and later, using the scientific advances made elsewhere in the field of induction, on electric power. Moreover, its labor shortage removed most institutional (such as Luddite), as well as economic, obstacles to a thoroughgoing exploitation of labor-saving technological opportunities. From textiles to metallurgy and to the many later applications of machinemaking in routinized mass production industries, the response to changes in the environment was usually rapid. Rosenberg points out that as the comparative advantage in cheap wood later dwindled, iron replaced wood, and coal and coke replaced charcoal as the primary source of fuel. The rapid overall pace of industrialization was accompanied by increasing capital intensity and associated scales. Increasing pressure for labor-saving technology in industry, together with the existence of a large, dependable, and expanding domestic market, propelled by the expansion of the railroad, provided the cornerstone for the "American system" of mass production. [7]

In U.S. agriculture, the favorable man/land ratio led to a mechanization trend, initially of the horse-drawn variety and later of the tractor type; both were labor-saving and land-using. The application of what Yjiro Hayami and Vernon Ruttan have called biological/chemical technology change did not seriously come into its own until after the closing of the frontier (circa 1890) seemed making such increased reliance on the resource-saving effects of science rather more warranted. [8]

But the patterns that evolved in Britain, the Continent, and the United States over time were due also to government policies that either facilitated or obstructed the system's accommodations to its changing relative endowments and capacities. In Britain, for example, a policy of substantial laissez-faire, which had been a liberating advantage in the eighteenth and early nineteenth centuries vis-à-vis the more mercantilist and still somewhat feudalistic countries of Europe, may have become a handicap later on. When technical education continued to enjoy a relatively low prestige and, as Cardwell puts it, the Indian Civil Service exam drew more attention than the Cambridge mathematics tripos, the British government, instead of leaning against this wind, chose to stand aside. With supremacy already having been achieved in textiles and metallurgy, and colonial markets safely protected, it did not feel the need to encourage scientific research or education. Delegations of businessmen visiting the United States in the 1850s could not convince the establishment at home that anything was amiss. It took World War I to bring a sharp realization of the extent to which Britain had become dependent on German science and science education and on U.S. machinery engineering accomplishments. It was only then that (belated) government action was taken. Cardwell, in fact, notes that even to this day, in spite of the new universities of the postwar era and the increase in defense-related R&D, Britain still finds itself in something of a "technology trap," with higher technical education something of a stepchild and routinized R&D not yet a major management tool.

Nineteenth-century Germany, on the other hand, represented, as Wolfram Fischer puts it in chapter 3, a case of "Smithian liberalism tamed by enlightened governmentalism." Spurred by the threat of British economic hegemony as well as by competition among the various German states, these governments of the latter generally did not question their responsibility to help—either through protective tariffs, as in the case of the important beet-sugar industry, or through the support of scientific research laboratories and scientific education, as in the case of the von Humboldt reforms. Prussia went so far as to set up costly state enterprises and to invite moneyed private parties to establish factories; but mostly, in contrast with the heavy interventionism of the French, the German effort was more indirect, through expositions, awards, subsidies, technical advice, and the establishment of a network of technical and scientific institutions at various levels that were to provide formal training in fields ranging from engineering and mechanics to manual arts and design. German government assistance to the institutionalization of private credit and the provision of public overheads, via the Crédit Mobilier type of mechanism, compared increasingly favorably with the inadequacies of the British private market for venture capital.

By contrast, even with respect to the acquisition of general cognitive skills by the population as a whole, Britain remained elitist and indifferent. In 1860, for example, only 50 percent of British school-age children attended elementary schools; in Germany, as a consequence of compulsory education laws, the equivalent figure was more than 97 percent. This was in addition to a longer period of schooling and a quality differential in favor of Germany. As Landes put it, "Once science began to anticipate technique—and it was already doing so to some extent in the 1850s—formal education became a major industrial resource."[9] While the British turned to enjoy their successes of the past in a gentlemanly fashion, exhibiting an increasing disdain for the (underpaid) scientist and the technically educated, German princes vied with each other in founding technical schools and research institutes as well as in becoming the patrons of individual scientists.

The role of government in the United States, while clearly more limited than in continental Europe, also served to facilitate the process of technological development. Rosenberg, in commenting on the growth of the multifaceted U.S. machinery industry, speaks of a surprising volume of public/private collaboration in "visiting one another's plants, sharing new technological knowledge and even occasionally borrowing one another's workmen." Other observers place heavy emphasis on the role of widespread general education that provided for a measure of technical literacy at lower skill levels, and for substantial empirical problem-solving capacity at higher levels. Even if most technology was borrowed, and even if there was no first-rate scientific establishment in evidence, Rosenberg observes that the United States was "highly discriminating in borrowing patterns and highly selective in the uses to which imported technologies were put." Clearly, the mechanical skills and ingenuity required for this task were considerable. And while the United States produced few contributions to frontier science until much later, the diffusion of labor-saving technology change and adaptations, from firearms to clocks and watches, to harvesters and typewriters, all part of the "American system," required engineers who had at least a grounding in science and its use, even if they were not active contributors to it.

Little wonder that the nineteenth-century U.S. attitude toward science and technology has often been called extremely pragmatic. While higher-risk basic science was neglected, technology was borrowed and improved. The continuing shortage of labor resulted in continuing bias toward labor-saving technology. Only agriculture was, to some extent, an exception; with private risks larger, so was the role of government. The unique institutional framework focused on the land-grant college system was able to generate substantial technology change that was tied to progress in the chemistry-related agricultural sciences and diffused widely after the turn of the twentieth century.

The nineteenth-century United States may thus be characterized as a frontier society controlling what seemed like unlimited natural resources, including fuel, and therefore, unlike Germany, not much inclined to invest heavily in basic science. Nevertheless, innovative activity, based largely on imported technology and assisted by public-sector action, especially in education and agricultural research and extension, proceeded at a very rapid pace, in association with rapid increases in per capita income. If we accept the notion, previously put forward, that all technology is likely to be (directly or indirectly) science-based, it is nevertheless true that this divergence in the historical paths taken by Germany and the United States is most instructive. It tells us that different combinations of endowments and policies may lead one country to participate in growth through basic science and science-intensive industries, while another borrows technology and uses a broadly based scientific literacy to improve and diffuse such technology.

THE LATE FOLLOWERS: HUNGARY AND JAPAN

Turning our attention to the two successful late followers in our sample of historical country cases, Hungary and Japan, we may note that, in terms of initial endowment and international "opportunities," the gap between them and the early arrivals was substantial, probably as substantial as that between the late followers and today's developing countries.

Hungary had a considerable disadvantage in terms of human and natural resource endowment relative to Britain, Germany, or the United States. While it shared a common European cultural heritage, it was not a full member of the elite inner circle of scientific/industrial exchanges through trade, migration, professional meetings, and industrial exhibitions, all of which were very common in Western Europe, especially after the middle of the nineteenth century. When we add to this the effects of 150 years of Turkish occupation, frequent wars, and the strong grip of feudalism, we should not be surprised that the transition to modern growth was delayed by at least half a century.

As Lajos Szántó, Péter Vas-Zoltán, and Mária Csöndes put it (see chapter 5), Hungary experienced a "second edition" of serfdom between the sixteenth and eighteenth centuries, while Western Europe was undergoing a major transition to mature growth, combining a free labor force and nationalistic governments to build, first, commercial and overhead, and later, the basic fixed industrial capital structure required. But perhaps Hungary's biggest handicap was that there was little possibility for an agricultural revolution to precede (and fuel) an industrial revolution; and when the possibilities for "catching up" finally existed, Hungary was assigned a position within the Habsburg Empire that did not permit their adequate pursuit. The quasi-colonial division of labor under the Austro-Hungarian regime called for Austria, Bohemia, and Moravia to provide the industrial base, with Hungary assigned to a largely agricultural role. While this might initially have been an appropriate static role, this colonial assignment of resource allocation deprived Hungary of a chance to move gradually into industrial activities of comparative advantage until much later.

Chapter 5, "The Role of Science and Technology in Hungary's Economic Development," reports that in the middle of the seventeenth century, only one university was in existence, and was in an area not under Turkish rule. By the end of the nineteenth century, a number of universities and institutes had been established and were contributing to both science and technology. However, in spite of increasing state support to redress the balance within the empire after 1867, through subsidies, improved conditions for attracting foreign capital, and enhanced support for scientific agriculture (in the interwar period), Hungary's development continued to be handicapped by political instability virtually until World War II.

Since then, under the socialist mode of organization, a substantial effort has been made to "catch up," mainly by the extensive use of R&D allocations. The latter have increased at a more rapid rate than gross national product (GNP) and are now reported at 3 percent of GNP, one of the highest on record. But as Szántó, Vas-Zoltán, and Csöndes acknowledge, in spite of the 1968 reforms that, inter alia, served to encourage R&D by making it chargeable as a current cost, "the incentives of production . . . do not [yet] seem to give adequate encouragement to the assimilation of newly developed technologies."

Japan represents the case of a small latecomer country that can today be even more definitively labeled "mature." Like Hungary, Japan before the Restoration in 1868 may be considered to have been feudal and poor, although it had the definite advantage of not having been subjected to colonialism and of having experienced substantial development of internal markets and of agricultural and human infrastructure during the Tokugawa period. While its "initial conditions" may thus be considered favorable relative to most contemporary developing countries, its successful transition to modern growth from initial endowments substantially closer to those of developing countries than to those of Western Europe or the United States has aroused unusually strong interest among development analysts and policy makers.

Most observers, including Shigeru Nakayama (see chapter 6), have detected the existence of important subphases in Japan's transition, during which the relationships among science, technology, and growth underwent considerable change. Partly because of the extraterritoriality treaties imposed by the West and partly because of the long seclusion period predating the opening to the West, Japan's initial efforts to support its industrialization drive through government intervention—what we would call import substitution today—were relatively mild. Nevertheless, as Nakayama also reports, we do find, between 1868 and roughly 1890, determined government efforts to "catch up with the West." With protective tariffs largely unavailable, government intervention took the form of public-sector participation in directly productive activities, subsidies, and other ways of influencing relative factor and product prices.

The Meiji government initially encouraged the large-scale borrowing of technology from abroad for both the agricultural and the nonagricultural sectors. As Nakayama points out, substantial errors were committed in the attempt to apply Western-style, land-abundant agricultural methods—mainly developed for wheat—to a small-scale, land-scarce rice economy. Similar errors were committed in industry by importing inappropriate "turnkey" technology for use in public-sector plants, some of which subsequently failed. Both the advice of the many invited foreign experts and the findings of the even larger numbers of Japanese sent abroad to seek the "right" country from which to borrow were frequently wide of the mark during this period.

While these facts often have been lost sight of by overly enthusiastic observers of the Japanese experience, it is perhaps more instructive to note that it took the Japanese relatively little time to recognize not only that imported technology had to be selected carefully but also that, for maximum effectiveness, it had to be substantially adapted to local conditions. In agriculture, for example (except for the northern island of Hokkaido, which had an atypical, almost United States-like, factor endowment), by the early 1880s attention had shifted completely from labor-saving (mechanization-oriented) Western methods to the diffusion of land-saving (fertilizer-and-cultivation-oriented) technology, using the experienced or "veteran" farmers and supported by government demonstration farms and extension efforts.[10] In industry, trial and error led to greater reliance on private-sector decisions regarding appropriate technology choice. In 1885, Nakayama reports, the Ministry of Technology was dissolved; and the gradual withdrawal (by sale to the private sector) of the government from all but some heavy industries was more or less complete by 1890.

The Japanese case has often been cited in support of the notion that a country can stay out of high-risk science and can concentrate instead on lower-risk technology imports. The Meiji government spent relatively little effort or resources on the advancement of pure science. Yet the quick, empirically based response of Japanese engineers and industrialists would not have been possible without a strong and well-dispersed educational base, both general and technical, which had been part of the Japanese scene from the beginning. To borrow wisely and to adapt, with an eye to differences in both the endowment and the demand patterns, requires, as Nakayama points out, intermediate-level scientific manpower, not "big science" or heavy R&D expenditures, which the Japanese generally sought to avoid (except, apparently, when the customer was the military). In this sense, we may detect a strong parallel with the nineteenth-century U.S. case: pragmatic borrowing of technology from abroad, plus extensive indigenous technology change supported by high, well-distributed levels of scientific and technical literacy.

Nakayama notes that the early public sector's import-intensive industrialization efforts were dominated by former samurai who had been displaced by the abolition of feudal rights. It is equally interesting to observe that many of the medium- and small-scale private entrepreneurs who later helped shift the center of gravity to indigenous technological experimentation and adaptation came from the ronin, the lower (and less well-connected) strata of samurai.

Production increases based on the diffusion of the "best" known agricultural technology naturally ran out of steam after some time; and, as Nakayama reports, chemical, science-intensive agricultural innovations became increasingly important after 1885. The trend quickly spread to the agricultural input industries, such as tools, seeds, and fertilizer, with substantial science-intensive (mostly German-oriented) technology change in evidence. This was also the time when government support of industry in general became more indirect than direct and, incidentally, more export-oriented (and thus of necessity more competitive) than during the import-substitution period.

With respect to industrial technology, the need to compete in international markets for silk, cotton yarn, textiles, and, later, rubber and electrical goods, provided an added impetus to the search for additional innovations and adaptations that would make intensive use of the relatively abundant unskilled labor force. [11] Nakayama cites the remarkable increase in patent applications during this period, the large majority of which were process- rather than product-related. A piece of interesting historical evidence that has come to my attention is the change from mule to ring spindles in the Japanese cotton-spinning industry in 1887. Rings, it turned out, accommodate much more unskilled labor per unit of capital, could produce a wider variety of yarn qualities, and could accommodate greater variations in the quality of raw cotton input. What is most interesting for our purposes is the almost instantaneous switch of all cotton textile mills in Japan, while Indian textile mills, supplied at the time mainly by the same capital exporter (Platt's of London), and facing an even more extreme surplus of unskilled labor, stayed substantially with the less efficient mule technology.

Some observers would place the approximate data of Japan's successful transition into modern growth shortly after World War I; others, after World War II. [12] What is more important for our purposes, however, is that in Japan, as in the United States, technology and industry became more directly science-based or science-intensive in the interwar period, with basic sciences receiving major attention for the first time; this was reflected in the growth of public and private research laboratories, the rise of sponsored research and university science departments, and the growing demands of the military. Nakayama considers the creation of Riken (the Institute for Physical and Chemical Research) in 1917 to be a landmark, for 85 percent of the funds came from industry rather than government. Another landmark was the creation in 1931 of the Japan Foundation for the Promotion of Science. At the same time Japan, which had never really opened its doors to direct foreign private investment,* experienced a new wave of foreign technology inflow embodied in joint-venture or licensing arrangements (such as the General Electric/Shibaura alliance, responded to by the agreement between Westinghouse and Mitsubishi). Also, an increasing number of indigenous innovations, beginning with the Toyoda loom but moving into chemistry and later to electronics and related fields, began to appear. Clearly, as had happened earlier in Germany, science had become increasingly viewed as an essential national instrument, especially in Japan, short of natural resources and thus increasingly dependent, once labor surplus had become exhausted, on the ingenuity and resourcefulness of its people.

THE DEVELOPING COUNTRIES: BRAZIL AND GHANA

Brazil and Ghana are the two developing countries represented in our study. As is so frequently the case when anyone attempts to generalize about the developing world, these two systems clearly have as many differences between them as they do with respect to a "typical" developed economy. While it might be useful to deal with our two cases within the context of a systematic typological framework that differentiates among "families" of developing countries by, for example, size, land/labor ratios, and/or human resource endowment, this would take us beyond the limits of this chapter. We will therefore attempt to draw what reasonable generalizations seem to emerge about the relationships among science, technology, and development in developing countries, and content ourselves with an

---

*The minor flows in the nineteenth century and the rather more substantial flows thereafter mainly took the form of loan capital. The absence of any marked colonial pressure surely precluded the appearance of some of the more customary manifestations of multinational business interests.

occasional comment concerning relevant differences among developing countries.

Both Brazil and Ghana began as colonial entities, with much of the observed pattern of resource allocation and growth dictated by the needs of that system. Jose Pastore (see Chapter 7) and Edward Ayensu (see chapter 8) report on the almost exclusive emphasis of science on the public sector during this period and the concentration on flora, fauna, and geological surveys—mainly aimed at the location and exploitation of primary raw materials. It was gold, then mainly cotton, sugar, and coffee in Brazil; gold, then cocoa, in Ghana. But the patterns were the same. In both cases we witness not only a neglect of industry—in fact, some destruction of artisan production by industrial imports—but also of food-producing, domestically oriented agriculture.* And in both cases the concern with the exportable cash crop was supplemented mainly by an interest in health, and the required adaptation of medical science to the overseas territory. Colonial governments thus clearly recognized that some indigenous scientific capacity was required in agriculture and health where local conditions, with respect to soil, climate, and disease potentials, are apt to vary substantially.

Political independence came much earlier to Brazil, which emerged from its "Iberian period" in the early nineteenth century, than to Ghana, which did not become independent of England until the 1950s. Nevertheless, as Pastore is at pains to point out, political independence did not alter the basic triangular colonial pattern of resource flows in Brazil until the 1930s. Both systems, Brazil's in response to the international vagaries of the Great Depression and Ghana's in pursuit of domestically oriented national development goals under Kwame Nkrumah, embarked on a fairly standard type of import-substitution industrialization strategy. Abstracting from the differences in the sizes of the two domestic markets, the sheer volume of natural resources available, the extent of regional diversity, and the educational base achieved, import substitution in both cases meant a rather determined effort to import developed-country industrial technology without much emphasis on indigenous science, on the one hand, or technology adaptation, on the other, and with a continuation—if for somewhat different reasons—of the policy of relative neglect of food-producing domestic agriculture.

---

*This, incidentally, differentiates the Japanese colonial system (as in Korea and Taiwan) from others. But that is because, once domestic Japanese agriculture slowed early in the twentieth century, it was food that the mother country wanted from its colonies in this case.

Since the agricultural hinterland remains the large and crucial sector in most developing countries, whether measured in terms of people, of output, or of the potential for the application of science-based technology, this continuation of colonial neglect under independent national governments is of great concern. Rice and maize research in Ghana received as scant attention as research on beans and rice in Brazil. Cocoa and coffee, on the other hand, continued to be viewed as the major source of fuel for the operation of the system and thus received most of the attention of agricultural research concerned with variety improvements, new fertilizer combinations, and resistance to plant disease. Pastore records some recent changes in Brazil in this respect; the situation is less clear in Ghana—certainly the institution of the state farm system there did not encourage cultivator pressure or receptivity.

With respect to industry, Pastore finds not only Brazilian technology but also the entire pattern of growth still heavily influenced by foreigners—now through the multinational corporation—even during the import-substitution subphase of development. He notes that "domestic technological and scientific establishments were not encouraged to innovate," and that a surprising 62 percent of industrial know-how still emanates from abroad, with half of large-scale Brazilian firms holding permanent foreign technical contracts. Science, until quite recently, apparently remained a highly individualistic Europe-oriented art form. The first science- or technology-oriented university-level training programs did not begin until the 1930s, and then still with an expressly abstract slant. It should not surprise us that the portion of Brazilian output growth not attributable to increases in physical inputs—which, with all its shortcomings, is called "technology change"—has been measured at about 20 percent, as opposed to around 80 percent for the advanced countries. Pastore is undoubtedly correct in concluding that the development of science cannot be left to laissez-faire forces if the requisite critical mass of human and physical resources is to become available.

On the other hand, we note that heavy government intervention in science (and technology), which certainly was the situation in Ghana during the Nkrumah period, by no means guarantees a more favorable outcome. Ayensu is kinder than I in his evaluation of the long-term developmental impact of Nkrumah's Science City, the Volta River project, and other large-scale government efforts aimed at forcing Ghana into modernity, without the benefits of a fully socialist institutional structure. He nevertheless recognizes that the increasingly heavy government participation in directly productive activities during the First and Second Plans could not solve (I would say probably worsened) the middle-level management capacity shortage in the country. In fact, there should be little surprise, given

Ayensu's own figures of only 1.4 percent of school-attending males and 0.7 percent of school-attending females getting even a modest technical or commercial education, that the Ghana Council of Scientific and Industrial Research (CSIR) has performed as badly as he reports.

Even where the effort is less extreme, and the economy remains more "mixed," there seems to be ample evidence that "in general, import substitution policy and full-scale protection of consumer goods industry have tended to promote a passive attitude to the utilization and development of indigenous R&D efforts, during the early phase of industrial development."[13] The distortions affecting output and technology choice, in terms of both relative prices and lack of competitive pressures, in favor of modern "engineering" and against appropriate "economic" choices, are well known. The relevant issue, rather, is how severe the import-substitution policies are and for how long they are maintained. For, while it is generally acknowledged that they have a valid and important role to play in the early postcolonial life of a developing country still lacking an industrial entrepreneurial capacity, the fact is that they as often convert themselves into ballast that later is politically difficult to discard. While Ghana remains in the fairly early stages of nondurable consumer-goods (primary) import substitution, it is, I believe, accurate to say that Brazil, except for a brief 1963-68 interlude, has intensified its import-substitution policies, moving into the technically more complicated (secondary import-substitution) industries: durable consumer goods, capital goods, and raw-material processing.

Such a shift is, if anything, yet more dependent on foreign technology and yet more dissociated from domestic scientific or adaptive technological ingenuity. I agree with Pastore that "a strong scientific establishment is necessary in order to understand trans-national knowledge, both in science and technology." I would only add that the dimension of "strength" involved includes the capacity to choose and to reject, to adapt and to diffuse; the contrast in performance between a relatively small, natural-resources, poor-island nation. Japan, which proceeded to turn outward after a period of relatively mild import substitution, and large, natural-resources-abundant Brazil, is bound to be instructive in these respects.

FINDINGS AND CONCLUSIONS

In these last pages I will attempt to record some of the findings and insights that seem to have emerged with respect to the many complicated questions raised, focusing mainly on those facets of historical experience that may serve to illuminate basic contemporary developing-country concerns. These are personal conclusions drawn from the seven country papers as well as other sources, and do not implicate the individual authors in any way. Almost all countries today accept the importance of the impact of technology on growth, as well as on distribution and other important dimensions of development. But they are profoundly uneasy as to how much of their technology can be, or should be, home-grown, imported, and/or imported and adapted. They are even more uneasy with respect to the volume of resources and energy they should commit to basic science as the underpinning for technology change. Waiving other motivations and considerations, in other words, they are concerned about the price of the "ticket of admission" to the community of science.

The analysis of the role of science and technology in the history of the now-developed countries led me to conclude that to divide technology into empirically-based and science-based categories is likely to be off the mark. Epochal technological change of the type to which we have become accustomed in the twentieth century is likely to be more directly related to major scientific breakthroughs than in the past. We need only think of electronics, plastics, the computer, or atomic energy, to make the point. [14] In the eighteenth and nineteenth centuries, on the other hand, the pace of science was slower—some would say "big science" had not yet arrived—and consequently any epochal technology change, such as the steam engine, equally based on a major scientific discovery, might yield its technological impacts and applications over a longer period and in more diffuse ways. This does not make the sum of such innovations less science-based but, rather, less science-intensive.

Second, keeping the United States and Japanese experiences particularly in mind, these systems admittedly were not pioneers in frontier science; but they developed a definite capacity to absorb science as a necessary basis for their own very substantial achievements in importing and adapting technology. As Kuznets has pointed out, [15] this capacity to use science wisely is more likely to be national rather than supranational. But it does not just happen. It is related to the educational system, to the national ethos, and to the types of interventions, either direct or indirect, practiced by governments. An educational system that imparts a modicum of scientific understanding to a substantial portion of the population, a pragmatic "catch-up" philosophy that accompanies "latecomer" status, and national governments' willingness to move away from dirigiste mercantile interventions can provide basic building blocks for this type of science capacity at a relatively early stage of a country's development. As the experience of both Japan and especially the United States also illustrates, the same country may later, in its modern growth phase, acquire the capacity to advance the international frontiers of science.

The "typical" contemporary developing country thus cannot afford to "sit back" and let the advanced countries incur all the expenditures attached to the trials and errors of international science—especially not now, in the twentieth century when the pace of science has much accelerated and the gap between it and technology narrowed. Yet it cannot afford, and should not try, to "show the flag" in every field of basic scientific endeavor; the developing world is strewn with scientific institutes and other expensive white elephants which contribute neither to science nor to technology. Most observers agree that the biggest waste of all is second-rate basic research. The "middle road" points in the direction of a broad enough spread of science and technical education and a flexible enough economic environment to permit appropriate scientific and technological choices as well as indigenous improvements and adaptations. International science is only slightly more a "free good" than technology; there are important search, identification, transfer, and assimilation costs involved.

Julian Engel (in a private communication) sees "little justification [in developing countries] for basic research except for sustaining a viable teaching effort and keeping your best brains at home." This is in general accord with the above position, except that it may go too far. There are fields of scientific endeavor that must be strongly represented within the developing countries because of their country- or region-specific character. The best examples are in agriculture and health. Without basic agriculture/science-oriented research on a country or at least regional basis, the recent chemical and plant genetics-based innovations that have gone under the name of the Green Revolution do not, as we are now finding out, have the necessary sustaining power and the necessary defense against specific local problems (such as pests and diseases). Similarly, in the field of health, few people would argue that one transnational science can really be equally responsive to the very different conditions around the globe. It is in such areas that the "puzzle"-solving capacity of science in response to technological problems clearly requires a first-rate scientific establishment.

Are there other areas in which the same criteria apply? This is perhaps the most difficult question of all. In one sense, all human activity is affected, to a larger or smaller extent, by the particular soil, climate, and other conditions under which it is carried out. It is, for example, relevant even in industry—think of the relationship of fertilizer and agricultural implements to the conditions of the soil, and of the importance of humidity and temperature conditions, and of natural fiber quality, to spinning and weaving operations. Where, then, does the need for individual developing-country basic research end, and do the caveats against a wasteful buckshop approach take hold?

This is by no means an easy matter on which to pontificate in an abstract way. I would insist, nevertheless, that the burden of proof be on those who would like to initiate advanced university training and basic research, including some obligation to demonstrate a flexible, time-phased relevance to technology changes—which, in turn, can be expected to affect the productive system. This may seem like the typical hardheaded, narrow economist's prescription. What about the importance of those many possible chance interconnections, decades apart, that may flow, in some entirely unpredictable way, from what looks like an unconnected intellectual pursuit? Without disparaging these possibilities in any way, we would respond—if we are indeed offended by the spectacle of open-heart surgery research in countries where malnutrition is prevalent—that science really should not expect to be entirely outside the realm of some flexible, sophisticated version of cost/benefit analysis. Such analysis must try to balance the potential benefits against the possible alternative allocations of scarce financial and (perhaps more important) human resources. The higher risks of science, due partly to the uncertainty of predicting future two-way interactions between science and technology, and partly to the likely inappropriability on a national scale of any such "returns," render this task unusually difficult. But analysis must still be done; an act of faith does not suffice.

In addition to placing the burden of proof on those who would like to have developing countries pay the "price of admission" in a given field of basic scientific endeavor, it might be possible, although admittedly difficult, to encourage much more specialization within, and possibly also among, those countries on a regional basis. This type of agreement has been reached, for example, in European atomic energy and ballistics research and African efforts to combat yellow fever and rinderpest—that is, where the required scale and the need to avoid expensive duplication were sufficient to overcome nationalistic jealousies. Although the record on similar agreements among developing countries in the field of common market investment allocations, for instance, has not been very encouraging, it has been somewhat better with respect to the use of regional training institutes and research organizations—whenever regionalism is not forced, but flows from the recognition of mutual self-interest.

If we agree that no country can really afford to be either a full-time borrower of science or an across-the-board contributor to it, what about technology? First of all, our historical forays seem clearly to indicate support for the Bernal-Kuznets position that technology gives rise to as many "puzzles" required for further scientific progress as the other way around. Consequently, much of what has been said applies to technology as well. When we speak about a society's national capacity to utilize and modify international science creatively, we are also referring to a kindred capacity to select appropriate technology and adapt it to differing environments. If we but keep in mind that contributions to human knowledge that break new ground and provide scope for major new technological break-throughs will, with few exceptions, remain the province of the leading mature countries, what can we say about the direction that new science-intensive and engineering-intensive technology change is likely to take?

The two elements that seem most responsible for this direction are changing resource endowments and public policy. The very different behavior of the natural-resources-rich, labor-scarce United States relative to a relatively capital-scarce England and a Germany that felt cramped for natural resources should be instructive in this respect. Engineering-intensive technology took a different, more capital-intensive path in the wide-open spaces of the United States than in England. And in Germany, metallurgical science responded to the demands of an iron ore with high phosphorus content; official encouragement of the entire chemical industry was based on the felt need to overcome, by artificial and synthetic shortcuts, the relative unkindness of nature. Japan, after first exploring its abundant labor resources—and taking an engineering-intensive route analogous to that of the United States, but capital-saving—has, with the disappearance of that labor surplus, tended to place more of its eggs in electronics and other high-technology baskets.

But, as has also been pointed out, while government policies cannot legislate away the basic endowment of a society, they can, if flexible and able to overcome narrow sectional interests, provide important assistance to the transition effort of a developing economy as its endowment changes with time. Analogously, if dominated by narrow vested interests and/or lacking in historical perspective, such policies can attempt to draw a veil over the endowment and lead the system into expensive scientific/technological dead ends and economic stagnation. While there is no rigid unidirectional sequence of phases that every developing country must follow on the path to mature growth, some attention to the changing roles of science and technology in terms of a changing resource endowment and, especially, changing human capacities is essential in all but the most unusual cases.*

---

*A country like Kuwait, for example, may be able to buy its way into the charmed circle with turnkey oil-oriented technology; but, even so, there is some doubt as to whether it qualifies as a mature

At the micro and institution-building levels, the appropriate role of government in the mixed economy context is not unrelated to the appropriability or nonappropriability of the new knowledge acquired. Investment in basic science carries a high risk, in part because of its, at best, indirect and long-term relationship with technology and growth, but also partly because it is generally an international good not even appropriable by a country, not to speak of any private party within a country. As we move from basic international science to changes in technology, risks are reduced and private appropriability becomes much more important. As the extent of appropriability rises, so, normally, does the level of private R&D expenditures.

Appropriability, of course, depends not only on how basic the research effort is but also on the overall state of competitiveness or noncompetitiveness of the system. This is partly a function of the overall policy environment; for example, during periods of intense import-substituting industrial protection and large unearned profits, there would seem to be less interest among industrialists in searching for the best technology; instead, satisfying behavior and the use of inappropriate (often prestige) technology frequently seem to displace maximizing behavior. [16] But, for any given industry or sector, the state of competitiveness also depends on conditions peculiar to the particular market, with respect to goods, information, or technology. Agriculture, for example, is typically the most competitive field, and therefore exhibits the least private appropriability possibilities and the least willingness (or capacity) by individual farmers to incur R&D expenditures. Consequently, not only basic scientific agricultural research but also the search for appropriate adaptive technology, and even its dissemination to individual farmers, usually represent activities (and costs) that fall to the public sector. The same is true of health—perhaps even more so.

This is fairly well recognized. But what is perhaps less well understood is the fact that there are other industries—again on a continuum moving through agricultural processing and input industries to light consumer goods, some services, and beyond—where similar characteristics abound: a competitive market structure, the relative absence of scale advantages, and thus the need for possible government involvement in R&D, education, and extension. If, in the absence of pronounced market imperfections, new technology can be borrowed selectively from abroad, the burden on high-cost domestic

---

economy. It certainly does not meet all the Kuznetsian stylized attributes of a system under modern growth.

R&D is reduced and a minimum of government support can lead to rapid diffusion of technology change. This certainly was the case in Japan, where the profound technology change associated with the switch from mule to ring cotton spindling in the late nineteenth century was diffused as rapidly as the agricultural practices of the "veteran farmers."

Whether or not, in mature market economies, competitive or noncompetitive industrial configurations yield relatively more private R&D activity remains an unresolved empirical issue. Competitive industries have more incentive but less capacity. With respect to the developing countries, it seems to me, any viable science and technology policy must begin with an examination of the extent of the overall competitive pressures felt by individual decision makers with respect to economic versus engineering choices. It must include sensitivity to differences in the market structures of specific industries, and consideration of selective government action in creating social overheads in science and technology. Such interventions may be addressed to ensuring that existing technological alternatives are known to all sizes of firms, or to helping expand the range of alternatives through the support of university or R&D institute activity. In either case it is, however, important to ensure not only that the areas of activity be selected with some of the (flexible) cost/benefit previously referred to in mind, but also that the specific activities supported within these fields carry built-in devices to ensure that the criteria of ultimate contributions to social and economic development, and not any exclusively internal criteria of the "invisible college," be addressed. One such device, frequently referred to, is that government subsidy of R&D institutes be set on a long-term declining basis, with private-sector contracts forced to fill the widening gap. Another is to concentrate on the more competitive "nonappropriable" sectors and, in fact, to ensure that access to information and to the required complementary inputs be relatively equal across firms.

One dimension of this general problem that has been mentioned only fleetingly is that of process versus product innovation. Economists, as this chapter demonstrates, spend most of their time discussing technique or process change, while industrialists and R&D allocators spend most of their energies and resources on product change. When technology change is of the cataclysmic or epochal type, such as the invention of the automobile based on the internal combustion engine, it is more of a semantic issue whether we call this a change in the transportation process or a change in product (from the horse and buggy); but when we are dealing with more common sequential and adaptive innovations, the distinction may be more real and related to the competitiveness of markets. It seems clear, for example, that product differentiation is of greater importance in less competitive markets and process differentiation more important in competitive markets. If we apply, loosely and briefly, the product-cycle idea, it is clear that patents and trademarks may represent one device to extend the period of quasi-monopoly position beyond what would be possible through simple process and price considerations. The contrasting role in today's developing countries of the Japanese multinational corporation, which is largely process- and price-oriented, and that of the U.S. multinational, which is largely product- and quality-oriented, is rather startling in this regard. It is no accident that the distribution of domestic patents between process and product innovations today is overwhelmingly in favor of process in Japan and of product in the United States. In an earlier day the United States was process-oriented relative to the product orientation of Britain. In recent years Japan has shifted toward product innovation. It thus appears, ceteris paribus, that the richer and more scale-dominated the mature economy becomes, the more important the relative role of product innovations.

This question of competitiveness is, of course, important for the developing economy subject to the blandishments of domestic and foreign technology salesmen, with respect to both process and product. Jacob Schmookler[17] long ago pointed out the importance of demand factors in technology change. Whether the change in product quality is real or imagined, and carries additional real benefits or not, is another question. The fact is that the absence of competitive pressures and adequate information often does not give consumers in developing countries an unimpaired choice, at realistic relative prices. Much of the (I believe quantitatively important) misallocation of the developing countries' resources on inappropriate (such as overspecified) goods and inappropriate technologies is related to the presence of proprietary and noncompetitive elements in areas not warranted by the basic scale relative to the size of the market, as in soft drinks and drugs. The evolution of modern appropriate goods for local markets, at prices that reflect quality differentials, is similar to the adaptive changes in processes arising from "blue collar" R&D, practiced on the factory floor and in the machine shops, in contrast with the more visible "white collar" variety carried on in corporate and university laboratories. It is similarly linked to empirical learning by doing and to experimenting processes. While some scientists, some economists, and many engineers may well disdain to call both of these related types of activity "technology change"—and may be especially reluctant to admit any relationship to science—I would argue not only that they are important for social and economic growth in developing countries, but also that they are only one step removed from the mechanical or engineering-intensive innovations of nineteenth-century labor-rich Japan, and two steps from those of the nineteenth-century labor-scarce United States.

It is, in summary, admittedly futile to attempt to manipulate basic science in any particular direction; the relationships and feedbacks are much too diffuse and complicated. But developing countries can, and must, exercise restraint as to the fields in which they decide to maintain a first-rate scientific establishment. They must make a serious effort to reorient their educational structures toward the ability to make these selections, through the achievement of a broadly based scientific literacy that not only conveys the ability to perceive where indigenous frontier capacity should be installed, but also guarantees the necessary access to the international networks.

With respect to technology, the task of public policy is perhaps easier, but by no means simple. Internationally, there exists a substantially larger number of borrowing options by country and by type than was formerly believed, but many remain obscured by a lack of information and other institutional impediments, some related to the public and private capital transfer mechanism. The options that exist with respect to indigenous or adaptive technology change, either new or derived, are much more numerous. Governments can do much at the aggregative level to ensure that the veil between relative prices and endowments that must sometimes be drawn is neither excessively thick nor kept excessively long; governments can help ensure that sufficient workable competition exists for entrepreneurs to be interested in finding the most appropriate technology, rather than being in a position to indulge their preference for prestige and the "quiet life." And, perhaps most important, as a complement to these aggregative measures there is the possible intervention of the public sector in institutional areas, in ensuring a freer flow of information on market, quality, and technology options, and in providing support to technical education and R&D, especially of the unspectacular adaptive, nonappropriable type.

The interactions among science, technology, and development are potentially of very great benefit to the transition effort of the developing country, which typically finds itself more restricted than the rich in terms of its ability to rely on the contribution of more physical capital and other conventional inputs. But the opportunities of participating within an interdependent global system will not be realized if national policies are not more realistically geared to gradually improving our understanding of the fundamental behavioral relationships involved. It is hoped that the work of this symposium has carried us a small step forward in this direction.

NOTES

1. Simon Kuznets, "Technological Innovations and Economic Growth," in M. Kranzberg, ed., Science and Society (Atlanta: Georgia Institute of Technology, 1977), p. 30.

2. See Organization for Economic Cooperation and Development, The Conditions for Economic Cooperation and Development (Paris: OECD, 1974).

3. See, for example, David S. Landes, The Unbound Prometheus (Cambridge: Cambridge University Press, 1972).

4. As documented by A. P. Usler, The History of Mechanical Inventions (New York: McGraw-Hill, 1929).

5. First pointed out to me by Simon Kuznets in private correspondence.

6. Landes, op. cit., p. 181.

7. See also H. J. Habbakuk, American and British Technology in the 19th Century (Cambridge: Cambridge University Press, 1962), for a detailed U.S./U.K. comparison.

8. Yujiro Hayami and Vernon W. Ruttan, Agricultural Development: An International Perspective (Baltimore: Johns Hopkins Press, 1971).

9. Landes, op. cit., p. 150.

10. Yujiro Hayami and Saburo Yamada, "Technological Progress in Agriculture," in Klein and Ohkawa, eds., Economic Growth: The Japanese Experience Since the Meiji Era (Homewood, Ill.: Richard D. Irwin, 1968) demonstrates that the diffusion throughout Japan of the best known agricultural practices of the day accounted for the most of the (substantial) growth of agricultural productivity during this period.

11. See Gustav Ranis, "Factor Proportions in Japanese Economic Development," American Economic Review 47 (September 1957): 52-57.

12. See, for example, John Fei and Gustav Ranis, Development of the Labor Surplus Economy: Theory and Policy (Homewood, Ill.: Richard D. Irwin, Inc., 1964); R. Minami, The Turning Point in Japanese Economic Development (Tokyo: Tokyo University Press, 1970).

13. Nam Kee Lee, "Technological Development and the Role of R&D Institutes in Developing Countries," I.L.O. Working Paper, p. 19.

14. A study, The Conditions for Success in Technological Innovation (Paris: OECD, 1971) noted: "Technological innovation is as old as man, but it is only in the 20th century that science, technology and industrial firms have come together to play such an important role in it."

15. Private correspondence with the author.

16. For more on this, see Gustav Ranis, "Appropriate Technology in the Dual Economy: Reflections on Philippine and Taiwanese Experience," presented at the International Economic Association Conference on Economic Choice of Technologies in Developing Countries, Tehran, Iran, September 1976 (Proceedings to be published by Macmillan, London).

17. Jacob Schmookler, Invention and Economic Growth (Cambridge: Harvard University Press, 1966).

2

SCIENCE, TECHNOLOGY, AND ECONOMIC DEVELOPMENT: THE BRITISH EXPERIENCE

D. S. L. Cardwell

It is generally agreed that Britain was the first nation to become industrialized, and that this dramatic change began toward the end of the eighteenth century. It follows that any appraisal of the role of science and technology in the economic and social development of the country must start from an earlier date than would be appropriate for any other nation. This said, I believe that in the final analysis most, if not all, of the significant features of British development can be paralleled elsewhere.

DEFINITIONS

Unfortunately, the definitions put forward by philosophers of science in their discussions of fundamental science take no account of the varieties of applied science. Accordingly, such definitions have little, if anything, in common with those used by economists and other social scientists concerned with technological innovation. Since the history of technology, particularly in its relations with science, is still an undeveloped study, anyone proposing a set of definitions that takes account of both fundamental science and technological innovation is unlikely to satisfy all sections of the community. Nevertheless, the task must be attempted.

Francis Bacon, writing in the early seventeenth century, recognized two forms of invention. The first, going back to the dawn of civilization, can be called pragmatic invention. It may be described as the production of something useful or desirable by the wholly novel arrangement of familiar components or well-known principles that have little or no scientific content; or if they have scientific content, that it is so well known that it can be taken as common knowledge. Obviously an infinite number of pragmatic inventions is possible, and this form of invention is as important today as it has always been; witness the number of minor mechanical devices in the average home. Nevertheless, pragmatic invention does not, of itself, give rise to institutionalized research and development (R&D). Since a pragmatic invention contains no new or unknown components, it could in principle have been made anywhere, at any time; systematic research is irrelevant.

The second form identified by Bacon can be called science-based invention. This includes all those inventions that can be made only if, and when, the necessary and appropriate scientific knowledge is available. Once again an infinite number of science-based inventions is possible. Familiar examples include the telephone, television, and refrigerators. But "science" here is used in a broader sense than philosophers would commonly allow. It is equivalent to new knowledge. Thus, Bacon regarded the cannon as a science-based invention, for it could not be made until the explosive properties of compressed black powder were known; a knowledge of the basic chemistry of the reaction was not required, and indeed was not available until very much later. One might say that in this and similar cases, the inventors take a short cut once they have a sufficient knowledge base.

Broadly speaking, technical progress is characterized by two distinct features: revolutionary invention and evolutionary improvement.[1] Thus, for example, the invention of the printing press in the fifteenth century was plainly revolutionary. Its subsequent improvement by detailed inventions, additions, and modifications was evolutionary.

During the eighteenth century the process of evolutionary improvement was put on a rational basis. John Smeaton's search for a hydraulic cement to be used in building a lighthouse can be taken as exemplifying this advance in sophistication. Smeaton collected samples of lime and cement from different sources, mixed them in carefully graded proportions, and submitted the resulting mortars to strict tests of setting and strength; ultimately he found an entirely satisfactory hydraulic cement.[2] The procedure is inductive and therefore can be said to be scientific. It is now a normal feature of technological progress. But as this and similar cases show us, the underlying chemistry, the basic science, is not important; what counts is the final, usable product.

The most recent mode of invention is what may be called research innovation. It is characterized by the industrial laboratory, staffed by graduate scientists and technologists; and its appearance had to await the demand for, and consequent training of, appropriate numbers of specialized and professional men. These conditions were not satisfied until the last quarter of the nineteenth century. Research innovation can take two forms. In the first the research itself leads to something new, such as a new dye, or plastic, or pharmaceutical product, or insecticide.* In this case the product cannot be envisaged without reference to the research. In the second the research is carried out to aid or to perfect the production of something conceived or invented independently of the research. We might call this heuristic research. The manufacture of an entirely new jet engine, for example, will necessitate a great deal of heuristic, or ancillary, research. This form of research innovation received a great boost during World War II.

Summarizing, then, invention is as old as civilization itself; but the incorporation of science, particularly in the strict sense of the word, into the innovative process is both recent and rather complex in its effects. It would be misleading to suggest that these different modes of invention are, in any significant way, mutually exclusive. In many, if not most, large industrial laboratories all may be practiced at different times; and research workers may be switched from one to another as occasion demands. In addition, in some of the largest and best-known industrial laboratories some basic research, of the same sort as that practiced in universities, may be carried out.

BACKGROUND

The Eighteenth Century

There are good reasons for believing that, technologically and scientifically, Britain lagged behind the more advanced European countries until the end of the seventeenth century, and in some respects continued to lag behind until well into the eighteenth century. Bacon (1561-1626) had been a powerful advocate of the social benefits to be obtained from technolgy;† and the establishment of the Royal Society in 1660 (charter in 1662), together with the achievements of

---

*Sometimes the research procedure, particularly in the two last fields, is inductive; a large number of likely compounds are tested until the one that gives the best result is found. Methodologically this is a laboratory version of "Smeatonian" improvement, for the basic chemistry and biochemistry remain unknown.

†The word "technology" was coined in the seventeenth century, but "technologist" did not appear until late in the nineteenth century.

Newton and his circle, marked Britain's entry on the world's scientific scene. However, the riddle of the universe having been solved by Newton (<u>Principia</u>, 1687), there seemed, to British men of science, little more that needed to be done, at least in Newton's chosen fields. Accordingly, the studies of planetary astronomy and—potentially much more useful—mechanics languished, being advanced much more effectively in France and in countries under French cultural influence. The Newcomen engine, effectively the first successful heat engine, was admittedly science-based in that its invention depended on knowledge of atmospheric pressure and the properties of the vacuum, discovered in the seventeenth century. But the development of this engine was substantially through the normal historical process of evolutionary improvement. It was not until the advent of James Watt (about 1765) and Smeaton that science once again influenced this all-important branch of technology. With Watt, indeed, an element of R&D as now understood made an appearance, but it was a transitory phenomenon because it depended very much on the genius of Watt himself.

Much has been written and, no doubt, remains to be written on why the Industrial Revolution began in England in the eighteenth century. Here it is sufficient to mention a few of the more obvious reasons: an abundance of suitable raw materials, particularly coal and iron; an effective patent system, established in the previous century; a high degree of personal freedom and an absence of internal barriers to trade and commerce; highly profitable overseas possessions; and a favorable geographical position for the development of overseas trade. Not without significance in the latter context was one of the few instances of state action in the field of technology: the British government offered the (then) truly enormous sum of £20,000 to the first person to produce a chronometer accurate enough to be used for navigation. The prize was won by John Harrison (1693-1776).

At the same time agriculture, a basic technology, was undergoing transformation. The "enclosures"—that is, the substitution of relatively large and efficient farms for the inefficient medieval pattern of small strip cultivation and "common" land—not only resulted in increased production but also released labor most opportunely for the mills and factories that were put up toward the end of the century. In addition, a number of important innovators helped to transform the theory and practice of farming. "Gentlemen improvers" like Lord ("Turnip") Townshend (1674-1738) and Thomas Coke (1754-1842) systematically developed their farms while Robert Bakewell (1725-1795) effected a revolution in the breeding and rearing of stock, particularly sheep and cattle.[3] Taken together, these changes not only released manpower for industry but also ensured an adequate food supply for the industrial population. It is, incidentally, worth noting that Bakewell's achievements and ideas were studied by Gregor Mendel: an interest example of the influence of technology on science.

The industry that, more conspicuously than any other, exemplified the economic growth associated with the Industrial Revolution—the textile industry—seems to have had practically no connection with science. The achievement of Richard Arkwright (1732-1792) and his fellow innovators lay in their success in breaking down the complex human processes of spinning and weaving into discrete, separate stages, each of which could be mechanized by using the materials and techniques freely available at the time.[4] The machines were made of wood. Gears, rollers, and weights were made by clockmakers whose craft went back to the Middle Ages. The "flyer" was taken directly from the fifteenth-century Saxon spinning wheel. The inventions were, in fact, pragmatic; and through them the entire industry was mechanized, from the opening up and combing out of the cotton bolls to the weaving of the fabric on a power loom. The motivation was demand, for a radical advance in productivity at one stage meant strong pressure for complementary advances in the previous and following stages. The process began with John Kay's invention of the fly, or flying, shuttle in 1733, which greatly increased the productivity of the weavers. The response was Arkwright's water frame (1769), the first successful cotton-spinning machine. Had these inventors been science-oriented, we might expect them to have examined cotton staples under the microscope, or to have experimented with different liquids to improve adhesion between the staples, or perhaps to have tried to cultivate cotton with very long staples. In fact they did none of these things. And it was characteristic, in more ways than one, that Arkwright began his career humbly enough, as an apprentice to a barber and wigmaker: an ideal training for anyone wanting to gain practical experience of natural fibers, but one wholly barren of science.

The later development of this important industry exemplifies the chain, or trigger, effect that a great increase in productivity with rapidly expanding markets can exert on related industries, whether suppliers or consumers. As textile mills grew bigger, floors had to be strengthened to take the weights of more and heavier machines, and fireproofing became more urgent. These requirements led to the construction of the first iron-framed mill at the beginning of the nineteenth century, an invention that was necessarily science-based, although the science in question was archaic. This was, interestingly enough, based on Galileo's theory of the strength of materials, formulated a century and a half earlier. The theory was inexact, but it was good enough to enable engineers to calculate the correct proportions for iron beams. It is always worth remembering that the scientific component of a science-based invention does not have to be the latest scientific knowledge, fresh from the research laboratory; scientific ideas will often lie fallow and unexploited for a long time before they are successfully applied to invention. Other aspects of the textile industry in which science gradually became more important were in the provision of adequate power; in the heating, ventilating, and lighting of mills; and in solving the problems of cleaning, mordanting, dyeing, and printing the woven cloth. Science, however, was not involved in the invention of the elevator, or "teagle"; all that was required, besides the requisite ingenuity, was an appropriate need and a multistory building with power available on the top story: in other words, a typical large textile mill with, perhaps, a corpulent owner.

Much more could be written under this heading. Certainly there were great reserves of skill and enterprise in the industrialized areas of Britain, resulting from the ingenuity and spirit of self-help of the people themselves. But this did not, and could not, add up to institutionalized R&D as understood today. Two salient points should make this clear.

In the first place, over the greater part of the eighteenth century only one branch of science—Newtonian mechanics and planetary astronomy—had reached an advanced stage of conceptualization or was "established"; all the others were in "progressive" or empirical stages.[*] The scope for systematic and institutionalized R&D was therefore very restricted. And when, at the end of the century, chemistry and branches of physics began to advance rapidly, it was France that took the lead. The Ecole Polytechnique, founded in 1794, was soon associated with a remarkable galaxy of able mathematicians, physicists, chemists, and engineers. Science was systematically reorganized in the wake of the Revolution—the now almost universal metric system is a scientific legacy of those days—and French technical publications rapidly achieved an unprecedented degree of professionalism. The scientific memoir of multiple authorship first appeared in France during this period and soon became common. At the same time the Republic and the succeeding Empire made serious attempts to apply science to industry and to national defense.[5] On the other hand, it may be that French technological achievements after the Revolution have, perhaps, been undervalued by English-speaking historians deeply impressed by the Industrial Revolution in Britain. The Leblanc soda process, the use of chlorine for bleaching, and the Jacquard loom were all key French innovations that were

---

[*]These expressions were coined by the historian and philosopher of science William Whewell.

rapidly adopted in Britain. Admittedly, a balanced comparative assessment of the science and technology undertaken by the two countries and their economic performances has not been carried out; but the evidence suggests French supremacy in academic* and state technology, and British excellence in industrial and production technology (with little or no interest in long-term R&D).

Second, and to complete the equation, there was no national system of education in Britain. There was nothing remotely comparable with the École Polytechnique, and consequently formal means of training R&D personnel were completely lacking. It appears, indeed, that a nation concerns itself with education insofar as it is anxious to preserve, clarify, or establish its national identity. The English—as distinct from the Scots—felt no need to do any of these things; thus they neglected education in a way that by the end of the nineteenth century had become something of a national scandal, and may well have been the ultimate cause of some of Britain's later troubles. As regards tertiary education, it is true that Cambridge University had developed a reputable mathematical school by the beginning of the nineteenth century. This was a consequence of the establishment of the greatly esteemed mathematics tripos examination; but it was a common complaint that Cambridge mathematical teaching was of little or no practical use, and it was in any case based on such long-outmoded techniques as Newtonian fluxional calculus. The vast majority of the top men, the "wranglers," sought careers in the law, the Church of England, or in politics.[6] In fact, during the whole of this period the two ancient universities of England were little more than finishing schools whose aim, not always realized, was the liberal education of Christian English gentlemen. There is abundant evidence of the decadence of Oxford and Cambridge during the eighteenth century. Here it is enough to quote an unimpeachable contemporary observer: "A young man of eighteen is not in general so earnestly bent on being busy as to resist the solicitations of his friends to do nothing. I was therefore entered at Oxford and have been properly idle ever since."†

On the other hand, it seems that those countries in which the state was economically active were the first to pay attention to higher technological education. The famous mining academies of Germany

---

*Advanced textbooks of mechanical and civil engineering published in France were incomparably superior to those published in England.

†Jane Austen, Sense and Sensibility. The young man in question is Edward Ferrar.

and, particularly, the Schemnitz Mining College in Austria-Hungary were direct consequences of state interests in mining. In France, too, the state interest in roads and bridges led directly to the foundation of the École des Ponts et Chaussées. In Britain, where the state was substantially inactive in economic matters, higher technological education did not begin to develop until the last decade of the nineteenth century.

Institutionalized R&D did not, and could not, develop under these circumstances in Britain. The solution of the problem of industrial innovation took the form of the symbiotic partnership of the businessman and the engineer, or technologist. It is epitomized in the famous partnership of Matthew Boulton and James Watt. As late as the twentieth century another famous partnership was formed when C. S. Rolls joined forces with the engineer Henry Royce.*

In summary, then, the economic success of Britain during the Industrial Revolution was due to individuals, or more commonly to symbiotic partnerships, responding to intense demands by effecting pragmatic and then increasingly science-based innovations. The lack of a national system of education, and of recognizable R&D institutions, did not inhibit effective innovation. The more "practical" eighteenth-century English engineers were trained as millwrights, the more "scientific" (the term is comparative), as instrument makers, a vocation that was correlated with the importance of navigation. In both cases training was by apprenticeship (as was also the case for the majority of eighteenth-century physicians and surgeons). Apart from apprenticeship and, for those who could afford it, education at a Scottish or Dutch university, the only other significant way of getting a scientific education, at roughly junior high school level, was by attending a "Dissenting academy." The Dissenters, those who refused to subscribe to the Articles of the Church of England, formed a large and energetic section of the community. Excluded from the English universities (no great deprivation) the Dissenters established their own academies, in which experimental science, modern languages, and other subjects of commercial and practical importance were taught. So good was the education at some of these

---

*The virtual disappearance of the symbiotic partnership presents an interesting problem. It is easy to see how the engineer partner could soon be transformed into the research or technical director presiding over a hierarchy of R&D personnel and so lose his function as the acknowledged source of innovation in the company. But besides such institutional changes, other factors may well have been important. The question needs further study.

academies that it was quite usual for members of the Church of England to attend. Most, if not all, of the academies had closed by the end of the century as the discriminatory laws against Dissenters were repealed or allowed to lapse. But it has been suggested that the ideals of the Dissenting academies were not without influence on the founders of institutions such as Massachusetts Institute of Technology.

At the entrepreneurial level, scientific information was disseminated by means of formal or informal societies that often functioned as social clubs. By far the best-known of the early groups was the Lunar Society, with which Matthew Boulton, James Watt, Joseph Priestley, Josiah Wedgwood, and many other pioneers were associated.[7] The best-known of the later ones was the Manchester Literary and Philosophical Society. The needs of artisans and mechanics were met by peripatetic lecturers who gave lecture-demonstrations in town and village halls, using simple apparatus to illustrate their subjects. The science was elementary—the law of the lever, Archimedes' principle, Galileo on the strength of beams—but it was, in many cases, sufficient for the ingenious, the industrious, and the ambitious. Set in the context of Britain's natural, political, and legal advantages, a little science went a long way in the early days of the Industrial Revolution. In retrospect, then, it seems that conditions in middle-to-late-eighteenth-century England were favorable for the practice of industrial technology and pragmatic invention, although not for institutionalized R&D, as it developed later in the nineteenth century.

Finally, we must remember that in spite of the hierarchical social structure of Britain, in spite of the great differences in rank and wealth, there was nothing to stop any man granted good health, a willingness to work hard, and a necessary quota of ability from making a fortune. In how many other eighteenth-century European countries could a man starting life as a barber's apprentice earn the honors and distinctions, as well as the vast wealth, that Arkwright did? Or, to take an example from another field of technology, the navigator James Cook was the son of farm servants and began life as a deck boy on a coal boat sailing from Newcastle. He became a captain in the Royal Navy and a fellow of the Royal Society. Indeed, there are grounds for believing that up to the middle of the nineteenth century, England may in some important respects have been socially more mobile than it later became.

In conclusion, mention must be made of the part the Dissenters played in the Industrial Revolution. The theme has been a familiar one ever since Max Weber opened the debate. Protestantism is said to be favorable for the development of capitalism and, by inference, of industrialism and industrial technology. The English Dissenters, quintessential Protestants, excluded from the universities and offices of state, which were the preserves of the Anglican, landowning classes, turned to other activities and helped to create the industrial system that by 1851 made Britain the "workshop of the world." It is, of course, true that many of the leading entrepreneurs, technologists, and scientists of the period were devout Dissenters; but there are certain difficulties with the theory, plausible though it seems, and further discussion would be unprofitable. What is clear, however, is that with the progressive removal of formal disabilities, many prosperous non-Anglicans, like other affluent members of the manufacturing classes, set new social targets for themselves and, more important, for their families. It was said that a typical Manchester mill owner would "rather you admired his mill than his mansion."[8] Later generations of the family preferred to forget the mill, a forgetfulness made all the easier by the rapid development of the railroads from the 1830s on. These enabled the mill owners to live far away from the industrial areas, and to commute to their offices by rail. We may infer that this was one of the causes of the social disasters and the class divisions that have marred British life during the twentieth century.

The Nineteenth Century

The growth of British technology during the nineteenth century was by extended "chain effects": the urgent demands of, and inventions in, some industries triggering innovations in other industries, and in some cases leading to the establishment of wholly new industries. The telegraph is a case in point. A system of visual telegraphy had been established in France during the Revolutionary Wars and had been immediately copied in Britain. By the end of the Peninsular War in 1814, the system was quite extensive in both countries, but it fell into disuse with the coming of peace. Evidently there was no demand for high-speed communication. With the coming of the railroads in the 1830s, however, it became urgent to devise a system of communication that was faster than a railway locomotive, which itself was faster than the fastest horse. The result was the electric telegraph, which had been foreshadowed in the eighteenth century and was theoretically feasible by 1800.

The pioneers had been Italians and Frenchmen, Germans and Russians; but the first effective working telegraph was installed in 1837 by two Englishmen, W. F. Cooke and Charles Wheatstone, along the railway line from Paddington (London) to Slough, a distance of about 40 kilometers. The system spread with the rapid growth of the railroads in the succeeding decades, and the intensive and fast transportation services of later years would have been impossible without it. A quarter of a century after Cooke and Wheatstone, the first great Atlantic cable was laid, an achievement that captured public imagination much more than the completion of the transcontinental land line to India. With the growth of electric telegraphy, a new profession and a new industry arose. The first electrical engineers styled themselves "telegraph engineers"; and the German pioneer of the electrical engineering industry, Werner Siemens, began his career as a military engineer concerned with telegraphy. But electrical engineering was to develop in response to a different stimulus: the invention and popularization of domestic electric lighting.

Another example was provided by the ubiquitous textile industry. The change from wood to iron in the construction of textile machinery was probably the greatest single factor in the rise of the machine tool industry, since most of the pioneer machine-tool inventors in Britain were, or at least began as, textile technologists.* By 1832 the manufacture of textile machines had reached such a stage of sophistication that Charles Babbage could report that machines commonly became obsolete long before they wore out, so great was the pace of innovation kept up by the machinery makers, particularly the firm of Sharp, Roberts. Three years later Andrew Ure could talk of the advent of the automatic factory, but this was premature.[9] In brief, the textile industry in England played much the same role as the firearms industry in the United States (see chapter 4). So, for example, Richard Roberts, maker of textile machinery and inventor of the self-actor mule, was also the inventor of the planing machine and the radial drill. Innovation in a key industry like textiles therefore breeds innovation in related and ancillary industries; the imbalance caused by the introduction of a new machine, process, or product necessitates further innovations to accommodate the newcomer satisfactorily and to restore, at least for the time being, a state of harmony in the system. The desideratum must be that the key industry is expanding.

Another industry that benefited from the change from wood to iron and the associated development of engineering technology was agriculture. As the catalogs of the international exhibitions from 1851 on indicate, a wide range of agricultural machines made of iron appeared during the first half of the century. Harvesting, which from time immemorial had been by hand, was mechanized. The first

---

*For example, Joseph Whitworth, Richard Roberts, William Fairbairn, James Nasmyth, Joseph Clement. The centers of the machine tool industry were Leeds, Manchester, Glasgow, and London—all, except the last, textile cities.

reasonably successful steam plow appeared in 1832; at the International Exhibition of 1862, held in London, no fewer than nine British firms displayed steam plows. But British agriculture was open to scientific as well as to engineering innovation. Humphry Davy and Justus Liebig had popularized agricultural chemistry. In 1838 the Royal Agricultural Society was founded with the specific aim of encouraging scientific agriculture. At about this time J. B. Lawes and J. H. Gilbert, both of whom had studied under Liebig at Giessen, began their collaborative researches.

The Rothamsted research station, which effectively began in Lawes's home, continues their work on a vastly wider scale. This, in short, was the era (1837-74) of "high farming," when a combination of technical and scientific advance with the growth of large industrial towns and the spread of the railroads resulted in great agricultural progress and prosperity. However, after 1862, "the tide of agricultural prosperity . . . ceased to flow; after 1874 it turned and rapidly ebbed."[10] There were several causes: the long recession that began in the 1870s and, more particularly, the repeal of the Corn Laws in 1846; the development of the vast agricultural potential of North America; and the appearance of efficient, economical oceangoing steamships from about 1855.[11] The last is a good instance of a significant innovation in one industry inhibiting progress in another. It took two world wars and the threat of starvation by U-boat blockade to restore progress and prosperity to British agriculture.

The mid-Victorian period is commonly, and no doubt correctly, thought of as the age of the steam engine, which even had its poet (a bad one). Water-power resources in Britain were under pressure by the end of the eighteenth century; accordingly the steam engine was developed for a wide range of manufacturing purposes. This, as Rosenberg points out in chapter 4, was in contrast with American development, where the emphasis was on transport. The stationary steam engine in Britain reached two separate peaks of specialized perfection: the Cornish mine-pumping engine and the textile-mill engine. The development of an efficient and economical steam engine for sea transport was delayed by inadequate scientific theory and by the understandable emphasis laid on safety and simplicity. After the introduction of the high-pressure compound expansive engine, through the efforts of John Elder, W. J. M. Rankine, and others, the economy of marine steam engines is said to have doubled in a decade, and to have doubled again in the following decade.[12] But large power units were not the only ones in demand. As Charles Babbage pointed out in 1851, echoing the remarkably similar words used by Benjamin Cheverton 25 years earlier,[13] the small workshop urgently needed an engine that could be turned on or off as required, and that would not consume fuel when not in use.

Babbage suggested the construction of a network of underground mains to supply water at high pressure. Customers connected to these mains could use the water to drive small engines of a few or fractional horsepower. This largely forgotten branch of the power-supply industry was developed to some extent in Europe and America in the second half of the nineteenth century.

But mains of another sort had already been laid down in towns and cities: these supplied the coal gas increasingly used for domestic and factory lighting. Coal gas can be used as a fuel for internal-combustion engines; accordingly, from about 1860 small Lenoir and, later, Otto gas engines were manufactured in large numbers, particularly for those using small amounts of power. History repeated itself, for with the rise of the electricity-supply industry, following the invention of the carbon filament bulb in 1879, extensive electricity mains were laid to meet domestic demand. These had the incidental effect of making the electric motor feasible as an alternative power source for the small, and then the big, consumer. So, in the cases of both the gas engine and the electric motor, the existence of a network of mains to supply domestic illuminants was a prior condition for successful development. Since the electric motor has now displaced both the high-pressure water engine and the gas engine, and since most electricity is generated by steam power—whether the fuel be coal, oil, natural gas, or nuclear—it would be correct to say that the steam engine has triumphed over all its rivals except the hydraulic turbine.

A marked feature of the mid-Victorian age was the immigration of talent to the new industrial areas. Once again the process was cumulative. Able young technologists were attracted to the centers of innovation, no doubt partly because their abilities would have the widest scope in such areas, and partly because they could hope to make a lot of money. The first wave of immigrants included a high proportion of Scotsmen, the second a high proportion of Germans. The common factor, apart from any genetic considerations, seems to be that while Scotland had the best universities in Europe at the beginning of the nineteenth century, by the middle of the century German universities were increasingly taking the lead.[14] Indirect confirmation of the last point is provided by the fact that more and more British students attended German universities as the century progressed. This was particularly true of scientists, especially of chemists.

Nevertheless, the British had some cause to congratulate themselves by the time of the International Exhibition at London in 1851. But thereafter a certain unease can be detected in some of their observations on industrial progress. The French were feare , and the German states plainly had some things to teach the world.

The significance of the latter became increasingly clear in the following decades, most notably through the development of the synthetic dyestuffs industry. The story is well-known, so only the briefest outline will be given here.[15] In 1856 William Henry Perkin, a youth of eighteen, synthesized the first of the aniline, or coal-tar, dyes while working in London under the direction of A. W. von Hofmann. Perkin immediately went into business as a manufacturer of the new dye, mauve, and soon made a fortune. It should be noted that, in Perkin's favor, coal tar was abundantly available from the growing British gas industry and that the market, the British textile industry, was not only the greatest in the world but was also by then well accustomed to technological innovation.

Perkin's discovery was soon followed by many others. The landmarks were Emanuel Verguin's synthesis of magenta; the synthesis of the azo dyes in 1865, which made possible practically unlimited numbers of new dyes; the synthesis of alizarin (the coloring matter in madder) in 1869; and, by 1892, the synthesis of indigo. By the beginning of the twentieth century, natural dyestuffs had been replaced almost entirely by synthetic dyes, with the consequent destruction of the extensive madder- and indigo-growing industries. Two features of this development make it particularly interesting. First, the dyes were the products of laboratory science; that is, they were synthesized in university, and later university-type, institutions of the kind so successfully pioneered by Justus von Liebig at Giessen. And second, by about 1900 something like 90 percent of the dyestuffs used in Britain, where the textile industry was still very large, were made in Germany. The industry begun in Britain had emigrated to Germany in the space of less than forty years. The relatively small German dyestuffs firms had been able to establish research laboratories in their works and to staff them with research chemists holding university degrees.[16] These were, beyond reasonable doubt, the world's first industrial research laboratories. From the time of their establishment there has been a steady growth and development of such institutions all over the world. It is reasonable to conclude that without these laboratories and the men to staff them, the German dyestuffs industry could not possibly have grown to the position of world domination it achieved by 1914.*

The chain effect, so far as Britain was concerned, was broken— even though its textile industry had from the beginning been willing

---

*German dyestuffs firms were said to have been patenting an average of one new dye per day by 1914. This could not have been achieved without a large staff of scientists and technologists.

to accept the new dyes and the raw material, coal tar, was abundantly available. Germany increasingly took the place of France as the rival to be feared, and with justification, for it soon developed a range of new, highly scientific industries. Closely following dyestuffs came pharmaceuticals, photographic equipment, camera lenses, optical glass, scientific instruments, and electrical equipment of all kinds. Many, perhaps most, of these industries stood in the same sort of chain relationship that the textile and related industries had had at the beginning of the century in Britain. The main difference was that the new industries in Germany had a much higher, and indeed institutional, scientific content. In other words, the R&D equation was now complete, albeit on a small scale, judged by modern standards.

The evidence of German success was too convincing to be denied in Britain. Royal commissions were set up to investigate scientific education and to suggest ways in which it could be improved. Groups such as the Society of Arts and the British Association carried out inquiries of their own. But it is doubtful that concern among the general public ran very deep; after all, few, if any, British fortunes had been lost as a result of Germany's successes, and there was little, if any, technological unemployment as a consequence. The industries were all new ones, apart from dyestuffs; and the natural materials they replaced were mostly grown abroad. It is likely that the passing of the Trade Marks Act in the last decade of the nineteenth century, which made it compulsory for each and every article to indicate its country of origin, had at least as deep an impact as the rhetoric of the scientists, for it revealed the extent to which Germany had penetrated British markets on a wide front as well as on the narrow R&D one.[17]

There was no systematic attempt to discover how and why Germany had been so successful in applying science to industry.[18] The excellence of German universities and technical high schools was acknowledged, and the large numbers of well-trained students that they graduated each year seem to have been accepted as self-evident causes of German superiority. In this the British observers were probably correct, but there was more to it than that. The consequent appeals to British manufacturers to "value" science, to "incorporate" it in their manufacturing processes, and to provide employment and incentives to young men who had scientific qualifications were misconceived and, in the circumstances, naive. The government response was to create a system of technical colleges where there had been none before,[*] to encourage the foundation of new universities, and

---

[*]The Technical Instruction Act, passed in 1889, empowered municipal authorities to impose a local tax to pay for technical colleges.

to reform secondary education. All this was overdue, perhaps by as much as 50 years; but the real question was simply whether it was not too late, at least as far as the "lost" industries were concerned.

It cannot be doubted that a supply of well-trained scientists and technologists had been a necessary precondition of Germany's success, and that its unintentional discovery of R&D[19] had been a consequence, not a cause, of this supply. German universities had been training appreciable numbers of chemists and physicists long before there was a science-based industry in the country, before German industry was capitalized.[20] German science graduates were destined for secondary or tertiary teaching, or for some branch of state service. As for the technical high schools, they began as junior technical schools; but the standards of the top ones were progressively raised until, at the end of the century, eight of them were elevated to the status of autonomous, degree-awarding universities. While the course of German educational reform is fairly clear, it did not follow that the same principles and practices would work in Britain. There were significant differences in national styles of education: the British undergraduate student was motivated by the written examination, too often a memory test, on which the class of his final degree depended, while the German student was inspired by the requirement to carry out adequate research for the Ph.D. degree. The latter, it may be inferred, was a better training for the would-be industrial researcher, the former better adapted for squeezing a modicum of work out of young gentlemen like Edward Ferrar.

Beyond these differences in practice lay significant differences in the social philosophy of higher education. Germany had inherited a number of universities from the independent states that preceded the formation of the German Empire. There were certainly variations in quality among them, but no one questioned their general excellence. In England, on the other hand, where national education had never been a matter of urgent priority, there were two ancient universities, now substantially reformed as regards the inclusion of science and other "modern" disciplines but still basically dedicated to the ideals of a liberal education and inaccessible—such was the paucity of scholarships, the cost of higher education, and the inadequacy of secondary education—to the children of the vast majority of the population. Besides Oxford and Cambridge there were a number of new, struggling, and underendowed provincial universities (including, oddly enough, London University) whose resources and prestige were markedly lower than those of the two ancient universities.

No less important was the question of timing. England, at the beginning of the twentieth century, was trying to create a comprehensive system of education that its main competitors had completed decades earlier. It would be reasonable to suppose, although it requires confirmation by further study, that Germany, deploying a vastly superior labor force, had already captured the highly scientific industries that were, and with one new addition still are, most susceptible to advancement through intensive R&D, and that therefore competition in these fields was no longer feasible, at least under normal circumstances. Control, from basic research through sales, seems to have been complete.[21] Effective challenge of this domination would surely have required appropriate tariffs and/or government subsidies.

The new technologies apart, however, there remained many opportunities for British firms in the more science-based branches of industry. But it is interesting to note that by this time, the end of the nineteenth century, many of the more scientific firms in Britain seem to have been controlled, if not by Germans, at least by people from countries that accepted the German educational system: Alsatians, Austrians, Poles, Russians, Swiss. For example, in the Manchester area important firms were founded by Charles Dreyfus, Ivan Levinstein, Ludwig Mond, Hans Renold, and Heinrich Simon, to name only a few of the leaders. All these men had been educated in German, or German-type, universities or technical high schools; and their employees frequently included German or Austrian technicians, draftsmen, and commercial clerks.

Such firms apart, a formally trained British scientist seeking an industrial post would generally find himself confronted by an array of established industries that would not, for a long time to come, be R&D-minded, and that were not, even in Germany, conspicuous employers of scientists. A few firms, notably in the iron and steel industry, possessed laboratories but these were invariably for testing and control purposes; the standard of scientific knowledge required was low and the work routine. For the rest, the situation was summed up by the historian H. A. L. Fisher when he remarked, "We are an old country, of old and small industries."

So far I have discussed mainly the most scientific, the most truly "research" aspects of R&D. If we broaden our considerations to include the more general range of science-based and pragmatic inventions, we find that the German lead is less well-marked. In the case of the heat engine, for example, British leadership had been unquestioned up to about 1860.[22] Thereafter the main innovations, with the possible exception of the steam turbine, came increasingly from the United States, Austria, France, and Germany.

The pace of innovation in production engineering and the invention of labor-saving machinery seems to have been set by the United States. H. J. Habbakuk has argued that due to the relatively high price of skilled labor and the competitive cheapness of land, the United States was practically forced into becoming a land of labor-saving machinery.

Certainly this was true in the important industry of agriculture. Barbed wire and the McCormick reaper, which aroused such great interest at the International Exhibition at London in 1851, are cases in point. On the other hand, the excellence of the American firearms industry was officially recognized in Britain only a few years after 1851, with the consequence that the Royal Ordnance Factory at Enfield was set up on the American model. In fact, the development of the American system is of particular interest from the British point of view. [23]

During the first decades of the nineteenth century, and particularly at the time of Babbage and Ure, the British textile industry had provided an object lesson in high productivity and in stimulating innovation in related industries. America also achieved a high standard in production engineering; unlike Britain it continued until, by the beginning of the twentieth century, it was the leader in mass-production technology. There were, of course, significant differences between the firearms and the textile-machinery industries. While the first catered to mass markets, the latter made highly specialized products for a very limited market. Moreover, textile technology seems to have reached a plateau of excellence about 1835; over a period during which the soldier's weapon changed from the muzzle-loading musket to the light machine gun, such textile machines as the carding engine, the self-actor mule, and the Jacquard loom remained virtually unchanged. The pressure for innovation that the textile-machinery industry had exerted on related industries must therefore have slackened after about 1835, the year in which the self-actor mule was patented.

Besides the structural and technical differences between the two key innovative industries, it should also be remembered that more general comparisons were unfavorable to Britain. For instance, some 27 years—a generation—elapsed before Britain enacted legislation remotely comparable with the Morrill Act (passed in the United States in 1862). The standard of science taught in the early land-grant colleges may not have been particularly high; but it seems likely that, as in eighteenth-century England, it went a long way.

An interesting consequence of American supremacy was that a number of firms engaged in mass-production industry in Britain were subsidiaries of American corporations. Indeed the adjective "British" before the name of such a firm indicated not its nationality but the fact that it was a British subsidiary of an American firm. Before World War I we had, for example, the British Westinghouse Company and British Thompson-Houston. [24] Another well-known American firm to establish a British subsidiary was the Singer Sewing Machine Company, whose factory at Bishopton, outside Glasgow, was said to be the most advanced mass-production plant in Britain, if not in Europe. A few years later the Ford Motor Company established a factory at Manchester and later at Dagenham, near London. These American firms were not, generally speaking, engaged in science-based industries. And they were not the only ones to establish subsidiaries in Britain before 1914: the Mannesman brothers had a tubemaking plant in South Wales; Nobel Explosives had a factory at Ardeer, near Glasgow; and the Siemans family had their works at Woolwich.

It would not be unfair to say, therefore, that just before 1914, Britain was importing a good deal of its advanced technology, either through skilled immigrants or by means of the subsidiaries of foreign multinational enterprises. This is what the country had done in the seventeenth century, with this difference: earlier it had been a deliberate government policy to import foreign technology, embodied in skilled individuals, whereas now it was an unintentional consequence of what may well have been a growing technological backwardness on the part of a still very wealthy country.

The facts that now seem obvious aroused some anxiety among scientists at the time; hence the foundation of the British Science Guild in 1905, mainly through the efforts of the astrophysicist Norman Lockyer. If we wonder that the anxiety was not more widespread, it is well to remember that R&D was a very new activity, hardly represented at all in Britain, and in any case affected only a small sector of industry. Perhaps, too, nineteenth-century economics had become unduly abstract and prone to see the businessman as motivated by strictly rational and narrowly defined considerations, unwilling to accept that apparently noneconomic factors such as science or education could affect economic activities. Perhaps the English, the first people to industrialize, had never really appreciated technology; perhaps they had taken it too much for granted, never having seen it, as the Continental nations saw it, as the way to national supremacy. As J. T. Merz pointed out at the time,[25] the British had been tardy in establishing a national policy, or a national philosophy, for science; and he hinted that, unlike the French and the Germans, they had still not succeeded in doing so by the beginning of the twentieth century.[26]

Besides these administrative, intellectual, and sociological factors, there was another feature of British policy that may well have helped to aggravate any technological lag. But here a caveat is necessary. As Karl Popper has pointed out, it is the easiest thing in the world to find confirmatory evidence for any particular view or prejudice. Doubtless, if as an exercise one selected the economically most successful nation in the world—however success is to be defined—it would be possible to draw up a plausible list of reasons why that country should not be successful. Nevertheless, with this warning

in mind, one possible contributory factor to Britain's relative technological decline must be mentioned: the movement called imperialism.

As preached by Joseph Chamberlain and as practiced by Cecil
Rhodes, imperialism was an unpleasant creed; "painting the map red,"
and the wild dream of a Cape-to-Cairo railroad built entirely on
British-ruled land, now seem more absurd than vicious.[27] But imperialism in theory and in practice may have had two adverse effects
on British technology. In the first place, it encouraged investment
in overseas ventures rather than in productive industry at home. It
is not denied that British manufacturing industry may well have
benefited (in the short run, at any rate) from overseas investment;
but the effect could well have been the creation of protected markets,
with a mitigation of those pressures of consumers on suppliers that
so often stimulated, and stimulate, technological innovation. It is
not necessary to enlarge on the consequences of complacency.[*] Of
course it must be proved that overseas development itself denied
capital for innovative, or potentially innovative industries at home.
But one need not assume that the investor played an entirely passive
role. It seems to have been true that many of the more energetic and
aggressive members of the investing public thought that their best
chances of making a fortune were in South African diamonds rather
than, say, in British fine chemicals. And if these investors and spec-
ulators could drive the country into a futile and immoral "Boer War,"
what might they not have achieved if their energies had been directed
into more productive channels at home?

The pervasive ideology of imperialism may have been no less
important, although its effects are harder to assess. The ideology
itself is fairly clearly revealed in countless stories, anecdotes,
plays, songs, and novels; and we can be sure that it had due effect
on education. The many public (that is, private) schools founded in
the nineteenth century to cater to the needs of the affluent middle
classes undoubtedly preached the gospel of empire, and of service
to the empire, to their impressionable charges. If in the 1840s the
Cambridge mathematics tripos was the most prestigious examination
in the country, by 1914 the entrance test for the Indian Civil Service
was the most exalted, the most admired, and the one most ardently

---

[*]A number of years ago a director of a British group that had
included in its activities the manufacture of locomotives for export
remarked to me, "When we had an Empire the colonies used to be
glad to buy their locomotives from us; now that they've all got their
independence they buy them wherever they like." The firm in question
no longer manufactures locomotives.

prepared for. The attractions of the Indian Civil Service were obvious and powerful. I suggest that, ceteris paribus, the mills of Lancashire and Yorkshire, the machine shops of Birmingham and Manchester, the chemical works of Widnes and Tyneside, could not offer careers that could compete with the mystique of empire and the more tangible attractions of, for example, the Indian Civil Service.

The cult, for such I think it must be called, of empire survived World War I, although in attenuated form. Its final demise was marked by Prime Minister Harold Macmillan's "Wind of Change" speech.

WORLD WAR I AND AFTER

The outbreak of war in 1914 and the immediate severing of trade links with Germany made painfully clear the extent to which the latter country had taken over British markets, particularly in goods with a high scientific content, many of which were vital for the war effort (such as magnetos for automobile engines, all made by Bosch of Stuttgart, gunsights for British artillery, tungsten for high-speed steel). Steps were immediately taken to overcome the most critical shortages; committees were formed to deal with the problems posed by the cutting off of supplies of drugs and pharmaceutical products, optical glass, and various chemicals. Plans were made to set up new, scientific industries, immune from German competition; in particular a national dyestuff corporation, British Dyes, was founded with state backing. In 1916[*] a radical initiative was implemented when the Department of Scientific and Industrial Research (DSIR) was set up under the control of the Privy Council. It was empowered to grant studentships and fellowships for research, the recipients of such grants to be nominated by responsible scientists or technologists, and not by politicians or officials. DSIR was expected to initiate researches on its own account when the national interest demanded,

---

[*]In the same year, a very interesting if little-known book was published: H. B. Gray and S. Turner, Eclipse or Empire? (London: Nisbet, 1916). The authors used a case-study technique, very similar to that used by John Jewkes, David Sawers, and Richard Stillerman forty years later in The Sources of Invention (London: Macmillan, 1958), to prove that the great majority of modern innovations were made and developed outside Britain. Needless to say, they prove their case. The title of the book is rather misleading: the empire is hardly mentioned, nor is it relevant to the argument.

and it took over the National Physical Laboratory (instituted in 1900 in imitation of the Reichsanstalt) from the Royal Society. It also took over certain museums and other establishments, and it set up the Radio Research Board, the Fuel Research Board, and the Forest Products Research Board.

At the same time a number of research associations were organized. These, which seem to have been copied from certain pre-war German organizations,[28] were associated with specific industries or branches of industries; the intention was that each should provide research facilities or carry out research for its associated industry. The state, through DSIR, participated by paying up to half the costs of running each association, the industry being required to find the remainder. It was hoped that within a few years the research associations would be self-supporting, but this hope has not been fulfilled; they are still dependent on the state for about 25 percent of their running costs. In fact, there seems to be a certain ambiguity about the functions of a research association (there are now over 40 of them), for if the relevant industry is highly scientific, the firms in it will hardly need the services of a research association; if the industry is inefficient, unresponsive to R&D, or by nature has few direct links with science, the firms concerned will probably fail to recognize the value of their research association. This is not a criticism of the research associations. It is simply an observation that they do not, in themselves, constitute a solution to the problem of ensuring a wider diffusion of science in less progressive industries.

The last consequence of World War I that deserves mention was the institution by British universities of the Ph.D. degree. The first was initiated in 1918 by Oxford University (where the abbreviation is, uniquely, D.Phil.). W. H. Perkin, Jr., son of the discoverer of the first coal-tar color and himself a graduate of a German university, was instrumental in securing this reform. He was at that time a professor of chemistry at Oxford.*

---

*A factor in this reform was the desire to be of service to the English-speaking academic world. Many young Commonwealth and U.S. soldiers who had passed through Britain during the war were graduates, and some had wanted to return afterward to carry out postgraduate studies. Unfortunately there were not, at that time, suitable postgraduate degrees available. The other main factor was, of course, the long-standing discontent on the part of scientists like Perkin with degree courses that made no provision for research training.

From a narrow, specialist point of view, the war was not without incidental benefit for Britain. It made possible the establishment of a domestic organic chemicals industry, as well as the fostering of other highly scientific industries; it precipitated the foundation of necessary state institutions, such as DSIR, for the advancement of applied and basic research; it confirmed the value of science in industry for those who might otherwise have remained skeptical; and in the peace that followed the armistice, possible German trade rivalry was set back many years. But in all other respects it was an unmitigated disaster.

The most obvious item on the debit side was the slaughter of irreplaceable scientific talent. Whether Britain, where conscription was not introduced until 1916, suffered more in this respect than other countries may be debated. But all the combatant countries suffered appalling losses that must have included many able young scientists, and even more potential young scientists. The death in combat of one particularly talented young man, H. G. J. Moseley, provoked his mentor, the influential Ernest Rutherford, to write an angry and bitter obituary in Nature. In all probability this had a lasting effect, for certainly the death of Moseley has a place in the folk memory of British science.*

And with the destruction of so many young scientists there went the less obvious dismemberment of the old and complex network of science in Europe that had endured since the Middle Ages and that even the Napoleonic Wars had hardly touched. British scientific achievements depended directly, sometimes very directly, on previous European achievements; the converse was also true, of course. Since by 1914 Germany was, by common consent, the leading scientific nation in the world,† it followed that British science was cut off from its most fruitful single source of outside inspiration. This continued to be the case after the armistice and the peace, for a vendetta against all things German continued until well into the 1920s.‡ As far as Britain was concerned, the United States took the place of

---

*Footnotes in textbooks on atomic and nuclear physics published in Britain during the interwar years refer sadly to Moseley's death at Gallipoli in 1915.

†Over the first quarter of the century, Germans received far more Nobel prizes for science than did citizens of any other nation. Even more weight can be put on the fact that by 1914 practically every professor of chemistry in Britain had a German Ph.D. degree.

‡It was not until 1927 that the first postwar German Rhodes Scholar arrived in Oxford.

Germany in due course; but this could not happen immediately, and the damage done to the European framework of science was incalculable.

Two generalizations that originated in World War I have passed into the mythology of British industrial science. The first, which has almost entered national folklore, arose from the startling discovery that Germany had, by 1914, taken over the most scientific industries; it was therefore to the effect that "we [the British] make the original inventions, but it is the foreigner that reaps the profit." Later this was modified into the quasi boast "We are very good at science, but no good at applying it." This unappealing combination of self-flattery and self-pity must be treated with considerable reserve. Scientific ideas cannot be applied anywhere unless there is a high degree of scientific competence in the firm or organization that effects the application. Further, and contrary to the common view, the actual first source of the vast majority of scientific ideas is extremely obscure. Almost invariably, and for reasons that are obvious, we are most aware of the contributions made by our fellow countrymen; the contributions made by foreigners, except the most illustrious, are taken for granted.* Finally, it must be remembered that other European nations have, at different times, claimed exactly the same quality (or lack of it) for themselves.

The other generalization, which on the face of it has more weight, arose in consequence of the first, unsuccessful attempt of the British government to launch British Dyes in 1915.[29] This is to the effect that only those firms with scientists and/or technologists on the board of directors tend to be technically progressive. The implication is that firms controlled exclusively by accountants, bankers, lawyers, stockbrokers, and (for respectability) a lord or two, will not be open to new scientific ideas. This is a reasonable proposition, although there are (and presumably always were) many notable exceptions; but it does tend to include an element of circular

---

*For example, although W. H. Perkin discovered the first of the aniline, or coal-tar dyes, he was one of the very few Englishmen working in a field of science (organic chemistry) that was dominated by Germans. Furthermore, he was working under the direction of a distinguished German chemist, A. W. von Hofmann. It is reasonable to assume that the roughly contemporary and quite independent discovery of magenta by Verguin would have had the same general effect as did Perkin's discovery of mauve. We must regretfully conclude that had Perkin never lived, it would not really have been necessary to invent him.

reasoning. A progressive firm, by definition, will be one that employs a number of scientists/technologists in research, design, and development. In due course some of these will be promoted through the management structure and eventually reach the board room. The argument therefore tends to the undeniable but unexciting conclusion that a progressive firm is a progressive firm. On the other hand, it is reasonable to assume that a scientist/technologist on the board of an unprogressive firm will find himself in the position of the only sighted man in the country of the blind; he will see where the others do not, but he will be quite unable to convince them of the validity of what he sees because they will lack his basic experience and understanding.

War, and the threat of war, have long been recognized as great stimulators of technical progress. Compared with what was achieved during the war years, the developments that took place during the peace that followed were of rather less importance. The Research Council of DSIR was complemented by the formation of the Medical Research Council in 1920 and of the Agricultural Research Council in 1931. While the last two councils have supported some extremely important researches,[30] their industrial and economic importance has necessarily been less than that of DSIR. The state also intervened in the cases of two twentieth-century innovations: commercial air transport and radio. Of the inauguration of Imperial Airways—the archaic title may be noted—one might say that rarely has a new technology been less happily conceived and applied than in the case of this airline.* On the other hand, it would be fair to say of the British Broadcasting Corporation that while its technological impact was indirect, it has proved its value in other fields: cultural, educational, and political. Curiously, these two enterprises were initiated at about the same time. The chosen vehicle for these state monopolies was the "public corporation," a form used by subsequent governments to "take into public ownership" such basic industries as atomic energy, coal, electricity, gas, and iron and steel.

---

*Imperial Airways provided an empire service. It favored large and comfortable, but somewhat ungainly and decidedly slow, airplanes bearing classical names, such as Hannibal and Heracles. The growth of commercial airlines serving mainly domestic routes led to the formation of a group, British Airways, in 1939. At the outbreak of war, the government merged the two to form British Overseas Airways. This state carrier has recently become British Airways, so that it may be said that all traces of Imperial Airways have now vanished.

During the same period, research laboratories were expanded and new ones were founded in the chemical industries, in electrical engineering, and in the new and rapidly growing electronics industry. British Dyes amalgamated with other chemical firms, particularly with the great alkali groups, to form Imperial Chemical Industries in 1926. But for the rest the country had its share, perhaps more than its share, of recession and industrial unrest. And the great industries of the Industrial Revolution and the nineteenth century sank into comparative stagnation.[31]

WORLD WAR II AND AFTER

There is no doubt that defense requirements, from about 1935 (the year in which the existence of the German Luftwaffe was publicly acknowledged), were the main motivating factor in the expansion of R&D in Britain. This is reflected in the allocation of resources. In 1928 expenditure on R&D amounted to 0.10 percent of the total national income; by 1938 it had risen to 0.25 percent, and by 1958 to 1 percent: a tenfold increase in 30 years.[32] This was a respectable performance, particularly since the national income itself had increased over these years. But it is not so impressive as the U.S. achievement, which showed a twentyfold increase between World War II and 1958.

The threat of the air weapon and the vulnerability of the enormous capital to air attack could not be ignored by the government. The development of radar (at first misleadingly called RDF for security reasons) and other electronic devices is too well known to need repetition here. The government, acutely conscious of the need to conserve and increase scientific manpower, had asked the Royal Society to draw up a Scientific and Technical Register as early as 1937. The register recorded all professionally qualified scientists, technologists, and engineers down to and including undergraduate students in their final college and university years. All were to be exempt from military service and to be directed into civilian work of national importance in the event of war. Presumably the lesson taught by Moseley's death had been well learned,[33] although it must, in fairness, be added that in the light of the experiences of World War I, a frantic rush to the colors, of the sort witnessed in 1914, was hardly likely, with or without a register. On the other hand exemption from military service may well have attracted into science and technology students who might otherwise have chosen different careers.

During a war basic research, in universities and elsewhere, comes to a halt and all facilities are turned over to the solution of more or less short-range problems. A nation at war, it has been remarked, lives on its scientific fat, on its store of basic scientific knowledge. There would be no point in attempting a brief assessment of Britain's scientific war effort. An enormous literature on the subject is readily available. Here, it is enough to record that the British emerged from the war well satisfied with the performance of their scientists and therefore, for the time being, convinced of the public importance of science. Not many, perhaps, realized that one of the greatest scientific and technological triumphs of World War II was the provision of a balanced and satisfactory diet for a large, densely populated country almost cut off by submarine blockade. One grumbled about the rations but was immensely impressed by the invention of radar.

At the end of the war the government acknowledged the importance of science by instituting the Scientific Civil Service, the (honors) graduate members of which (scientific officers and above) were ranked with the administrative grade, the senior branch of the Civil Service, while the less well-qualified (experimental officers and above) were of executive grade status. In 1947 the Advisory Council on Scientific Policy replaced the old wartime arrangement. Two years later a new research council, the Nature Conservancy (now the Environmental Research Council), was established; and in the same year the National Research Development Corporation (NRDC) came into being. The latter was charged with aiding, or making possible, the development of scientific ideas or inventions coming from various government laboratories, or brought to it by private individuals or organizations. It was felt at the time that Britain was not particularly efficient at turning scientific ideas to practical and profitable account, and that there were many good ideas awaiting development. Five years later an act of Parliament enabled the National Research Development Corporation to initiate research likely to lead to new inventions.

Further changes in the 1960s included the setting up of the Department of Education and Science with responsibility for the research councils—including the Science Research Council, the scientific component of DSIR—as well as for the University Grants Committee. The Ministry of Technology (later the Ministry of Trade and Industry) retained responsibility for industrial research that had formerly been undertaken by DSIR and for the research carried out by the United Kingdom Atomic Energy Authority. The various defense departments maintained their own independent research facilities. Broadly speaking, then, government research in Britain today is divided between academic, supported by the Department of Education and Science; industrial, by the Department of Trade and Industry; and defense, by the Ministry of Defense. Over all of these is the

Central Advisory Council for Science and Technology, the chairman of which is the chief scientific adviser to the government. The Central Advisory Council has a small, part-time membership and a small nucleus of officials, most of whom have other responsibilities. An impartial commentator has remarked: "It may therefore be doubted whether, although the United Kingdom took the initiative in nationwide scientific co-ordination, it has yet achieved—despite several changes of organization in its science policy—a unified decision-making system in this field."[34]

Whether innovation can be economically and efficiently initiated by government decree is extremely doubtful; we shall return to this question later. For a number of years after World War II the emphasis in Britain was on "pure" science, the theory being that modern innovation represents the practical consequences of "pure" research. This theory was probably based on observation of the success of the German dyestuffs industry before 1914, and on the development of radar and the atomic bomb during World War II. In these and other particular cases, the theory is correct; but it is not the whole truth, for innovation is a complex and many-sided activity. It has been pointed out that few farmers, for example, carry out R&D, much less indulge in "pure" research. But in Britain, as in the United States, they are very willing to innovate and to accept suitable inventions, whether science-based or otherwise.

SCIENCE AND BRITISH INDUSTRY

Since 1945 Britain has consistently trained more scientists and technologists, and has spent more money on industrial R&D, both absolutely and as a percentage of GNP, than has any other European country (excluding Russia and East European countries, for which comparisons are difficult). It is therefore quite reasonable to ask why its economic performance has been disappointing compared with those of its neighbors. Can the fault, or faults, lie with its administrative arrangements, with its industrial structure, or, deeper still, with its social philosophy and way of life? Or perhaps, as some would argue, is R&D not so important for economic growth?

It might be profitable to examine briefly the first major commitment of the NRDC. During World War II, America, Britain, and Germany developed specialized computers for a variety of short-term purposes. After the war a group of academics at Manchester University who had had experience in government laboratories and who had complementary interests in mathematics and in electronics set about developing a universal computer.[35] By the end of 1948 the project had aroused the enthusiasm of the government chief scientist, who initiated a government contract with a local electronics firm. From the time it was set up, NRDC gave the Manchester project full support; thus, when the first machine, the Ferranti Mark 1, was delivered in February 1951, it could be claimed that not only was it the first commercially available computer but also that it represented an object lesson in university-industry-NRDC-government cooperation. Although the Manchester group went on to design and develop larger computers, the NRDC was "not altogether successful" in its efforts to encourage an alert and competitive computer industry in Britain; it seems that "industry was too cautious in those days."[36]

This moderate assessment apparently confirms the common impression that British industry is unenterprising. But, if that is so, it is difficult to understand why important sections of it have adopted R&D so willingly since the war. Significantly, too, the pattern of distribution among industries closely resembles that of the United States and other advanced countries. In Britain, as elsewhere, R&D tends to be concentrated in large firms; in 1955, for example, 309 firms accounted for 80 percent of the manpower engaged in R&D.[37] And the distribution of intensity of effort, measured by relative expenditure and manpower concentration, also resembles that of the United States. First come aircraft and defense industries, followed by electrical engineering and electronics, chemical and allied industries (including drugs), precision and scientific instruments, and mechanical engineering. Apart from aircraft and defense, the leading scientific industries are much the same as those that were conspicuous in Germany at the beginning of the century. The profile of R&D has not, it seems, changed much since about 1900, although the scale is now vastly greater.

In the 1950s an intensive research effort was mounted in Britain, in an effort to throw some light on whether British industries were backward in applying new scientific ideas.* Led by two academic economists, C. F. Carter and B. R. Williams, the group had a research staff of eight and could draw on a wealth of expert and specialized advice. The core of the effort was a large number of case

---

*The Science and Industry Committee, sponsored jointly by the British Association for the Advancement of Science, the Royal Society of Arts, and the Nuffield Foundation, functioned from 1953 to 1958. Its main publications are C. F. Carter and B. R. Williams, Industry and Technical Progress, Investment in Innovation, and Science in Industry (Oxford University Press, 1957, 1958, 1959 respectively). I was a member of the research staff from 1954 to 1956.

studies of firms in a wide range of industries. The case studies, which involved loosely structured interviews and the examination of documentary material, revealed, as might be expected, a wide range of abilities, from the efficient and progressive to the inefficient and complacent. But the characteristics of the progressive British firms, as they were established, turned out to be much the same as those revealed by studies in other countries.

R&D, Carter and Williams argue, represents investment in creating opportunities for further investment.[38] To this end, R&D personnel are essential in any firm, for besides carrying out R&D, they can assess the possibilities in new scientific ideas. Preferably the R&D department should include people with as wide a range of interests as possible, so as to increase the possibilities of cross-fertilization. Carter and Williams observe that there is

> . . . a clear distinction . . . between firms that have research (or design) and development departments, and firms that lack them. The former are more prepared to venture; the reason is partly the confidence gained from research and development results, and partly that where the firm employs scientists and technologists it is more likely to take the view that, because a new process is based on sound principles then any teething problems can be overcome. Because the firm possesses technical and scientific staff, it will have a basic confidence in the innovation that the firm without such staff will lack.[39]

The effective use of R&D depends on management practice, for R&D is essentially a tool of management. But R&D cannot be organized rigidly, hierarchically.[40] Of its very nature it raises new problems that would pose serious difficulties for an established, inflexible organization. From the nature of their work, scientists and technologists must be encouraged to form links with outside bodies, to attend scientific conferences, and to belong to professional associations. In this respect an R&D department should be organized rather like a large university science or technology department. This is not entirely surprising, since the objectives and the procedures have a good deal in common. On the other hand, the final purpose of industrial R&D is not the advancement of knowledge; the firm exists to make the maximum possible profit consistent with certain social restraints. From this it follows that scientists and technologists engaged in R&D should be cost- and profit-conscious.*

---

*Perhaps, too, basic science might benefit by being more cost- and profit-conscious. It is difficult to accept that the principle

In order to make innovation effective, therefore, there should be the closest possible cooperation between the R&D department and the production, finance, and marketing departments. A system of informal committees, with strong horizontal links between departments, seems to be the appropriate solution.

It may be that this conclusion has some bearing on the history of the British nuclear power program. In 1966 a leading British physicist, P. M. S. Blackett, suggested that the British program had been a mistake.[41] Atomic energy R&D should, he argued, have been contracted out, as in the United States, and not been undertaken in government laboratories. Certainly the study in question indicated that the more direct, immediate, and personal the contacts between R&D and other departments, the better. And from this point of view it is difficult to see any advantage in a formal separation of research, to be carried out in government laboratories, and production, to be the responsibility of individual firms. But in the case of nuclear energy this may be too short-term a view of things.

Tension between R&D, on the one hand, and production, marketing, and finance departments, on the other, has often been commented upon. It is a very old story. During the eighteenth and nineteenth centuries, the conflict of interest between the perfectionist engineer and his partner, the profit-conscious businessman, sometimes caused the partnership to break up. This was not necessarily a disaster, for new and more fruitful partnerships could always be formed. Tension, in fact, is not always a misfortune; it can be a stimulus to progress, and with good management, the more abrasive and destructive kinds can be reduced or eliminated. It appears that personal tensions may be most common when new R&D departments are grafted onto old, established firms in more traditional industries. Here the newcomers may challenge, or seem to challenge, the know-how and the skills of the established staff, and so to threaten their status and positions. This is a difficulty that time will resolve, although effective management can reduce the time required. Institutional tensions can arise from the conflict of interests between R&D and production departments with their stringent requirements and strict schedules. These kinds of tension can probably be reduced, if not eliminated, by adequate use of informal, interdepartmental committees. Finally, there may well be people in R&D who want a degree of intellectual freedom that is not possible, given the aims and objectives of the firm. These are academics manqués, and the only solution in their case is to seek more congenial employment.

---

that knowledge can be advanced only by employing ever more expensive machines embodies any law of nature.

Apart from the effective management of R&D, there are a number of other factors that govern innovation. They include adequate market research, budgetary control, and the proper costing of new projects. There is additionally the need to recruit and retain first-class people. All these are essential, for failure is self-perpetuating.[42] There is, as it were, a "technology trap." An unsuccessful firm in a static or declining industry will not attract good people, will not earn the profits needed to support R&D, and therefore will be at the mercy of whatever (unforseen) changes occur. It is likely that it was this "technology trap," this self-perpetuation of failure, that made it impossible for British firms to compete with more scientific German firms in scientific industries at the beginning of the twentieth century.

Studies therefore confirm that firms efficient at applying R&D exist in Britain as elsewhere.[43] The question is whether there are too few efficient firms, and if so, why? There are identifiable shortcomings. A relative abundance of scientists has highlighted a shortage of production engineers and, more particularly, of technicians. Carter and Williams observe: "We believe that in Britain there is an unreasonably high ratio of unconsidered projects, due to poor organisation of research and development, to there being too little of it in private industry, and to the shortage of trained men and women."[44]

In addition there is the likelihood that too many British firms deploy their scientists in an inefficient manner, concentrating them in R&D and having too few in production: "British industry employs far more of its scientists and engineers within special departments than American industry does; we think that this concentration is sometimes a source of weakness."[45]

It seems likely that further insights and understanding will require extended studies of the performances of British firms and industries in comparison with their counterparts in other countries.

SCIENCE AND BRITISH GOVERNMENT

During the eighteenth century the scope for state intervention in industry was limited by the absence of an effective bureaucracy and by the inclinations of government itself. The sponsoring of a few scientific expeditions; the facilitating of a few major projects, such as the construction of canals; and an occasional intervention in patent procedure, such as the extension of the patent granted to James Watt in 1774, were the most that could be expected. The Royal Society, nominally the government's scientific advisory body, was substantially a club for gentlemen, the nonscientific fellows considerably outnumbering the practicing scientists. As the Industrial Revolution got under way and its practical benefits became apparent, inclination and economic doctrine combined happily under the banner of laissez-faire. But the triumph of the manufacturing interest during the first two or three decades of the nineteenth century could only be short-lived. The rapid and unsupervised growth of the industrial towns, with the accompanying influx of displaced, and therefore desperately poor, agricultural laborers soon posed health problems that could not be evaded. At the same time humanitarian sentiment was offended by the spectacle of children working many hours a day in mine and factory. The period of intervention began.

Intervention and humanitarianism led inevitably to the extension of the franchise; that is, to democracy. Some members of the ruling class, Robert Lowe for example, were worried by this; they need not have been. The insistence by the progressives on the written examination as the sole test of merit, and therefore of fitness for office and professional qualification, meant that it was possible to judge absolutely impartially between the claims of (say) a duke's son and the son of a millionaire. But the great majority of the population could never aspire to such tests; the cost of education, particularly university education, ensured that. It may well be, therefore, that midnineteenth-century reformers, in their zeal for impartiality, actually restricted social mobility in some key sectors of society. And the extreme tardiness with which Britain introduced a national system of education ensured that the values of the gentleman amateur, the classical scholar manqué, persisted for a long time in the higher Civil Service. Some engineers and scientists believe that they persist today.

Civil servants, classically educated or otherwise, cannot be blamed for the excessive centralization that characterizes modern Britain. To some extent the country has been the victim of circumstances over which it had little control. The development of the railroad system and geography dictated that all routes should lead to London. Two world wars tended to concentrate economic administration in London. And, when the first post-World War II Labor government nationalized certain basic industries, few, if any, queried that the headquarters of the new public corporations should be in London. Even the National Coal Board has its headquarters in London, though very little coal indeed is mined in southeastern England. Under these circumstances it is hardly surprising that the large private corporations increasingly followed suit and moved their head offices to London. The net result has been the gradual impoverishment of the provinces and serious social problems in the distended capital city.

It is in principle difficult to see how innovation can flourish when more and more decisions have to be referred to London. As we have seen, it is essential that if innovation is to be effective, there should be close and continuing collaboration between the departments responsible for R&D, finance, sales, production, and sometimes personnel management. In general, then, the involvement of a possibly remote, certainly separate, authority or authorities, implicit in the concentration of power in London, may well retard the process of innovation. Recognition of this is one of the causes of the movement toward devolution in Britain today.

If these arguments are correct, it follows that at no stage would a coherent and strong policy for R&D have been sufficient to ensure the competitiveness of British industry. Before World War I it would have been pointless, in the absence of a parallel and complementary policy of raising the standards of production, sales, and other technical staff. After World War I it would have met the difficulties discussed above. R&D, we may conclude, is, like patriotism, not enough.

SUMMARY AND CONCLUSIONS

Britain's industrialization, which began in the eighteenth century, owed little directly to science. It was made possible by an extremely favorable geographical position, abundant supplies of raw materials (particularly coal), a benign political system that ensured a high degree of personal liberty although it denied full political rights to a large section of the community, a plentiful labor force, a highly efficient agricultural system, and, finally, perhaps as a degree of native ability and the existence of suitable skills in the working population.

The system was remarkably successful for about a century—so successful, in fact, that any substantial modification was unthinkable. In consequence only the minimum of necessary reforms were made, and then only in response to intense pressure. This meant that Britain did not have an educational system appropriate to the requirements of newly emergent scientific industries toward the end of the nineteenth century. Accordingly, the class of professional scientists and technologists appeared later in Britain than in its main European rivals and in the United States. A distorting factor that arose at the end of the nineteenth century was the doctrine of imperialism, preaching a national philosophy in which science and technology had little or no part.

Failure is self-perpetuating. Two world wars gave Britain chances to make good its failures, and in the main this was done.

But it may well be wondered whether enough was achieved, whether just catching up was sufficient. It has been said that British industry gets its fair share of top scientific talent, and this may well be so; but the question is whether enough of the ablest people seek careers in technology, as distinct from science, and in industry in general? No authoritative answer is possible. It may be sufficient to point out here that while statesmen and other public figures sometimes preach the virtues of technology and the urgent need for more and better technological education, they rarely practice what they preach. British education has still its irrational elements. The university with the greatest prestige in Britain is world-famous for its schools of history, philosophy, literature, chemistry, and physics; but it is in no way renowned for technology. When the first of the post-World War II universities was founded in 1949, technology was excluded from the syllabus, although the university is very close to a large industrial area. The most recent university founded in Britain, the University College at Buckingham (1972), also makes no provision for technology in its syllabus, although this may be an unavoidable consequence of its deliberate repudiation of state support (faculties of technology are extremely expensive). Against these rather depressing facts may be set the more hopeful one that a number of new universities that do not spurn technology were founded in the 1960s, including some nine technical colleges that were raised to university status. The full benefits of these long-overdue reforms have yet to be realized.

Industry still has a bad image in Britain. It has been bitterly criticized by eloquent historians, novelists, playwrights, and commentators of all sorts, none of whom has ventured to suggest alternative means by which a densely populated country could earn its daily bread.* Not since the days of Charles Babbage, Samuel Smiles, and Andrew Ure have the virtues of technology and productive industry been extolled by popular writers. Lack of enthusiasm for industry and technology is not difficult to understand, however, when one recollects the slums, the pollution, and the smogs that have only recently been conquered.

It used to be said that Britain gained its empire in a fit of absentmindedness. It would be equally true, and today much more significant, to say that Britain embarked on the Industrial Revolution without knowing what it was doing. The British, having been first in the field, have never really appreciated or understood that efficient

---

*Particularly influential have been the historians R. H. Tawney and J. L. and Barbara Hammond.

industry is essential if a nation is to prosper and to have influence in the world. This failure of understanding, particularly at the top level, was noticed by Charles Babbage in the 1850s, by J. T. Merz at the beginning of the twentieth century, and most recently by Carter and Williams:

> We doubt if it can be said that a government policy on the application of science really exists. The facts on which such a policy should be based have never been collected or assessed. Substantial research is needed; the right answers will not be reached by eminent scientists or economists in a few days work in a government commit-tee, nor will they naturally arise from the present organisation of the departments concerned. This is because the problem is both technical and economic and it requires extensive and continued study by a group containing both scientists and economists. We do not think that such a group exists at present; and without it policy making tends to go no further than a statement of generalities.[46]

British scientific industry is therefore the product of a long historical development, a fact that must be kept in mind in any discussion of British industrial performance today. Many different things have contributed to make it what it is, and a deeper understanding will come only with further research and scholarship. The desideratum, as far as I can see, is an adequate set of definitions of the processes involved in R&D. It cannot be satisfactory to treat such a complex activity as if it were a set of routine and simple operations. How many different kinds of research are practiced in industry? how do these relate to the different types of industry? to what extent is development a systematic procedure? When these and other questions are more fully answered—and I suggest that philosophers as well as scientists, technologists, and economists may have something to contribute here—we shall surely have a better understanding of the nature, the problems, the powers, and the limitations of R&D.

NOTES

1. S. C. Gilfillan, Inventing the Ship (Chicago: Follett, 1935), and S. C. Gilfillan, The Sociology of Invention (Chicago: Follett, 1935).

2. Arnold Pacey, The Maze of Ingenuity (London: Allen Lane-Penguin, 1974), p. 207.

3. Lord Ernle, English Farming, Past and Present (London: Longmans Green, 1936), pp. 176-89.

4. R. L. Hills, Power in the Industrial Revolution (Manchester: Manchester University Press, 1970), gives an excellent account of the development of textile machinery.

5. F. B. Artz, The Development of Technical Education in France, 1500-1850 (Cambridge, Mass.: M.I.T. Press, 1966); M. P. Crosland, The Society of Arcueil (London: Heinemann, 1967); and Robert Fox, The Caloric Theory of Gases from Montgolfier to Regnault (Oxford: Oxford University Press, 1971), give excellent accounts of the development of education and science in France at this time.

6. For an account of the development of the Cambridge mathematics tripos, see D. S. L. Cardwell, The Organisation of Science in England (2nd ed.; London: Heinemann, 1972).

7. R. E. Schofield, The Lunar Society of Birmingham: Provincial Science in Eighteenth Century England (Oxford: Oxford University Press, 1963), gives an excellent account of the organization and work of this important society.

8. W. Cooke Taylor, Notes of a Tour in the Manufacturing Districts of Lancashire (London: Duncan & Malcolm, 1842).

9. Charles Babbage, The Economy of Machinery and Manufactures (London: Charles Knight, 1832); Andrew Ure, The Philosophy of Manufactures (London: Charles Knight, 1835). Ure's book was to some extent a copy of Babbage's work.

10. Ernle, op. cit., p. 377.

11. D. S. L. Cardwell and Richard L. Hills, "Thermodynamics and Practical Engineering in the Nineteenth Century," History of Technology 1 (1976): 1.

12. Ibid.

13. I am indebted to Keith Hutchison for calling my attention to Cheverton's suggestion. See Benjamin Cheverton, "New Gas Power-Engine," Mechanics' Magazine 5 (1826): 326.

14. On the supremacy of Scottish universities in the eighteenth century, see S. D'Irsay, Histoire des universités (Paris: A. Picard, 1935), II, 290 ff. The later supremacy of German universities is confirmed by the notes, speeches, lectures, autobiographies, and other writings of the many English-speaking scientists who, in their youth, attended them.

15. This is dealt with in standard histories of chemistry in the nineteenth century. See in particular W. V. Farrar, "Synthetic Dyestuffs Before Perkin," Endeavour 33 (1974): 149.

16. J. J. Beer, "Coal Tar Dye Manufacture and the Origin of the Modern Industrial Research Laboratory," Isis 49 (1958): 123.

17. See E. E. Williams, Made in Germany (London: Heinemann, 1896; repr. 1937), with a preface by Austen Albu. This book caused a sensation when it was first published in the 1890s.

18. Apart from the institution of royal commissions to inquire into science, the teaching of science, and technical education, the main action taken by the government was to instruct British consuls in Germany to report on the state of the German chemical industries. See, for example, F. Rose's report in Diplomatic and Consular Reports, Cd 430-16 (London: H.M.S.O., 1901); and Supplementary Report, Cd 787-89 (London: H.M.S.O., 1902).

19. Whitehead once remarked that the greatest invention of the nineteenth century was the method of invention, referring to German R&D. But there is no evidence that this innovation was deliberate. See Beer, op. cit.

20. J. H. Clapham, The Economic Development of France and Germany, 1815-1914 (4th ed.; Cambridge: Cambridge University Press, 1948).

21. D. S. L. Cardwell, "The Development of Scientific Research in Modern Universities: A Comparative Study of Motives and Opportunities," in A. C. Crombie, ed., Scientific Change (London: Heinemann, 1963), pp. 661-77. The selling of German synthetic dyes was so well organized that the dyer had comparatively little to do.

22. For W. S. Jevons' confident assertion that the most efficient steam engines in the world were made in Cornwall, see The Coal Question (London: Macmillan, 1865).

23. See Nathan Rosenberg, ed., The American System of Manufactures (Edinburgh: Edinburgh University Press, 1969).

24. John H. Dunning, American Investment in British Manufacturing Industry (London: Allen & Unwin, 1958).

25. See his monumental History of European Thought in the Nineteenth Century, 4 vols (London: 1896-1914), I.

26. For an account of contemporary debates on the national importance of science, see Cardwell, The Organisation of Science in England, op. cit.

27. The classic critique of imperialism is by J. A. Hobson. There have been many works on this subject since his day. Most of them have discussed the injustices suffered by the peoples subjected to imperialism and/or the failure of imperialism to bring home the advantages claimed for it by its advocates. As far as I know, no one has yet suggested that it might have had a stultifying effect on the technological efforts of the "imperial" nation. The subject merits further study.

28. D. S. L. Cardwell, "Science and World War I," Proceedings of the Royal Society A, 342 (1975): 447-56.

29. Ibid.

30. For example, much of the research work on the structure of the DNA molecule was supported by the Medical Research Council. See R. C. Olby, The Path to the Double Helix (London: Macmillan, 1974).

31. For a critique of science in Britain just before World War II see J. D. Bernal, The Social Function of Science (London: Routledge, 1939).

32. C. F. Carter and B. R. Williams, Science in Industry (Oxford: Oxford University Press, 1959), p. 9.

33. According to Albert Speer, scientists were not exempted from military service in Germany until 1942. See his Inside the Third Reich (London: Weidenfeld & Nicholson, 1970).

34. Jacques Spaey et al., Science for Development (Paris: UNESCO, 1971).

35. Simon Lavington, History of Manchester Computers (Manchester: NCC Publications, 1975).

36. Ibid.

37. Carter and Williams, op. cit. There are other broad similarities between the structures of American and British R&D. For example, according to Edwin Mansfield, 82 percent of all American industrial R&D in 1969 was concentrated in five industries: aerospace, electrical equipment and communications, chemicals and drugs, machinery, and motor vehicles. Furthermore, as in Britain, R&D is directed mainly toward new products rather than new processes. See Mansfield's "The Contribution of R&D to Economic Growth in the U.S.A.," Science 175 (1972): 477. According to Richard R. Nelson, Merton J. Peck, and Edward D. Kalachek, only 4 percent of the total R&D effort is devoted to basic research. The figure for Britain is slightly greater, but still under 5 percent, according to K. Pavitt and S. Wald, the authors of The Conditions for Success in Technological Innovation (Paris: OECD, 1971). Nelson, Peck, and Kalachek also note that in 1968, 85 percent of industrial R&D was due to only 384 firms, while 7 percent of the remainder was accounted for by 260,000 firms. See their Technology, Economic Growth and Public Policy (Washington, D.C.: Brookings Institution, 1967).

38. C. F. Carter and B. R. Williams, Investment in Innovation (Oxford: Oxford University Press, 1958), p. 17.

39. Ibid., p. 76.

40. Carter and Williams, Science in Industry, pp. 47–50, 172.

41. Christopher Layton, European Advanced Technology (London: Allen & Unwin for PEP, 1969), p. 60.

42. Carter and Williams, Science in Industry, p. 105. See also their Industry and Technical Progress (Oxford: Oxford University Press, 1957).

43. Quoted in Pavitt and Wald, op. cit., p. 94 ff. See also the Research Reports of the Department of Liberal Studies in Science, Manchester.

44. Carter and Williams, Investment in Innovation, p. 85.

45. Carter and Williams, Science in Industry, p. 33. This characteristic of British industry has also been noticed by Pavitt and Wald, op. cit., p. 119; and by R. C. O. Matthews, The Role of Science and Technology in Economic Development, UNESCO Science Policy Studies and Documents (Paris: UNESCO, 1970).

46. Carter and Williams, Investment in Innovation, p. 103.

3

THE ROLE OF SCIENCE AND TECHNOLOGY IN THE ECONOMIC DEVELOPMENT OF MODERN GERMANY

Wolfram Fischer

An evaluation or even a measurement of the impact of science and technology on economic and social development, and vice versa, requires the application of concepts, techniques, and data from such a variety of fields that it seems impossible to arrive at satisfactory results given our present state of knowledge. To be sure, important and broad questions can easily be asked and general statements made as to the nature of these relationships; but only small and probably not so important questions can be put in an empirically testable fashion and definitely resolved. In this dilemma I chose to approach my subject at the middle level and to concentrate on problems that are sufficiently general and interesting to be explored and that, while not being operational in a strict sense, may nevertheless allow some conclusive arguments and empirically founded answers. They will be arranged in loosely chronological order insofar as characteristics of the period of "proto" or early industrialization, ranging from the later eighteenth century to around 1850, will be discussed first while the main part of the paper deals with the later nineteenth and the earlier twentieth centuries.

EARLY INDUSTRIALIZATION

There is widespread agreement, particularly among English economic historians, that during the first phase of the Industrial Revolution, technological change had little to do with sciences, that its impact on improvements was negligible, and that nearly all the major technological achievements of the Industrial Revolution were demand-induced and found by practitioners in their respective fields.[1]

A. E. Musson and E. Robinson have taken issue with this view. They have maintained (and documented) that there were manifold relationships between mathematicians, physicists, and chemists, on the one hand, and practicing sea captains, instrument makers, iron founders, and machine builders, on the other, and that these personal and intellectual relations actually led to many small and some significant impacts of scientific knowledge and methods on the production process.[2] Nevertheless, conventional wisdom remains very much on the former side and tends to expand this view to a kind of general statement: The first phase of industrialization nowhere is particularly connected with science and certainly is not very much dependent on scientific inputs.

No similar discussion, neither in size nor in depth of argument, exists for Germany. There it has always been taken for granted that science was important (as was government), though perhaps less so at the beginning of industrialization than later, in the 1870s, when the synthetic-dye industry began systematically to apply achievements of organic chemistry to the production process. The reason for this assumption is that innumerable contemporary sources emphasize the importance of science in helping the craftsman and farmer to improve their knowledge and, thereby, their businesses. There was a widespread belief during the Enlightenment that better knowledge leads to a better society in a moral as well as in a material sense. There is also ample proof that some of the best scientists of the time, foremost chemists and "technologists"[3] but sometimes also mathematicians, devoted a large part of their energy to the solution of practical questions. A recent study confirms that the numerous German academies of science and scholarly societies that were founded during the eighteenth century, beginning with Leibniz's Prussian Academy of Sciences in 1700, tried to promote technological progress by offering prizes for the best solution to agricultural improvements and technical devices.[4]

Very little, however, is known about the effect of such endeavors, and until now most historians who have written about them seem not to be aware of this problem. The existence of a general belief in the usefulness of scientific research and advice in technological matters, and the proven response of scholars to this challenge, certainly indicate a possible relationship between progress in science and practical improvements; but they are no proof of actual results. This proof cannot yet be given on a macroeconomic or macrosocial level because our knowledge of the structure and movements of the German economy in the later eighteenth and earlier nineteenth centuries is still incomplete. We have only informed guesses about the most fundamental data, such as population growth, employment (overall and by sectors), social product (GNP or national income), and productivity. They rest on the early private and semipublic statistical publications and on monographic research covering single sectors and regions.

These data show an economy that was by no means "traditional" in the sense of being either "static" or purely "agricultural." It was an economy that seems able to grow enough to accommodate the population increase (of 58 percent in 70 years, an average annual growth rate of 0.7 percent) and even to improve, however, slightly, the income per capita. This economy showed, as is to be expected, a marked decline in the percentage of the labor force employed in the primary sector (agriculture, forestry, and fishery) and an about even increase in the percentage employed in manufacturing and the services. Also, the structure of the manufacturing sector changed visibly. In 1800 textiles, produced mainly in the putting-out system (cottage industry), dominated the sector, whereas by 1850 metal producing and metal-working and mining, glassmaking and chemistry, wood and paper production, and printing increased their combined share from 36.2 percent to 42.6 percent. Also the share of food processing industries had grown. Textiles, textile products, and leather, while remaining by far the largest single industry group and experiencing a growth of about 500,000 employees between 1800 and 1850, nevertheless lost relative to all the other industries. This is to be expected in a slowly industrializing economy. There is nothing sensational about it: neither revolutionary changes or leaps, nor big spurts or breakthroughs, nor insurmountable blocks to development.

To be sure, for the problem considered here the employment data alone do not reveal much. As long as they cannot be compared with other relevant data in the same industries—capital formation, output, productivity—the question cannot be answered by scrutinizing macroeconomic figures. The relative growth of employment in mining and metalworking can be due to technologically induced growth as well as to a lack of such change (which would make a higher labor input necessary), and the relative losses of textiles and related fields may have resulted from a particularly high rate of labor-saving technical change in these industries.[5] The same may be true for agriculture, where the additional labor input is particularly small, while the output seems to have grown at about the same rate as the population.[6]

we want to obtain more precise results, we have to use a microeconomic approach, either concentrating on single firms or distinct products, or on production processes, or on smaller regions with better-than-average data bases. Even a whole industry might be too large a unit to reveal relevant change.

The most complex sector was, of course, agriculture; it was also the most important one. Despite its relative loss of employment, it remained the largest sector of the economy. It was also one of the most diversified in terms of products, production processes, size of production units, and regional differences throughout the whole period. There can be no doubt that educated landowners (noblemen, civil servants, businessmen-turned-farmers, country clergymen) invested much thought and experiment in improvements of agricultural methods. Soil conditions, crop improvements and rotation, cattle feeding and breeding, drainage, and the use of implements were discussed throughout Europe in journals, in sermons, and in travelers' reports. The academies of sciences invited competitions for the best solutions.[7] We know that the competition for the prizes in agriculture was much harder than for any other one. While questions on mathematics, physics, astronomy, or trade technology often attracted no response, many of the agricultural prizes were sought by about 40 competitors. We also know that some of the answers, even those receiving prizes, gave solutions that later knowledge proved wrong. Furthermore, most of the competitors relied mainly on their own experience and experiments, thus exemplifying the trial-and-error method.

Thus, step by step, better chemical and botanical insights were obtained. By the 1830s knowledge of mineral fertilizers was fairly well established.[8] But we know next to nothing about the impact of such knowledge, the spread of innovations, their effect on yields or on crop selection and rotation. Certainly there has been some demonstration effect of agricultural improvements, that is, some innovation. But even for the best known one, the introduction of clover, which brought the winner, Christian Daniel Friedrich Schubart, fame and an imperial knighthood, our knowledge has been spurious. We know now that clover culture was well established before the prize was announced in 1783.[9] Although the intellectual influence of that prize was tremendous throughout Europe—Schubart received numerous visitors and hundreds of letters, was invited to Russia, and saw his treatise go through nine editions during his lifetime (it probably was illegally reprinted even more often)[10]—we do not know what acreage of land was actually devoted to clover as a consequence, nor how this influenced agricultural productivity.[11]

The same holds true for the major prize questions of the Prussian Academy on stall-feeding (1788), rotation of crops (1792), and fertilizing (1800).[12] We know the prize winners: a Protestant clergyman from Pomerania; two civil servants (one a Prussian war counselor, the other an administrator of a royal demesne); and an estate manager/clergyman. We know their state of knowledge and sometimes, as in the case of the prize-winning expert on fertilizing, we do have productivity results for his own land-holdings; but we know very little about the spread of their knowledge and its results.[13] Somewhat later the classic authors of German agricultural science, A. D. Thaer and J. H. V. Thünen, provide examples of the impact of scientific and technological innovations.[14] But more than examples can be found neither in them nor in the most recent histories of German agriculture.[15] Even one of the most knowledgeable experts on late eighteenth-century German agriculture, the East German historian Hans-Heinrich Müller, concludes his latest book with the somewhat vague statement: "Difficult as it may be to quantify the 'scientific agriculture' in 18th century Germany, we are fully entitled to say: science had already knocked at the doors of agriculture."[16]

If the impact of science cannot be stated clearly for agriculture as a whole, at least one crop was definitely advanced by scientific research: the sugar beet. This is possibly the earliest, and certainly a classic, case of science-led growth of an industry with discernible backward linkages to agriculture. It also is a classic case of a demand-induced innovation caused by a serious political distortion of the market forces, in this case by the Napoleonic Wars and the Continental blockade. The chemical identification of the sugar content of the beet with the sugar of the cane had been established by the director of the mathematical-physical class of the Prussian Academy of Sciences, Andreas Sigismund Marggraf, as early as 1747. Marggraf, who is said to have been the first chemist to use a microscope in his research, had systematically searched for a substitute for sugarcane among indigenous plants. He reported his findings, which were the results of work in the Academy's laboratory, to his colleagues and expressed his confidence "that this sweet salt can be prepared in our homeland as well as in the regions in which sugar cane is grown" and that "economic advantages" would result, particularly for the peasants.[17]

Marggraf's successor at the Prussian Academy, Franz Karl Achard, a gifted inventor and researcher, continued Marggraf's experiments, turning the only laboratory of the Academy into what his colleagues indignantly called the "Royal raw sugar factory." At the same time he experimented with plants at his estate near Berlin, pushing himself deeply into debt. When he had depleted his funds he asked the Prussian king in 1799 for a credit "out of deep love for the Prussian fatherland, in order to create a new branch of European industry," and received it. With this money he bought an estate in Silesia where he established the first beet-sugar factory in 1801-02 and continued his agricultural experiments, striving for beets with higher sugar content. When Napoleon announced his Continental blockade, the hitherto uneconomic process suddenly became competitive with cane sugar; Napoleon transplanted it to France in 1811. After 1815 the competition of cane sugar strongly affected the new industry, but did not extinguish it. As soon as favorable fertilizing, large plantations and processing plants, and cheaper transportation (by railway) developed, the beet-sugar industry could hold its own against cane sugar, at least as long as it enjoyed the advantage of tax exemption. In the 1830s and 1840s the biggest beet-sugar factories, such as the one at Waghäusel, Baden, employed more than 1,000 people and transformed a whole region around it to sugar-beet agriculture. Silesia and the region of Magdeburg became the main centers of sugar-beet growing and processing. Until well into the later part of the nineteenth century, strong competition forced the industry to improve its productivity through systematic experiments in plant cultivation and seed production, and improvements in the manufacturing process.

Eventually beet sugar became a noticeable part of German agriculture and food processing. The land area devoted to sugar beets was 176,000 hectares in 1878 and 569,000 hectares in 1913 (0.7 percent of the total land under cultivation in 1878 and 2.1 percent in 1913). The yield per hectare grew from 25 tons in 1878 to 30 tons in 1913, and the total production of sugar beets from 23,000 tons in 1846 to about 15 million tons in 1913.[*] The sugar industry was the fastest-growing branch of food processing in Germany before 1913.[18] A part of this growth, which was much greater than the increase of sugar-beet harvests, stemmed from increases in the sugar content of the beets through continuous selection. Marggraf found a sugar content of about 2 percent, and Achard raised it in his best beets to about 5.5 percent; but generally it remained around 5 percent as late as 1838. By 1913 it had been raised to 15.5 percent.[19]

The economic results of this early science-led innovation became apparent only after a trial period of several decades. They were not demonstrated statistically before the second half of the nineteenth century. This should not mislead us, however, to assume that the decisive technical improvements (and scientific explorations were made only in the later period. Basically, the problem was solved by Achard around 1800.

Exceptional as this may have been, it points to a more general problem: the gestation period of inventions and innovations. It may

---

[*]All weights are in metric tons. Unfortunately the physical production of beet sugar was not counted by official statistics directly. Statistics on cultivated land were not compiled until 1878, and did not include beets until 1893. See W. G. Hoffman, Das Wachstum der Deutschen Wirtschaft seit der Mitte des 19. Jahrhunderts (New York: Springer, 1965), pp. 270 ff., 378.

be shorter or longer, but it nearly always exists. Whether it tends
to become shorter in more modern times, as is often suggested, or
whether its length depends on individual circumstances remains a
matter of debate. [20]

The beet-sugar industry also may have been exceptional because
it is a fairly clear-cut case. Other industries relying at least partly
on knowledge and techniques derived from chemical observations and
research seem, however, to follow similar patterns. This pattern
was not first developed, as is often assumed, by the coal-tar-dye
industry after Perkins' synthesis of the first coal-tar dye, mauveine,
in 1856. Although it constitutes the most outstanding example of an
industry moving toward reliance on scientific progress, it was neither
the first to establish industrial laboratories nor the first to employ
scientifically trained chemists for research and development. [21]
There were many forerunners. The close alliance of scientists and
entrepreneurs in chemical engineering already existed in mercantilistic France, where since the second half of the seventeenth century
chemists either managed their own chemical manufacturing plants
or were appointed directors of state-owned enterprises. In Germany
similar cases were found by the later eighteenth century; by around
1800 a number of scientists were involved in economic enterprises
and, even more astonishing, some scientifically trained men with
experience in industry were appointed to professorships at universities. An interesting case is the apothecary Johann Wolfgang
Döbereiner, who, after producing pharmaceuticals and chemicals
in his own small establishments and running bleaching and dyeing
firms, breweries and distilleries, was in 1810 appointed professor
of chemistry at the University of Jena because the Grand Duke of
Saxony-Weimar wanted a man who would combine "originality in
science with a practical tendency." [22]

We do not yet have a full knowledge of the German chemical
industry before the 1860s, but from scattered evidence one can conclude that the pattern was laid before the three coal-tar-dye giants
were founded in the 1860s. [23] Johann Friedrich Gmelin used the term
"technical chemistry" as early as 1786 to indicate the possibility of
technological application of chemical knowledge. Friedrich Wilhelm
Hermbstaedt, another chemist of the Prussian Academy, may serve
as an example of such application. He held at one time or another
such public appointments as royal apothecary and professor at the
Collegium Medico-Chirurgicum (from 1790), member of the Prussian
Administration of Public Health (from 1794), member of the Administration of Manufactories and Trade (from 1796), apothecary to the
Prussian Army (from 1798), war counselor and medical counselor
(1804 and 1806), and, from 1811, professor of chemistry at the
newly founded University of Berlin. But before entering public

service—during which period the Prussian government built him a "service house" that combined living quarters for his family and servants with laboratories, chemical and mineral collections, and an auditorium—Hermbstaedt had been working at a chemical enterprise in Berlin and probably built a laboratory there. Later he was not only sought as an adviser by chemical industrialists but also devised and installed at least two chemical plants, one for sugar processing, and one for tar production and carbonizing bones. These were not large undertakings, but the pattern is clear.[24] Another case is Karl Weltzien, the first professor of chemistry at the Polytechnical School at Karlsruhe, who advised nearly every chemical firm established in the upper Rhine Valley and in Württemberg from the 1820s to the 1840s. He seems to have been a partner in some and to have sent his students to work for them.[25]

Justus Liebig, who established his laboratory at the University of Giessen in 1825 and educated some of the outstanding chemists of the nineteenth century (including A. W. von Hofmann), is the most famous of these chemistry professors who became seminal for the establishment of chemical industry. However, he certainly was not the first to establish a laboratory at a university, as is often assumed. Even if we disregard the laboratory at the tiny University of Altdorf, near Nuremberg, established in 1682, we find many examples before 1825—for example, at the mining schools in Chemnitz and Freiburg before the middle of the eighteenth century. It is true that Jöns Jakob Berzelius found the establishment in Freiburg "small and inadequate" in 1819, but so would a chemist of the later nineteenth century have judged the famous laboratories of Liebig at Giessen and of Robert Bunsen at Heidelberg. Most of these early research and teaching laboratories were financed by the professors themselves, some of whom, like Karl Gottfried Hagen at Königsberg, were appointed to their posts because they possessed such a laboratory. And some of the early teaching, mainly in pharmaceutical chemistry, took place outside the universities, in private schools established by research-oriented pharmacists in connection with their pharmacies. At least four such private institutions were founded before 1800.[26] It was in this tradition of chemical-pharmaceutical research that the first known research laboratory was set up in an industrial enterprise by Emanuel Merck, at Darmstadt, in 1826.[27] If all such cases of cooperation between theory and practice were collected, I am sure we would find a prototype of science-oriented chemical engineering. The structure of Badische Anilin- und Sodafabrik (BASF), Hoechst and Bayer, as developed in the late 1860s and 1870s, will then appear as a more developed form of long-established patterns.

THE NINETEENTH CENTURY

By 1830 Germany was not a leader in technical change; but neither was it as backward as often is believed. If it was a latecomer to industrialization, it was, like Belgium, Switzerland, France, and the United States, a very early latecomer and, like them, had the potential to catch up with the leader, Great Britain, within one or two generations. Chemistry is only one field in which Germany did so quickly; others are the crucial sectors of iron and steel and machine building. Certainly German manufacturers relied heavily on English (as well as Belgian and French) experts during the first third of the nineteenth century, and even longer. As in other Continental countries the first spinning jennies and steam engines were introduced into Germany late in the eighteenth century. Soon afterward indigenous construction began, but failed in most cases. Men like the ingenious Swiss mechanic Johann Georg Bodmer, who established the first machine shop in Germany (combined with a textile mill and a gun factory) in 1809, were exceptions.[28]

Much of the new British technology was both known and mastered by the 1830s; and some German inventions such as the weaving machine of Louis Schönherr, were among the most technologically sophisticated. Still, when the first railways were built, Germany had to obtain locomotives almost exclusively from foreign suppliers. Until 1843 more than 90 percent of the locomotives on Prussian railways were English and the rest, except one, Belgian. Only ten years later the picture had changed drastically. Of the 729 locomotives running on Prussian lines in 1853, about 70 percent were made in Germany, and 54 percent in the Berlin shop of August Borsig. German production began to be competitive as early as 1842, when more than 25 percent of the newly acquired engines came from firms within the country; three years later more than 50 percent were German-made; and after 1850 only a few new foreign locomotives were bought, mainly from Belgium.[29] Interestingly, for three years in the early 1840s American locomotives also were bought (from W. Norris in Philadelphia). In 1843 these amounted to 23 percent of all purchases, but altogether only 11 American engines were acquired over the three years. (The figures for the Zollverein show a somewhat higher early German share because southern German firms like Kessler in Karlsruhe and Maffei in Munich, which began to build locomotives in the same year as Borsig (1841), received early orders from the state-owned railways in Baden and Bavaria.)

A similar pattern can be seen in the purchase of rails. Until the end of 1844 about 89 percent of the rails of German railway lines were produced in Britain or Belgium; by the end of the decade the figure was still about 75 percent, but during the 1850s German rails were quickly substituted for English and Belgian ones. At the end of that decade Germany began to have a favorable balance of trade in rails.[30] Through the microeconomic study of H. Wagenblass and the macroeconomic study of R. Fremdling, it has become quite clear that railways, with their backward linkages into the iron industry and into mining, were not only instrumental in generating indigenous growth of these sectors, but also that these sectors accepted and adapted the best technology available. It took only 10 to 15 years to make them independent of the more advanced nations.

The same is true of the transfer of the new steelmaking processes from the time of Henry Bessemer. It took several years to adapt the process to indigenous conditions. Alfred Krupp, who tried unsuccessfully to receive a patent for Prussia in 1856, mastered the process in 1862; a few years later, however, several German producers used it successfully.[31] As in France and Belgium, German steelmakers derived advantages from the Siemens-Martin open-hearth process and particularly from the Thomas-Gilchrist process, which allowed them to use their phosphorus-bearing iron ore; both these processes were adopted quickly.

Similar patterns of slowly, then rapidly, advancing transfer and adoption can be documented in many other technological processes. Just one other example, connected with ironmaking and steelmaking, must suffice here: the substitution of coke for charcoal in smelting pig iron. As late as the 1850s, when the Belgians already smelted most of their iron with coke, only 10 percent of German iron production came from coke-fed blast furnaces. In 1840 such furnaces were used only in Silesia; then they took hold in the Saar; and only at the end of the decade were they introduced into the Ruhr. But by 1860 75 percent of the pig iron blasted in Prussia (and a somewhat lower percentage in the whole Zollverein) was smelted by coke; three years later the figure for Prussia was 89 percent and that for the Ruhr was 98.7 percent.[32]

The evidence presented thus far seems to confirm W. W. Rostow's view of a long, drawn-out preparatory period in which the preconditions were laid, followed by a sudden takeoff into self-sustained growth during some two decades. This trend is not as obvious in many industries with less visible technical improvements. Different chronologies and overlapping time spans for different sectors or production techniques also are apparent, so that for the economy as a whole the model of a slow, but steadily accelerating, progress fits much better.

Nevertheless, if one tries to explain the transition from slow to quick responsiveness to technical progress, some puzzles are left.

They stem partly from the fact that sometimes the same reasons are given for the slow and frustrating part of the learning process as for the quick and successful one. Among these are certain features in the social or institutional structure of Germany, distinct behavioral patterns of the German people, and specific forms of government tutelage. We will restrict our discussion to two related aspects that often are stressed as causes of the final success: a deliberate government policy of promoting science, technology, and industry, and a high degree of literacy and formal schooling.

Prior to that, however, I should like to emphasize a more basic reason that is not restricted to Germany, but can be applied to most parts of Western and Central Europe: the close economic and cultural interchange within Europe, particularly among its scientific and technological elite. The exchange of ideas, whether philosophical or mathematical, or of a more pragmatic nature, was well established throughout the centuries preceding the Industrial Revolution. Cross-fertilization was not the exception but the rule. The tendency to think in terms of nation-states and in national achievements is fairly recent in European history and tends to obscure the fact that, despite different languages and a host of regional idiosyncrasies, the fundamental principles of cultural and economic life were basically the same. The migration or rotation of cultural elites never stopped, and the communication among them was close. One has only to note how many books and pamphlets that were seminal in advancing knowledge or technology were immediately translated from one language into others, sometimes within the same year. Whatever could be learned from another country usually was adopted. Moreover, for a considerable length of time technological transfer was not a one-way-street. The overwhelming dominance of Britain in the decades after the Industrial Revolution seems to have been an exception rather than the rule. Even then, Britain did not stop learning from the Continent.[33]

Government participation or leadership in this exchange was an established pattern throughout Europe, particularly in fields crucial to military strength, such as iron founding, cannonmaking and gunmaking, fortification, and shipbuilding. If the Continental governments sent technical experts to Britain after 1770 to spy on its recent achievements, this was not unusual. However, government involvement in economic affairs took different forms in the nineteenth century. In Germany it slowly changed from the mercantilistic supervision of all activities to more indirect assistance and sponsorship of private enterprise. The transition was gradual and lasted until the middle of the nineteenth century.[34] Help in acquiring new technology belonged in this category of indirect assistance. Exhibitions, publications, and the sale or lease of imported machinery or of models were arranged. Technical education on a limited scale and on a lower-to-middle level was offered.

The sums spent on such promotional efforts were, however, minimal compared with those spent on the military or on general administration, and with public expenditure on general education or physical infrastructure (such as road construction, river improvements, and canal construction).[35] It has been argued that in Prussia the effects of a permanent budget surplus, which was deliberately created to enable the state to repay the debts contracted during the Napoleonic Wars, far outweighed the modest financial outlays for the promotion of trade and agriculture.[36] And in most of the other German states, whose combined deficits almost equaled the Prussian surplus, the possible growth effects from deficit spending were counterbalanced by the continued restrictions on trade, particularly for craft shops. If government promotion was effective, it was mainly in the backward regions of Prussia (including Silesia). In the more developed areas like the Rhineland and Saxony, private enterprise developed, beginning in the later eighteenth century, through the lack of efficient government control.[37]

By 1830 the economic philosophy of the Prussian civil service could be called as Smithian liberalism tamed by enlightened governmentalism. By 1850 most of the traditional direct government controls were gone; even the Prussian Mining Administration was beginning to surrender its right (and duty) to manage the mines throughout the state.[38] The impact was not everywhere so spectacular as in the Ruhr, where the abandonment of government management coincided with the emergence of the region as one of the leading industrial centers of Europe; but it is safe to generalize that the rise of German industry took place while government tutelage and direct intervention receded. At the same time it must be stressed that government supervision never disappeared totally; it remained strong, for instance, in the technical supervision of safety regulations in mining. In fact, government activity grew stronger in economic legislation while many state governments retained considerable property in mining, forests, and heavy industry.*

---

*The share of government participation varied considerably, remaining strongest in forestry and mining. The Saar mines were fully owned by the state; in Upper Silesia the state (or king) was one of the biggest entrepreneurs; and the Ruhr was under private ownership. The governments of Bavaria and Württemberg remained in the field of iron production, while Baden sold its older iron works or

It cannot be discussed here in detail how far other instruments of economic policy, such as customs, were used to foster technological adaptation and economic growth of domestic industries. Though much has been written about motivation of customs policy, we do not yet have an overall evaluation or calculation of its economic effects.[39] As we have seen in the case of beet sugar, some industries were effectively shielded from foreign competition during their infancy; but, generally speaking, the Prussian and the Zollverein customs were low between 1815 and 1879, particularly in the crucial period after 1830, as compared with France, Austria, Russia, and the United States, though they were higher than those in England, Belgium, Switzerland, and the Netherlands. Thus it cannot be said that the German drive to technological and industrial maturity took place behind a protective customs wall. The pressures of competition, rather than protection, urged a constant drive for technological improvement.[*] And certainly the technologically most advanced German industries—chemical, electrical, and mechanical engineering—were not among those that were particularly protected.

If fiscal and economic policy cannot be regarded as crucial in the developmental process during Germany's transition to a leading industrial economy, perhaps educational policy played a decisive role, German governments recognizing the importance of general, vocational, and scientific education, fostering their schools and creating special institutions for educating engineers and scientists. To answer this question we have again to distinguish between proclaimed goals and publicized motives, on the one hand, and actual

---

discontinued them. In Spandau, near Berlin, Prussia retained one of the largest weapons factories in the kingdom.

The railways were state-owned from the very beginning in Baden and Hesse. In Bavaria and Württemberg the main lines were built and run by the government; Prussia began to buy the private railways within its borders in the 1870s. A move by the Prussian government early in the twentieth century to buy the second largest company in the Ruhr (the Hibernia), in order to ensure enough coal production in wartime, failed because of concerted counteraction by private interests.

[*]Both international and inter-German competition tended to become more effective during the creation of the Zollverein and the corresponding abolishment of customs between German states. Some non-Prussian territories feared competition from Prussian businesses more than that from other European countries.

solutions and effects on the other. Again, it is much easier to document the first than the second. There can be no doubt that, when means for improving economic performance and social welfare were discussed, education played an important role. This can be traced back not only to the Enlightenment, when practical education ranked high among public goals, or to Luther and the Reformation, but also to the thriving commercial cities of the later Middle Ages. Literacy among the burghers of the German towns was traditionally high, while the countryside lagged considerably behind. But by 1830 or 1850, some elementary education was widespread even there.

However, one must not overestimate the standard of this education or its economic and social effects. The bare knowledge of the three Rs did not suffice for more sophisticated jobs in an industrialized economy. On the other hand, the traditional education of the urban middle classes (including craftsmen and traders) prepared enough people to fill positions in a gradually industrializing economy. One should not, therefore, put too much emphasis on global literacy figures. But if one does, Germany certainly ranked above average. By all measures employed by C. M. Cipolla—census data, illiteracy among recruits, newlyweds, or immigrants into the United States—Germany had a lower rate of illiteracy than any other major country in Europe. Only smaller Western and Northern European nations—the Netherlands, Switzerland, the Scandinavian countries, and Scotland—had an equal or higher rate of literacy between 1850 and 1914.[40]

If elementary education was above average in nineteenth-century Germany because of traditions and continuous government effort, it remains a matter of debate as to what effect government promotion of higher education had on economic development. It is true that the Prussian university reforms, promulgated mainly by Wilhelm von Humboldt, set an example for the combination of research and teaching at the highest level that was soon followed in other German states, and was much admired and sometimes adopted in other countries during the century. But this reform deliberately aimed at a scholarly education freed from all practical purposes. It was to be an end in itself. Its purpose was progress of scholarship and pure knowledge, not its application to technology or the economy. Many of the foremost German scholars of the century despised such application as undignified. The humanities, not sciences, stood at the core of this institution of learning. Useful knowledge was regarded as second-rate, belonging to intermediate rather than to higher schools. Vocational training was kept separate from the universities. It can be argued that this was a fertile atmosphere for breeding pure sciences—mathematics, theoretical physics, and chemistry—in one type of institution and applied sciences and practical knowledge in other types, and that both flourished; but it certainly was not an optimal system for interchange of ideas and careers, and of development of new ones.

Variety may have been an asset of the German educational system, but flexibility certainly was not its strength.[41] Variety was also the most interesting feature of government expenditure for scientific research and higher learning: the smaller German states spent a much higher percentage of their budgets for higher education than either Prussia or the Reich (which did not support universities, except one at Strassburg after 1871), and expenditure in the smaller states was also higher in relation to GNP and to population. If one takes the average expenditures, outlays for scientific purposes actually declined as a share of public expenditure during the 1880s and 1890s, thanks to the great weight of Prussia and Bavaria, where the share fell in the 1880s and the 1890s, respectively, and recovered, at least in Prussia, only slightly before World War I. On the average, expenditures for science remained around 1 percent of the government budget during the whole period, rising only to about 1.5 percent at the peak in the 1870s.[42]

The reason for the differences in the relative sizes of expenditures for science and higher learning cannot be found so much in different conceptions about higher learning as in the weight of the existing institutions. Baden, for example, the smallest of the five states under consideration, during the Napoleonic era inherited two ancient universities that had been outside its territory, Heidelberg and Freiburg. And since it, like the other states, founded a polytechnical school, which later became a technische Hochschule, its responsibility to fund institutions of higher learning was relatively high. Prussia, on the other hand, concentrated its efforts on relatively few newer institutions, such as the universities of Berlin and Bonn, together with some older ones, including Halle, but refused to found new institutions in the region of highest population growth (besides Berlin), the Ruhr (where universities were established after decades of debate only in the 1960s). In addition, two of the richest recently acquired cities, Cologne and Frankfurt, were left without financial support of the government when they established universities shortly before World War I. Such local or regional efforts often were decisive in the development of institutes of higher learning during the second half of the nineteenth century. When the Prussian government, after several decades of competition between Cologne and Aix-la-Chapelle for a technische Hochschule in the Rhineland, finally decided in favor of Aix-la-Chapelle in 1863, the reason was, besides some political considerations, essentially that the business enterprises and the city of Aix-la-Chapelle were capable of raising more funds for the new school than was Cologne.[43]

The decentralization of German public, economic, and cultural life furthered German education more than central government decisions. Where the urban middle class was fairly numerous and many towns were spread throughout the countryside, favorable conditions existed. Moreover, Germany had many capitals. In education, as in sciences and the arts, this meant competition for rank and prestige. Universities, low as their standards may have been in the eighteenth century, Latin schools, military academies, small engineering institutes, and medical schools existed in many places. This decentralization continued into the nineteenth and twentieth centuries. Whenever a new type of school was established in one state, others soon followed. In the third quarter of the eighteenth century, mining academies were founded in Austria-Hungary (Chemnitz, 1750), Saxony (Freiburg, 1765), Prussia (Berlin, 1770), and Hanover (Clausthal, 1775). All four schools drew students not only from German-speaking countries but also from the rest of Europe and from America. When, early in the nineteenth century, polytechnical schools were established in Austria (Prague, 1806; Vienna, 1815), others again followed suit: Baden (Karlsruhe, 1825), Bavaria (Munich, 1827), Saxony (Dresden, 1828), Württemberg (Stuttgart, 1829), Hanover (the city of Hanover, 1831), Brunswick (the city of Brunswick, reorganization of an older school, 1835), Hesse (Darmstadt, 1836), and, much later, Prussia (Aachen, 1870). Earlier (1821) Prussia had combined two older technical schools in Berlin to form a trade academy with similar, though somewhat less ambitious and more vocation-oriented goals. Toward the end of the century, again in a competitive move, one polytechnical school after the other received university status.[44]

Probably this competition was more effective than the competition with England, which played such a large role in the rhetoric of German industrialization and has been stressed by later observers.[45] Here a comparison with centralized France reveals more, particularly since France, not England, was the country to which German states looked when they established new educational institutions. In the eighteenth century they were modeled after the École des Mines and the École des Ponts et Chaussées; in the nineteenth century the École Polytechnique was the primary example. Actually, the German polytechnical schools developed quite differently. They were less mathematically and theoretically oriented, and catered not to the military but to the civil sector. Only very late in the century did they approach the academic level and prestige of the École Polytechnique. What first seemed to be their weakness proved to be a strength: they educated engineers and technicians for a variety of middle-level jobs in industry and government (such as safety inspectors). Of the 100 students trained at the Berlin Trade Academy before the middle of the nineteenth century whose careers could be traced, nearly two-thirds worked in industry and government agencies. While among the students of the School of Construction about 75 percent eventually started their own businesses or took over a construction firm, 91 percent of the students in the metalworking department and 68 percent of those in textile engineering did not become self-employed entrepreneurs.[46]

Whether this holds true for the students at the polytechnical schools as well remains to be explored.[47] There are indications that they had higher goals than they actually achieved. Throughout the 1800s the emerging engineering profession complained about its low social status and wanted to raise it by means of higher education, one of its persistent goals being the equality of the polytechnical schools with universities.[48] The leading professors of machine engineering—Jakob Ferdinand Redtenbacher, Franz Grashof, Karl Karmarsch, and Franz Reuleaux—tried to develop machine design and machine building into a systematic science founded on mathematics, only to be accused by some of their colleagues at the end of the century, when university status was obtained, of having headed in the wrong direction: Prospective engineers needed experience with real machines in industry more than mathematical and physical knowledge, those around the Berlin professor Alois Riedler maintained.[49]

The discussion as to a more theoretical or practical orientation of their education also took place among the chemists and physicists in industry. While many foreign observers believed that in Germany university chemists, industrialists, and governments "worked smoothly together for their mutual good," German businessmen complained that the universities educated the chemists too long and too arduously, and that their research was of no practical value. Universities complained that governments did not provide the necessary atmosphere and funds for the adequate education of scientists.[50] It is difficult to evaluate what would have been adequate; but recent research in the history of German science policy and administration has revealed an astonishing degree of reluctance, if not resistance, particularly on the part of the Prussian government (mainly because of fiscal reasons), to endow scientific research and teaching. All the highly praised innovative institutions for scientific research, the Physikalisch-technische Reichsanstalt (1887), the Institutes for Applied Physics and Chemistry at the University of Göttingen (1896 ff.), and the Kaiser-Wilhelm-Gesellschaft (1910), had to be endowed by businessmen like E. W. Siemens, Alfred Krupp, and the Berlin private banker Leopold Koppel. In each case it took 15 to 20 years for scientists to convince industrialists and governments that such institutions would have more than prestige value for the nation.[51] The same obstinacy had prevented the establishment of a Prussian

École Polytechnique in the first half of the century; even the continuing advocacy of Alexander von Humboldt had not been enough to overcome the fiscal scruples.[52]

Something like an "American pattern" of research promotion can be detected in the way these institutions originated and were funded. Not government policy, but private initiative, was the moving force. Indeed, in Germany the endeavors to promote education and research for industrial technology have been labeled "Americanism." As early as 1830 a Heidelberg professor of philology complained in a letter to the founder of the Polytechnical School in Karlsruhe, C. F. Nebenius, about "the spreading Americanism, Polytechnicism or however one wants to call the material tendencies." Throughout the century the humanistically oriented cultural elite opposed this trend; but among those responsible for the education of engineers, American technology was considered the most interesting challenge, and after the world exhibitions in Philadelphia (1876) and Chicago (1893), self-criticism and proposals to follow American examples in technical education abounded. Riedler, the influential professor of engineering at the Technical University of Berlin who fought against the preponderance of theoretical and mathematical studies in the education of engineers, wrote a report to the Prussian minister of education about the role of laboratories in American technical education. The Association of German Engineers took up the matter, and only a year later the minister announced that the financing of laboratories for the Prussian technical universities would be forthcoming.

The fact that Karl Linde, the pioneer of refrigerator technology, had established such an institute at Munich already in 1875 had not sufficed. But now, again in a competitive move, the other German technical schools received similar laboratories within a few years.[53] This American influence may seem surprising during a period in which American educational institutions increasingly sought German examples to follow. But it only confirms what has been said about the exchange of innovative ideas between European nations: it was seldom a one-way street, but a cross-fertilization. Though Germany around the turn of the century was doubtless one of the leading countries in science as well as in technical education, and probably also in the methods of adapting scientific innovations to practical needs, it did not possess absolute leadership, and recognized and adopted developments successful in other countries.

That American patterns were followed mainly in technological fields close to the needs of industry does not come as a surprise, since the United States had become the fastest-growing and most innovative economy in the last decades of the nineteenth century. Leading German firms saw in their American counterparts the most

formidable challenge in the world markets; and when they tried to
modernize their plants, they often introduced American machinery
and American principles (mainly specialization) in the organization
of the production process. Siemens departed from the "manufactural"
stage when he introduced an "American assembly room" in 1872,
after having sent a worker to learn about the arrangements from the
sewing-machine producer Ludwig Loewe, who was one of the first
to use American production methods in Germany. In 1906 the leading
producer of agricultural machines in Germany, Heinrich Lanz in
Mannheim, incited the first strike in his plant when, after a visit to
the United States, he too quickly introduced American methods and
machinery.[54]

Government promotion of scientific research for economic
purposes did not follow American examples in the same degree or
with the same consciousness. But despite all hesitations and obstacles,
it grew considerably from 1850 to 1914. Frank R. Pfetsch has tried
to calculate the share of government expenditure for sciences devoted
to general, military, and economic purposes. Though the ascription
of certain expenditure to the one or the other category is not beyond
doubt, it seems clear that those areas oriented toward promotion of
industry and agriculture gained considerable influence throughout
the period, while those related to military purposes became less
important after 1875. While the state governments were, and remained,
responsible for the general promotion of sciences, mostly through
maintaining the universities (except Strassburg), the expenditure of
the empire was devoted first more to military research (63 percent
in the 1870s) and then, from the change of its economic policy in
1878-79 more and more to economy-related work, mainly in the
form of research in technology and applied sciences (including agriculture), which amounted to about 50 percent of its outlays for
science in 1910-14. Together, the empire's and the states' economy-
related expenditure grew from around 11 percent of their science
budgets in 1850-55 to 36 percent in 1910-14.[55]

How this compares with public expenditure for science in other
countries is difficult to establish because of the many differences in
the organization and financing of institutes of higher learning and
research. For one segment, the outfitting of universities with physics
laboratories and the number of assistants and students working in
them, a comparison among England, France, and Germany has been
attempted. If we can trust the figures, taken over by Pfetsch from
contemporary sources, France and Germany in 1870 each had four
physics laboratories in which scientific education took place; England
had one. In 1880 Germany had nine, a remarkable advance over
France, which still had four, and England with two. In 1890 Germany
still led with 15, but England had 11 while in France the number had

increased only to six. But in 1900 France had caught up again, boasting 24 such laboratories, the same number as Germany, while England held a close third place with 21. In the number of assistants employed in these laboratories, France and Germany equaled each other in 1870 and 1880 with fewer than ten; but German institutes expanded faster, employing 62 assistants in 1900. England, which had a late start, came close with 60, while France had only 27. A similar picture develops in regard to the "output" of these institutions. At the beginning (1870), France and Germany educated about the same number of people: 26 and 25, respectively. In the next decade Germany developed a marked lead and kept it in comparison with France until 1900, when 2,029 students worked in the German physics laboratories, but only 750 in the French ones. Again England, starting late, overtook Germany as early as 1890. In 1900, English laboratories counted 2,297 students. [56]

INDUSTRIAL CONTRIBUTIONS

Whatever the contribution of public authorities to financing pure and applied sciences, most of the actual technical progress in the decades before World War I seems to have been achieved by R&D within industry. For chemistry some figures are now available. Frank R. Pfetsch has compared the employment of chemists in three large chemical enterprises (BASF, Hoechst, Bayer) and in universities. While in 1865 only three chemists were employed in these firms (which just had been founded), 72 chemists taught at universities. During most of the period up to World War I, employment of chemists in business firms grew more quickly than in institutes of higher education. In 1875-80 the number of industrial chemists in the three enterprises surpassed the number of university chemists; and in 1910 the firms had 651 chemists on their staff, as against 360 at the universities (now including technische Hochschulen). The number of chemists employed also grew much more quickly than general employment in the chemical industry. [57] Most of the chemists worked, of course, in production of coal-tar dyes and pharmaceuticals. In 1896, for example, 712 of a total of 1,191 chemists in chemical engineering (60 percent) were in this sector; each of the plants in this sector had, on the average, nearly 15 chemists on its payrolls, while the chemical industry as a whole employed about six chemists per establishment. For every 27.8 employees in organic chemical firms there was one chemist available, while in the chemical industry as a whole the relation was 37.5:1.*

---

*In the oil industry it was 84.7:1, in heavy chemicals 67.6:1, in artificial fertilizers and explosives, about 60:1. Note, however,

That firms hired their own chemists did not mean, however, that cooperation between them and the university chemists was discontinued. The close cooperation between industry and university or research institutes can be observed throughout the period. From the beginning the major competing chemical firms tried to stake out "claims" on certain professors and universities to which they could turn for scientific advice and for recruiting the most promising students for their own R&D.[58] Later some of the greatest scientific assets to industry were the Nobel Prize-winning bacteriologists Robert Koch, Emil von Behring, and Paul Ehrlich, who developed vaccines against tuberculosis, cholera, tetanus, diphtheria, and other diseases, and laid the foundations for chemotherapy, which became an important branch of the pharmaceutical industry in Germany (and elsewhere). Koch and Behring cooperated with Hoechst, and Paul Ehrlich with Cassella, both leading pharmaceutical firms. In the realm of coal-tar dyes the cooperation between industrial chemists and university scientists like A. W. von Hofmann and Adolf von Baeyer continued to bear fruit.

Fertilizers became another, more recent, mass product relying partly on scientific research. Though they were not developed in the laboratories to the same extent as synthetic dyes, the recognition of the fertilizing value of by-products of the Thomas process or of mineral resources like the Stassfurt potassium salts resulted from chemical observations and research. One of the leading chemists of the time, the Dutch Nobel Prize winner Jacobus van't Hoff, devoted most of his research as the salaried chemist of the Prussian Academy of Sciences (from 1896) to the chemical properties of the Stassfurt deposits. The most spectacular result of close cooperation between research chemists and chemical engineers came immediately before World War I, when Fritz Haber, professor of chemistry at the Technical University of Karlsruhe, and the industrial chemists of BASF in Ludwigshafen, headed by Carl Bosch, developed the ammonia synthesis that made Germany's wartime production of explosives independent of imports. Haber, who in 1911 had become the first director of physical chemistry at the newly established Kaiser-Wilhelm-Institut, in 1917 found himself running a research factory with 1,500 employees instead of the five he had had at the outbreak of the war, and a research budget 50 times higher than in 1914.[59]

---

that the total employment given by F. R. Pfetsch (p. 158) is 138,000 for 1895. Pfetsch does not care to explain this difference, which must be due to a more narrow definition of the chemical industry in his source for Table 4: F. Fischer, Das Studium der technischen Chemie an den Universitäten und technischen Hochschulen Deutschlands (Brunswick: 1897).

Aside from chemistry the main industrial benefactor of scientific research probably was the optical industry. Ernest Abbe, professor of physics at Jena, provided much of the basic research that his friend, the mechanic Carl Zeiss, turned into profitable production of optical and photographic instruments beginning in 1867. A dozen years later the association of the chemist Otto Schott made possible the production of heat-resistant glass in another firm, Schott & Genossen, which was started with a credit from the Prussian government. It may be noted that Abbe retained his chair of mechanical and experimental physics at the university even when, after the death of Carl Zeiss, he became the sole proprietor of the Carl Zeiss works.

The electrical engineering industry exemplifies the difficulty of sorting out scientific and empirical inputs into technical progress. All pioneers in electricity, whether trained physicists and mathematicians like Alessandro Volta, André Ampère, and H. C. Oersted, or experimenting inventors with some scientific background like Michael Faraday, Thomas Edison, and E. W. Siemens, contributed to technical progress mainly by advancing knowledge through trial and error. The main discoveries were made in several countries and quickly exchanged. If the United States and Germany achieved a lead over the other nations as far as technical and commercial application is concerned, this seems to stem from the fact that both countries possessed inventive and innovative businessmen like Edison, George Westinghouse, Siemens, and Walter Rathenau who employed professional mathematicians or physicists and engineers at an early stage.

Siemens, perhaps more than others, fostered a close cooperation of scientists and engineers, since he was somewhat of both (in addition to being a successful businessman), boasting among his many achievements membership in the Prussian Academy of Sciences.* He was certainly one of those who most influenced the scientific policy of later nineteenth-century Germany, taking the initiative for and helping to finance some of the innovative research institutions set up in Germany. The foremost of these was the Physikalisch-technische Reichsanstalt (1887), which he actually "forced upon" the government by providing the funds. However, like other businessmen he was later disappointed that, under the directorship of the physicist

---

*Siemens' first scientist was the founder of the firm himself, though his formal training had been limited. Schuckert in Nuremberg, whose firm later merged with the Siemens company, hired electrical engineers and established a laboratory in 1882. See S. Franck, "Georg Hummel (1856-1902), ein Pionier der Elektrotechnik," Technikgeschichte 38 (1971): 237 ff.

Hermann von Helmholtz, it devoted most of its research efforts more to basic problems of physical science rather than, as planned, its technological applications.[60] While German electrical engineering in its early years certainly profited from his unique combination of talents, in the 1880s Siemens' stubbornness and some wrong management decisions contributed to a relative stagnation of his firm. Only the aggressive competition of the German Edison Company under the leadership of Rathenau forced him to move into the quickly developing field of high-voltage technology. Scientific and technical expertise of an aging and family-oriented businessman could, as shown by Siemens' insistence that all new applications of electrical engineering, such as electric bulbs and power stations, be developed within his own firm, sometimes hamper the innovation of new technological developments. While he pioneered the electrification of public transport by installing the first electric tram at Lichterfelde, near Berlin, in 1881, it was Rathenau's firm that first introduced it commercially; and while Siemens allowed his chief engineer to conduct futile experiments on a Siemens-developed system of electrical lighting, his competitor bought the Edison patents and carried the market.[61]

Taken together, the German electrical engineering industry certainly was, and remained, second only to that of the United States. This was possible, however, only because a firm with strong technological and inventive orientation did not monopolize the market but was forced to make quicker adaptations by a more commercially oriented firm that transferred technology whenever it could profitably do so. A decade later Siemens regained some of its lost strength. In the long run its scientific-technological orientation paid off after some necessary corrections in the organization and financing of the firm had been taken. Thus the history of German electrical engineering teaches several lessons, one of them being that technological excellence is not a sufficient condition, though sometimes it may be a necessary one, for economic growth and business success.

Whether this technological excellence could be achieved and maintained through reliance on academic research or through R&D in the firms became a matter of debate in the last decade of the nineteenth century. For chemistry the answer was clear: the cooperation of both had been the optimal solution, and a division of labor had developed in which the more basic research was left to academia while the applied research was taken on by the firms. Many variations of this pattern were possible. Outside chemistry, businessmen were rather skeptical about the value of academic research for their immediate business purposes, partly because university laboratories of applied physics and engineering were newly founded and therefore had not had a chance to prove their usefulness. The physics laboratory founded at Karlsruhe in 1853 had remained small and was used mainly for teaching. Only the technische Hochschule in Munich had established a mechanics laboratory (1868) and a laboratory for research in refrigerator technology (in the 1870s).

When the Göttingen mathematician Felix Klein tried to find sponsors for institutes of applied sciences at his university in the 1890s, leading firms like Siemens, Krupp, and Maschinenfabrik Augsburg-Nürnberg (MAN) answered reluctantly; and Eugen Langen, Otto's associate in the development of the internal combustion engine and himself a trained engineer, stated bluntly that industrial research was more useful, more rational, and cheaper. At this time the leading firms possessed a generation's experience with their own research laboratories.* If Klein finally mustered quite an impressive list of industrial sponsors, it was not because they hoped to benefit directly, but because they felt obliged to be munificent.[62] That the technical universities around the turn of the century were equipped with laboratories not only of physics but also of engineering was not so much the result of private munificence, however, as of successful lobbying at the government level by the professors at these schools, who argued for the superiority of the American way of teaching engineering by using large laboratories in which not only models of machines but also real machinery could be demonstrated and analyzed. The argument that German industry would lose in international competition in both the long run and the immediate future proved to be a successful one.

---

*For example, Krupp had hired a young chemist as early as 1857 and had given him additional practical polytechnical training before putting him to work. In 1862 Krupp opened a chemical research laboratory at his steel plant for the analysis of raw material, fuels, exhaust gases, and products. Over the years chemists were also attached to other plants of the firm; and when Krupp's son Friedrich Alfred had finished his education as a chemist and metallurgist, an additional central research laboratory for the improvement of steel technology "in independent and original research" that employed both chemists and physicists was established. H. Bummert, "Die Entwicklung neuer technischer Methoden unter Anwednung wissenschaftlicher Erkenntnisse im Bereich der deutschen Schwerindustrie . . .," in W. Treue and K. Manuel, eds., Naturwissenschaft, Technik und Wirtschaft im 19. Jahrhundert (Göttingen: Vandenhoeck & Ruprecht, 1976), I, pp. 335 f.

EFFECTS OF TECHNOLOGICAL IMPROVEMENTS

To measure the relative importance of technological improvements or, even more difficult, the scientific contribution to technological improvements in single industries or the economy at large is not possible with an adequate degree of certainty for the period 1830–1914. One indicator, however crude, is the distribution of patents between the industries and their correlation with economic performance. Before 1877, because there was no common patent law, we must examine the data of individual German states. Prussia may serve for the purpose.[63] The absolute number of patents is not a good indicator, since Prussia used its patent ordinance of 1815 restrictively; in the 1850s and 1860s it refused about 80 percent of all requests for patents, in comparison to about 25 percent in the United States at the same time.[64] Between 1815 and 1845 only 779 patents were granted, an average of 25 per year. While the annual numbers varied considerably during the first decades, a definite increase is discernible after 1838, when 40 patents were issued. In 1840 the number was 53, and in 1845, 72. From 1850 to 1875 Prussia granted an average of 82 patents a year. But this average was reached only three times in the early 1850s and three times during the 1860s.

The number of patents grew suddenly after 1873, when, under the impact of economic depression, Prussian authorities became less hostile toward patents. But even then the number of Prussian patents was not much higher than those issued in tiny Baden or in much smaller Bavaria, and only one-third to one-half the number of patents issued in Saxony under different regulations. How few inventions passed the criteria of the Prussian patent authority for novelty and originality comes to full light only in a comparison with countries where patents did not have to pass a test, but could be registered for a fee: between 1870 and 1875 Prussian patents amount to 6.4 percent of the Belgian ones, 4.4 percent of the British, and less than 1 percent of the American.[65] When the patent law of the German empire became effective in 1877, the absolute numbers immediately jumped to more than 4,000 per year and stayed at that level until 1890. When the law was revised in 1891 to allow both production processes and finished products, particularly chemical ones, to be patented, the number rose to 5,500 and remained between 5,000 and 6,500 until the end of the century. Then it took another leap. In the last five years before World War I, an average of 12,338 patents were issued.

The distribution of patents by industry after 1877 may tell more than the absolute numbers. Over the period up to World War I, the lion's share did not go to chemistry or electrical engineering, which generally are regarded as the most innovative industries, but to metalworking, mainly machine building. In 1877–78 metalworking (except steam engines, vehicles of all kinds, and electrical implements) accounted for 39 percent of all patents issued in the German empire, while chemistry received only 4 percent and electrical implements less than 1 percent. In 1913 metalworking still led with 32.5 percent, while chemistry had risen to 11 percent and electrical engineering to 8.4 percent.[66] Nevertheless, in relation to other indicators of the relative size of industries in Germany, such as the numbers employed, the number of patents received by chemistry was far above average, since the industry employed only 1.4 percent of the industrial labor force in 1877-78 and 2.5 percent in 1913. But metalworking (now including steam engines, vehicles, and electrical implements) also had a relatively small labor force, accounting for 12.3 percent in 1877-78, compared with a share in patents of 46.5 percent (including the above-mentioned branches) and for 17.4 percent in 1913, compared with 46.2 percent of patents.[67]

The insights gained by comparing such crude structural figures are limited. More can be revealed by looking at the growth rates of specific industries. Industries with strong scientific inputs and a high degree of technical change can be found among those with the best performance. At the top is one branch that took its innovative inputs mainly from improvements developed in other industries: the public utilities (gas, water, and electricity supplies), with an annual increase in production of 9.7 percent between 1870 and 1913. Other industries with growth rates above 6 percent include paper manufacturing (6.9 percent), steel production (6.3 percent), and chemicals (6.2 percent). Unfortunately, electrical engineering is not separated in the work by W. G. Hoffmann, but is included in the large category of metalworking, which grew 5.3 percent per annum—still well above the average of all industries and crafts (3.7 percent).[68] Altogether, the relations between the number of patents and the sectoral growth rate are not very clear. In order to obtain better results, one would have to divide the units of observation into smaller groups, include a time factor to allow for the lag between inventions and their economic effects, and probably also devise a method to weigh the economic importance of a patent. One also would have to determine whether the Imperial Patent Office continued to have a bias against granting certain types of patents, for example, products and procedures in chemical or metallurgical technology, as against such other types as mechanical devices.[69]

TECHNOLOGICAL PROGRESS AND ECONOMIC GROWTH

The methods of measuring technological progress as a residual component of economic growth—after determining the contribution of additional factor inputs—developed mainly in the United States by Robert Merton Solow, John W. Kendrick, Edward F. Denison and others have recently been used by a few German economists.[70] According to these studies, "technical progress" contributed 42 percent to the average growth rate of the German economy of 2.6 percent per annum between 1850 and 1913. The average growth rate would have been 1.5 percent if only additional factors of production had been put into the production process without technical change. D. André tried to break down the "residual" further and came up with a calculation according to which an annual growth of 0.1 percent can be attributed to "structural effects," that is, to the moving of factors of production from sectors of lower productivity to sectors of higher productivity, and 0.2 percent to a prolongation of education (and therefore to an improvement in the quality of labor inputs). Stated a different way, if the distribution of the factors of production had remained static, total factor productivity would have grown only 1.0 percent per annum. In addition, if there had been no improvement in the quality of labor through longer (that is, better) education, it would have been only 0.8 percent.[71]

There are, of course, sectoral differences in the growth rates and their composition. The highest rate is in the secondary sector (industry), and the lowest one in the primary sector (agriculture, extraction), while the tertiary (service) is closest to the average.

The share of "technical progress" is greater in agriculture than in manufacturing and the services. This can be explained by the fact that additional inputs of labor and capital into this sector were much smaller, so the share of "progress," since it is a residual, is bound to be higher. Moreover, it could be argued that this may reflect the relatively early endeavors to rationalize agriculture. Data show, therefore, some payoff of scientific and technological inputs that began late in the eighteenth century. A third possible explanation is that the slack in the agricultural economy was so high that even small improvements in organization and techniques led to fairly visible results, that small additional inputs were needed to raise productivity. On the other hand, the growth rate in agriculture was so much lower than that of the other sectors that even a share of more than 50 percent led only to an annual growth below 1 percent. Thus the impact of technical change in absolute terms was smaller in agriculture than in industry.

Of course, there are pitfalls in the data base and the techniques employed in such measurements, but we cannot discuss them here in detail. There are many which the workers in this field are well aware of, but others they do not seem to recognize clearly. These have to do mainly with the distorting effects of "heroic assumptions." The numbers that have to go into such calculations are, in the German case, quite tenuous because they have been derived mainly by inter-polations and extrapolations of a few censuses in which data crucial to measuring technological progress were not gathered. But even if "heroic assumptions" abound in all these studies, they are the best available informed and methodologically controlled estimates we have, and constitute now a body of "conventional wisdom" that did not exist prior to 1965. They enable us to delineate the development of the German economy between 1850 and 1913 more sharply than before. We can now identify periods of faster and periods of slower advance, structural changes, and imbalances.

Overall, the impression of a continuously accelerating economic growth prevails. Railways, with their strong backward and forward linkages, can be identified as the most important source of growth during the 1840s through the 1860s. Later the metal-producing and metalworking industries gained a momentum of their own, providing the necessary linkage back to coal mining and forward to the con-sumer industries and services.[72] As the patent statistics suggest, this was also the sector of the economy in which technical innovation was most visibly clustered: in 1907, the last census year before World War I, half of all patents were taken out by metal-producing and metalworking firms; and within this sector 20 percent of the industrial labor force produced about 23 percent of industrial goods.[73] If we add chemistry, we can say that nearly a third of the industrial production and about a quarter of the labor force were in the sectors of "high technology" in pre-World War I Germany.[74] Throughout the whole economy less visible changes also occurred.

We find a higher population increase than in the first half of the century (up to 1 percent on the average, compared with 0.7 per-cent before 1850), and an even more marked increase in employment (1.2 percent annually), resulting in a 1910-13 labor force double that in the middle of the nineteenth century. Most pronounced is, however, the growth of national income, which, in constant prices, rose 2.6 percent annually and 1.6 percent per capita. Thus per capita income in Germany in the years preceding World War I was 2.75 times higher than in the middle of the nineteenth century. If we take into account the fact that "technical progress" seems to have accounted for around 40 percent of German economic growth, we can say that the per capita income of the German population more than doubled thanks to improvements in education, organization, and the tech-niques of production between 1850 and 1913.

GERMANY SINCE WORLD WAR I

Twentieth-century political development in Germany was more discontinuous than in any other major industrial country. Three principal changes in the political regime and several territorial contractions, expansions, and divisions make it difficult to provide any continuous analysis of social and economic developments, including the impact of science and technology. Though international comparisons of long-term trends since the 1950s and 1960s seem to confirm that German patterns fit general patterns fairly well, short-term fluctuations tend to be more irregular and violent. For the war and immediate postwar years adequate data are lacking or distorted. It took about five years after World War I and four years after World War II before reliable statistical series resumed, and a somewhat longer period was necessary to overcome the wartime distortions.

In our context the number of patents awarded may serve as an interesting example. In 1919 it had fallen to 7,766 from an average of 12,338 in the five years preceding the war. In the following year the number almost doubled. However, the number of patents dropped quickly to around 15,500 for the period 1925-29, only to rebound during the Great Depression to more than 26,000 in 1932, then fall during the Nazi regime to around 15,000 in 1938. In 1950 it was down to a mere 2,383 due to Allied restrictions; but only one year later, with the beginning of postwar recovery and the reinstitution of a legal basis, the number was up to nearly 28,000. It was more than 37,000 in 1952 and 1953. As after World War I, inventions seemed to have piled up for several years, and did not start to appear without delay until 1954.[75]

As an indicator of technological improvements, the number of patents seems to be even less adequate than before 1914. If we disregard the short-term fluctuations and look at the structure of inventive activity as expressed in the percentage of patents issued for different sectors of the economy, however, we find a continuation of pre-World War I trends. Over the whole period between a third and a quarter of all patents were in metalworking. Machine-building remained in the forefront of inventive activity, while electrical engineering, which counted for around 12 percent in the 1920s, enlarged its share to almost 20 percent in the 1930s and more than 20 percent in the 1950s. Chemical engineering followed the same trend, lagging about 5 percent behind electrical engineering. Germany remained an economy with a comparative strength in investment industries.[76]

Looking backward, this seems to follow naturally; but looking ahead from, say, 1919, it could not be taken for granted. During the entire interwar period (excluding Hitler's drive to military superiority) and again after World War II, under the impact of reparations, dismantling of industries, and production and research restrictions, Germans feared exclusion from major developments. Indeed, the loss of international patent rights during and after the wars meant the loss of world supremacy for the German chemical industry after World War I.

By 1923 inflation had destroyed all private foundations, and the tight financial restrictions of the state's budgets after the consolidation of public finances prevented an adequate expansion of public assistance for R&D. In 1920 the Notgemeinschaft der Deutschen Wissenschaft was founded, on the initiative of scientists, civil servants, and industrialists, to collect and concentrate public and private means for research. Under the able leadership of the former Prussian minister for cultural affairs, Friedrich Schmidt-Ott, it served as the main agency of research funding until it was taken over by Nazi party members in 1936.[77] For some time also private funds provided by, among others, a Japanese industrialist, American foundations, and a research council of the electrical engineering industry helped to finance not only applied but also basic research, such as that in theoretical physics.

One of the amazing facts of the interwar period is that German physics not only kept the position gained in the decades preceding World War I, but even improved it. Under the sponsorship of established leaders like Max Planck, Albert Einstein, Max von Laue, and Max Born a new generation of physicists grew up in Germany and advanced research in such fields as quantum mechanics, under abominable economic conditions. When Walter Gerlach, at a demonstration before the German Physical Society in 1921, used an X-ray tube he had made himself, he spoke of the <u>physica pauperum</u> to which at least the experimental physicists in Germany seemed to be condemned. And Werner Heisenberg, while serving as an assistant to Max Born on the work that later earned him the Nobel Prize received the equivalent of $23 a month from the Notgemeinschaft; the sum was raised to $34 in 1924.[78] That highest standards could be kept in spite of such relative poverty throws some doubts on the hypothesis that there is a necessary direct link between ample funding of basic research and its quality. Despite the unfavorable circumstances, Germany was able to continue its lead in the number of Nobel Prize winners in the 1920s and 1930s: 16, compared with 14 British and 12 American ones.[79]

Chemistry remained one field of German preeminence; and after some years of adaptation to new conditions, the German chemical industry continued to be the most research-intensive of all industries. In 1925 its main firms formed I. G. Farbenindustrie A.G., a gigantic industrial complex with a capital base of more than 1 billion marks and several research laboratories, in which pioneer developments were carried on, most notably the production of synthetic fuel through hydrogenation of coal. Since the Haber-Bosch ammonia synthesis, a "technological momentum" had developed[80] that carried the German chemical industry far ahead of its competitors in the use of hydrogenation and high-pressure processes; but during the Great Depression it also carried the industry deep into a financial crisis.

The products derived from the new technologies were so expensive that they could not compete with standard products like gasoline, particularly since the prices fell rapidly during the Depression. It was under these circumstances that I. G. Farben sought state subsidies, contacting republican governments and later Hitler, who finally guaranteed success by protection and rearmament.

The alliance of research, business, and political-military interests made the chemical industry crucial to National Socialism and gave I. G. Farben chemists like Karl Krauch an eminent position in the Four Year Plan bureaucracy. It is not necessary in this context to repeat how science, technology, and armament were amalgamated during the preparation and conduct of World War II; but it must be stressed that promotion of science and even technology, massive and hectic as it was, was restricted to those parts considered to be of immediate use. Not only basic research, but also well-developed and powerful technologies were abruptly halted when Hitler did not see any use in them. As in economic policy, the National Socialist government did not possess a coherent idea, or plan, or clear insights regarding necessities. It could not, therefore, systematically promote science or technology, even where they would have served its political goals. This meant, on the other hand, that some established scientists and areas of research could continue fairly undisturbed even during the war, so long as nobody considered them enemies of the movement. One basic government aim was of course to purge science of Jewish persons and influence. This led to the denial of Einstein's theories (even by some other Nobel Prize-winning physicists). However, the majority of the German physicists did not abandon it, and at a conference held at Munich in 1940 even forced a type of coexistence upon the representatives of German physics.[81]

The main loss was on the human side. By 1936, 1,600 scientists had been removed from their positions; others had left them "voluntarily." About 100 physicists emigrated to the United States, most of them under 40 years of age.[82] There was also a considerable loss in terms of morale; many forms of resignation and "internal emigration" developed. Finally the war took its toll. Germany suffered much heavier losses of lives, and therefore also of talent, than during World War I. Only a few scientists had been classified as indispensable and freed from military service; in engineering the proportion was higher, and aircraft engineers were among the few privileged categories that expanded during the war. Altogether, scientific-technological research was the "stepchild" of the system, as one historian has observed,[83] and an original source of human aspirations basically opposed to a dictatorial government that sought to subdue all independent utterances.

After World War II the two German states that emerged were quickly integrated into the two main political and economic blocs of the world. Recovery, particularly in West Germany, was much quicker and steadier than after World War I. As far as the relation of science to technology and the economy is concerned, West German developments paralleled those of other West European nations. It is a matter of perspective whether one emphasizes this return to "normality," which is indeed impressive in most statistics, or whether one stresses the long-term effects of the Third Reich. There are many indicators: the smaller number of Nobel Prize-winning scientists; the continuing negative balance of patents; the lasting dependence on American technology in fields like computers, aircraft, and space; and, until very recently, the relatively low expenditure for R&D in relation to public expenditure, GNP, or population. The economic performance certainly is one of the most impressive in the world, and the capability to transfer and adapt new technologies seems to be at least better than average. The chemical, electrical engineering, and machine-building industries are of lasting strength, except in some of the most innovative fields. Of 491 major technological innovations between 1953 and 1973, only 33 (less than 7 percent) originated in West Germany, compared with nearly 65 percent in the United States and 17 percent in Great Britain.[84]

Correspondingly, technical progress seems to have contributed only around 50 percent to economic growth in the later 1950s and 1960s.[85] As the negative balance of payments for technological innovation (licenses for use of patents, for instance), the transfer of management techniques from the United States, and the expansion of multinational firms with headquarters outside Germany (mainly in the United States but also in Canada, Britain, the Netherlands, and Switzerland) indicate, a considerable part of this progress is imported. Unlike other countries, West Germany can easily afford such imports because they represent only a small fraction of its permanent surplus in the export of goods, mainly investment goods. "High technology" products have been scarce among these, though recently reactor technology, which for years was limited by Allied restrictions, has gained some prominence. The bulk of German exports is represented by sophisticated traditional investment goods: machines of all kinds, automobiles, electrical equipment, chemical and pharmaceutical products. The strength of the German technology seems to rest not with a few top products but with the variety and flexibility of its supply, which allows German industry to compete successfully in the industrialized West, in socialist countries, and in the Third World.

This success has been achieved with only an average input of funds and human resources into R&D. In this respect Germany has done less than the United States or Britain, with relatively higher success. Such a comparison may be distorted, however, since the higher research efforts of the three nuclear Western powers stem mainly from government expenditure on military and (in the American case) space research, the spill-over effect of which seems to have been less than expected. In the 1960s the awareness grew that in the long run Germany may fall behind because of underinvestment in R&D. During the second half of the 1960s and the early 1970s the growth rate of R&D and, particularly, public expenditure for higher education advanced faster than before and faster than the general growth rate of the economy. In 1963, R&D expenditure in West Germany was only 1.5 percent of GNP, compared with 3 percent in the United States, 2.3 percent in Great Britain, 1.9 percent in the Netherlands, and 1.7 percent in France; by 1971 it had risen to 2.1 percent, compared with 2.6 percent in the United States, 2.4 percent in Great Britain, 2.0 percent in the Netherlands, and 1.8 percent in France.[86] The per capita R&D expenditure more than quadrupled between 1962 and 1972 in current prices, and nearly trebled in constant prices.[87]

How efficient these additional inputs are is a matter of debate. A distinction should be made between the two-thirds of R&D funds that are spent in the private sector* and the third that is spent in the public sector, mainly at universities and research institutes. In the public sector the marginal efficiency of the additional outlays may have been rather small, since they serve partly to expand the administration of science and the participation of every employee in that administration or the material well-being of researchers through higher salaries. At the universities, at least, costs per student jumped considerably without tangible result, such as higher standards of education.

In the long run, however, the broadening of the system of higher education may lead to a mobilization of reserves of talent that formerly lacked higher education, particularly in regions traditionally underserved by institutions of higher education: the Ruhr, eastern Westphalia, northern Bavaria. At present there are signs of an oversupply of more educated people, particularly in fields where one would not expect it: chemistry and civil engineering. On the other hand, science teachers are still in short supply, in contrast with those in the humanities and social sciences. At least for the immediate

---

*This percentage is about the same as in the United States, Great Britain, Belgium, and the Netherlands, but smaller than in Switzerland (85 percent) and higher than France (50 percent) or Canada (40 percent). Bundesminister für Forschung und Technologie, ed., Fünfter Forschungsbericht (Bonn: Federal Government, 1975), p. 98.

future a conflict has arisen between the goal of broadening higher education and the ability of the economy to absorb more people with higher education. The critics of this trend toward more theoretical, university-oriented education maintain that the corresponding loss of prestige of the practice-oriented, intermediate technological schools, which were certainly an asset in nineteenth- and early twentieth-century Germany, may be greater than the gain through expanded higher education. Since the end of World War II the share of the population with university education jumped from about 2 percent to around 20 percent. It will be interesting for future social scientists to find out how this change will effect the performance of the economy and the social structure.

NOTES

1. For a concise statement of the interrelations see P. Mathias, "Who Unbound Prometheus? Science and Technical Change 1600–1800," Yorkshire Bulletin of Economic and Social Research 21 (1969); an improved version was reprinted in P. Mathias, ed., Science and Society 1600-1900 (Cambridge: Cambridge University Press, 1972), pp. 54-80.

2. A. E. Musson and E. Robinson, Science and Technology in the Industrial Revolution (Manchester: Manchester University Press and Toronto: Toronto University Press, 1969).

3. Technology was an all-embracing field that flourished at some German universities, mainly Göttingen, from the last third of the eighteenth century well into the first two decades of the nineteenth century. See U. Troitzsch, Ansätze technologischen Denkens bei den Kameralisten des 17. and 18. Jahrhunderts, vol. 5 in the series Schriften zur Wirtschafts- und Sozialgeschichte, W. Fisher, ed. (Berlin: Duncker & Humblot, 1966).

4. H. H. Müller, Akademie und Wirtschaft im 18. Jahrhundert. Agrarökonomische Preisaufgaben und Preisschriften der Preussischen Akademie der Wissenschaften (Versuch, Tendenzen und Überblick), vol. 3 is the series Studien zur Geschichte der Akademie der Wissenschaften der DDR, H. Scheel, ed. (Berlin: Akademie-Verlag, 1975).

5. That this indeed was the case, at least in cotton-spinning, is demonstrated by G. Kirchhain, Das Wachstum der deutschen Baumwollindustries im 19. Jahrhundert (Ph.D. diss., Münster, 1973).

6. F. W. Henning, Die Industrialisierung in Deutschland 1800-1914 (Paderborn: Ferdinand Schöningh, 1973), p. 51. The figures that W. G. Hoffmann, Das Wachstum der deutschen Wirtschaft seit der Mitte des 19. Jahrhunderts (New York: Springer, 1965), pp. 270-

323, provides for German agriculture before 1850 (mainly since 1816) suggest, however, a far greater growth rate particularly in meat production, which, according to Henning, could not keep pace with population growth, resulting in a diminishing of its share in food intake from 15 percent of the caloric value per capita in 1800 to 11 percent in 1850. Hoffmann (Berlin, Heidelberg, New York: Springer, 1965, pp. 297 ff.) finds an increase in the stock of cattle of about 50 percent between 1816 and 1850 and a similar increase in the average weight (the case was similar for swine and other types of meat). This is not irrelevant to our subject, since the feeding of cattle was one of the major topics of "scientific" agriculturalists, beginning in the later eighteenth century, and since scientific inputs through veterinary medicine, which seem to have been considerable, can account for much of this increase.

7. For a list of prizes on agricultural problems offered by the Prussian Academy of Sciences between 1769 and 1847, see Müller, op. cit., pp. 126 ff.

8. M. Trenel, "Zur Frühgeschichte der Agrikulturchemie," in Berliner Forschung und Lehre der Landwirtschaftswissenschaften (Berlin: Akademie-Verlag, 1956); H. W. Schütt, "Anfänge der Agrikulturchemie in der ersten Hälfte des 19. Jahrhunderts," Zeitschrift für Agrargeschichte und Agrarsoziologie 21 (1973): 83-91.

9. G. Schröder-Lembke, Die Einführung des Kleebaues in Deutschland vor dem auftreten Schubarts von dem Kleefelde, vol. 10 in the series Wissenschaftliche Abhandlungen, Deutsche Akademie der Landwirtschaftswissenschaften zu Berlin, ed. (Berlin: Akademie-Verlag, 1954).

10. Müller, op. cit., pp. 149-78.

11. W. Abel, Geschichte der deutschen Landwirtschaft vom frühen Mittelalter bis zum 19. Jahrhundert, vol. 2 in the series Deutsche Agrargeschichte, G. Franz, ed. (2nd ed.; Stuttgart: Eugen Ulmer, 1967), p. 309, cites data from five large estates where the percentage of acreage devoted to clover around 1800 fluctuated between 2.8 and 6.5 percent.

12. Müller, op. cit., pp. 187-260.

13. Some valuable regional studies about agricultural innovations are H. H. Müller, Märkische Landwirtschaft vor den Agrarreformen von 1807 (Potsdam: Ueröffenthchungen des Bezirksheimat-Museums, 1967). G. Schröder-Lembke, "Die mecklenburgische Koppelwirtschaft," Zeitschrift für Agrargeschichte und Agrarsoziologie 4 (1956): 49-60.

14. A. D. Thaer, Geschichte meiner Wirtschaft zu Möglin (Berlin and Vienna: 1815); and Grundsätze der rationellen Landwirtschaft, 4 vols. (6th ed.; Berlin: 1868); J. H. V. von Thünen, Der isolirte Statt in Beziehung auf Landwirtschaft und Nationalökonomie (Rostock: Leopold, 1826; 2nd ed., 1850; 3rd ed., 1863).

15. Such as Abel, op. cit.; H. Haushofer, Die deutsche Land-
wirtschaft im technischen Zeitalter, vol. 5 in the series Deutsche
Agrargeschichte, G. Franz, ed. (Stuttgart: Eugen Ulmer, 1963); or
W. Abel in H. Aubin and W. Zorn, eds., Handbuch der deutschen
Wirtschafts- und Sozialgeschichte, I (Stuttgart: Union Verlag, 1973),
for the period before 1800; and G. Franz, in ibid., II (Stuttgart:
Klett, 1976), for 1800-50.

16. Müller, Akademie und Wirtschaft, p. 258.

17. Quoted in ibid., p. 129. For a short discussion see
Haushofer, op. cit., pp. 87-90. The older literature includes F. C.
Achard, Die europäische Zuckerfabrikation aus Runkelrüben (Leipzig:
Hinrichs, 1809); F. Fischer, Rede zur Eröffnungsfeier des neuen I.
Chemischen Instituts der Universität Berlin am 14. Juli 1900; E. O.
von Lippmann, Geschichte der Rübe als Kulturpflanze von der
ältesten Zeit an bis 1809 (1925); and Geschichte des Zuckers seit
ältesten Zeiten bis zum Beginn der Rübenzucker-Fabrikation (Berlin:
Springer, 1929); A. Hildenbrandt, Die Entwicklung des deutschen
Zuckerrübenbaues mit besonderer Berücksichtigung der Acker-
bautechnik (Ph. D. diss., Landwirtschaftliche Hochschule, Berlin,
1927).

18. E. O. von Lippmann, Die Entwicklung der deutschen
Zuckerindustrie von 1850 bis 1890 (Leipzig: 1900). Hoffmann, Das
Wachstum der deutschen Wirtschaft, pp. 380 ff., gives only an index
with 1913 = 100. The sugar production in 1838 was then 0.26; for
breweries, it was 18.7; for butcheries 19.6; and for distilleries,
55.6. Even if we take into account that the earlier figures underestimate the production, the growth rate remains considerable.

19. Haushofer, op. cit., p. 89; Hoffmann, Das Wachstum der
deutschen Wirtschaft, p. 378.

20. See G. Mensch, "Zur Dynamik des technischen Fort-
schrittes," Zeitschrift für Betriebswirtschaft 41 (1971); F. R. Pfetsch,
Zur Entwicklung der Wissenschaftspolitik in Deutschland 1750-1914
(Berlin: Duncker & Humblot, 1974), pp. 161-66. A great variety of
case studies is provided by J. Jewkes, D. Sawers, and R. Stillerman, The Sources of Invention (London: Macmillan, 1958); and by
the National Bureau of Economic Research, The Rate and Direction
of Inventive Activity: Economic and Social Factors (Princeton:
Princeton University Press, 1962).

21. As is suggested by J. J. Beer, "Coal-Tar-Dye Manufacture
and the Origins of the Modern Industrial Research Laboratory," Isis
49 (1958): 123-31; and The Emergence of the German Dye Industry
(Urbana: University of Illinois Press, 1959).

22. See E. Schmauderer, "Die Stellung des Wissenschaftlers
zwischen chemischer Forschung und Industrie im 19. Jahrhundert,"
in W. Treue and K. Mauel, eds., Naturwissenschaft, Technik und

Wirtschaft im 19. Jahrhundert (Göttingen: Vandenhoeck & Ruprecht, 1976), II, pp. 622 ff.

23. The pharmaceutical industry, however, has been studied thoroughly. See W. Vershofen, Die Anfänge der chemisch-pharmazeutischen Industrie. Eine wirtschaftshistorische Studie. 2 vols. (Berlin-Stuttgart: Aulendorff, 1949-52).

24. I. Mieck, "Sigismund Friedrich Hermbstaedt (1760 bis 1833). Chemiker und Technologe in Berlin," Technikgeschichte 32 (1965): 325-82, esp. 363 ff.

25. W. Fischer, Der Staat und die Anfänge der Industrialisierung in Baden, I (Berlin: Duncker & Humblot, 1962). For the period after 1848 this is documented in two studies by Peter Borscheid, Naturwissenschaft, Staat und Industrie in Baden, 1848-1914 (Stuttgart: E. Klett, 1976), pp. 50 ff., and "Widerstand und Fortschritt in den Naturwissenschaften. Die Chemie in Baden und Württemberg 1850-1865," in U. Engelmann, V. Selin, H. Stuke, eds., Soziale Bewegung und politische Verfassung Beiträge zur Geschichte der modernen Welt (Stuttgart: E. Klett, 1976), pp. 755-69.

26. D. Pohl, Zur Geschichte der pharmazeutischen Privatinstitute in Deutschland von 1776 bis 1873 (Ph.D. diss., Marburg, 1972); Pfetsch, op. cit., pp. 155 f.; E. Schmauderer, op. cit.; Müller, Akademie und Wirtschaft, p. 22.

27. W. Treue, "Unternehmer, Technik und Politik im 19. Jahrhundert," in Treue and Mauel, op. cit., I, p. 226.

28. On Bodmer's activity in Germany, see W. Fischer, "Die Anfänge der Fabrik von St. Blasien (1809-1848). Ein Beitrag zur Geschichte der Frühindustrialisierung," Tradition 7 (1962): 59-78; and Der Staat und die Anfänge der Industrialisierung in Baden.

29. R. Fremdling, Eisenbahnen und deutsches Wirtschaftswachstum 1840-1879. Ein Beitrag zur Entwicklungstheorie und zur Theorie der Infrastrucktur, vol. 2 in the series Untersuchungen zur Wirtschafts-, Sozial- und Technikgeschichte (Dortmund: Gesellschaft für Westfälische Wirtschaftsgeschichte, 1975), pp. 74 ff.

30. H. Wagenblass, Der Eisenbahnbau und das Wachstum der deutschen Eisen- und Maschinenbauindustrie 1835 bis 1860, vol. 18 in the series Forschungen zur Sozial- und Wirtschaftsgeschichte (Stuttgart: Gustav Fischer, 1973), pp. 65, 86, 171 ff.

31. A Heggen, Erfindungsschutz und Industrialisierung in Preussen, 1793-1877 (Göttingen: Vandenhoeck & Ruprecht, 1975), pp. 82 ff.

32. D. S. Landes, The Unbound Prometheus, Technological Change and Industrial Development in Western Europe from 1750 to the Present (Cambridge: Cambridge University Press, 1969), pp. 216-18.

33. The German influence on British technology in the seventeenth and eighteenth centuries has been stressed by Hans-Joachim Braun, Technologische Beziehungen zwischen Deutschland und England von der Mitte des 17 bis zum Ausgang des 18. Jahrhunderts (Düsseldorf: Pädagogischer Verlag Schwann, 1974). The opposite view is presented by W. Kroker, Wege zur Verbreitung technologischer Kenntnisse zwischen England und Deutschland in der zweiten Hälfte des 18. Jahrhunderts, vol. 19 in the series Schriften zur Wirtschafts- und Sozialgeschichte, W. Fischer, ed. (Berlin: Duncker & Humblot, 1971). Also see W. O. Henderson, Britain and Industrial Europe. 1750-1870: Studies in British Influence on the Industrial Revolution in Western Europe (Liverpool: Liverpool University Press, 1954; 2nd ed.; London: Frank Cass, 1969).

34. I. Mieck, Preussische Gewerbepolitik in Berlin 1806-1884, in the series Veröffentlichungen der Historischen Kommission zu Berlin (Berlin: de Gruyter, 1965); Fischer, op. cit.; and "Government Activity and Industrialization in Germany (1815-1870)," in W. W. Rostow, ed., The Economics of Take-Off into Sustained Growth (London: Macmillan, 1963), pp. 83-94; U. P. Ritter, Die Rolle des Staates in den Frühstadien der Industrialisierung. Die preussische Industrieförderung in der ersten Hälfte des 19. Jahrhunderts (Berlin: Duncker & Humblot, 1961).

35. K. Borchard, Staatsverbrauch und öffentliche Investitionen in Deutschland 1780-1850 (Ph.D. diss., Göttingen, 1968); W. R. Ott, Grundlageninvestitionen in Württemberg. Massnahmen zur Verbesserung der materiellen Infrastruktur in der Zeit vom Beginn des 19. Jahrhunderts bis zum Ende des Ersten Weltkrieges (Ph.D. diss., Heidelberg, 1971).

36. R. Tilly, "The Political Economy of Public Finance and the Industrialization of Prussia, 1815-1866," Journal of Economic History 26 (1966): 484-97; and "Fiscal Policy and Prussian Economic Development, 1815-1966," in Third International Conference of Economic History, 1965, I (Paris and The Hague: Mouton, 1965), pp. 771-78.

37. M. Barkhausen, "Staatliche Wirtschaftslenkung und freies Unternehmertum im westdeutschen und im nord- und südniederländischen Raum bei der Entstehung der neuzeitlichen Industrien im 18. Jahrhundert," Vierteljahrsschrift für Sozial- und Wirtschaftsgeschichte 45 (1958): 168-241; H. Kisch, "The Textile Industries in Silesia and the Rhineland: A Comparative Study of Industrialization," Journal of Economic History 19 (1959): 541-64; and Variations upon an 18th Century Theme: Prussian Mercantilism and the Rise of the Crefeld Silk Industry (Philadelphia: American Philosophical Society, 1968).

38. H. D. Krampe, Der Staatseinfluss auf den Ruhrkohlen-
bergbau in der Zeit von 1800 bis 1865, in the series Schriften zur
Rheinisch-Westfälischen Wirtschaftsgeschichte (Köln: Rheinisch-
Westfälisches Wirtschaftsarchiv, 1961). I have dealt with the reform
of the Prussian mining law in 1851-1865 in three essays reprinted in
W. Fischer, ed., Wirtschaft und Gesellschaft im Zeitalter der
Industrialisierung. Aufsätze-Studien-Vorträge, vol. 1 in the series
Kritische Studien zur Geschichtswissenschaft (Göttingen: Vandenhoeck
& Ruprecht, 1972), pp. 138-78.

39. One of the best more recent studies is K. W. Hardach, Die
Bedeutung wirtschaftlicher Faktoren bei der Wiedereinführung der
Getreide- und Eisenzölle 1879, vol. 7 in the series Schriften zur
Wirtschafts- und Sozialgeschichte, W. Fischer, ed. (Berlin: Duncker
& Humblot, 1967).

40. C. M. Cipolla, Literacy and Development in the West
(Harmondsworth: Penguin, 1969), pp. 113 ff.

41. For a discussion of the relative merits and shortcomings
of the German system of higher education in comparison with the
French, English, and American, see J. Ben David, The Scientist's
Role in Society. A Comparative Study (Englewood Cliffs, N.J.:
Prentice-Hall, 1971), pp. 108-38.

42. F. R. Pfetsch, Zur Entwicklung der Wissenschaftspolitik
in Deutschland 1750-1914 (Berlin: Duncker & Humblot, 1974), p. 47.

43. H. M. Klinkenberg, ed., Rheinisch-Westfälische technische Hochschule Aachen 1870-1970 (Stuttgart: Oscar Beck, 1970),
esp. pp. 31 ff.

44. The older schools are listed in K. H. Manegold, Universität,
technische Hochschule und Industrie. Ein Beitrag zur Emanzipation
der Technik im 19. Jahrhundert unter besonderer Berücksichtung
der Bestrebungen Felix Kleins, vol. 16 in the series Schriften zur
Wirtschafts- und Sozialgeschichte, W. Fischer, ed. (Berlin: Duncker
& Humblot, 1970). Also see the well-documented Festschrift for the
centennial of the Rhenish-Westphalian Technical University in Aachen:
Klinkenberg, op. cit.

45. R. Tilly, "Los von England. Probleme des Nationalismus
in der deutschen Wirtschaftsgeschichte," Zeitschrift für die gesamte
Staatswissenschaft 124 (1968): 179-96. Critical of later judgments by
Gerschenkron, Rostow, and others is K. Hardach, Nationalismus—
Die deutsche Industrialisierungsideologie?, no. 26 in the series
Kölner Vortrage zur Sozial- und Wirtschaftsgeschichte (Köln:
Forschungsinstitut für Sozial- und Wirtschaftsgeschichte, 1976).

46. P. Lundgreen, Techniker in Preussen während der frühen
Industrialisierung. Ausbildung und Berufsfeld einer entstehenden
sozialen Gruppe, vol. 16 of the series Einzelveröffentlichungen der

Historischen Kommission zu Berlin (Berlin: Colloquium Verlag, 1975), p. 191.

47. Only the careers of early students of the Polytechnical School in Hanover are known. See K. Karmarsch, Die polytechnische Schule zu Hannover (2nd ed.; Hannover: 1856), appendix.

48. Manegold, op. cit.; G. Hortleder, Das Gesellschaftsbild des Ingenieurs. Zum politischen Verhalten der technischen Intelligenz in Deutschland (Frankfurt-am-Main: Suhrkamp, 1970).

49. Manegold, op. cit., p. 145.

50. P. Hohenberg, Chemicals in Western Europe. An Economic Study of Technical Change (Chicago: Rand-McNally, 1967), pp. 69 ff.

51. F. Pfetsch, "Scientific Organization and Science Policy in Imperial Germany, 1871-1914: The Foundation of the Imperial Institute of Physics and Technology," Minerva 8 (1970): 557-80; and Zur Entwicklung der Wissenschaftspolitik in Deutschland, 1750-1914; W. Ruske, Reichs- und preussische Landesanstalten in Berlin. Ihre Entstehung und Entwicklung als ausseruniversitäre Forschungsanstalten und Beratungsorgane der politischen Instanzen, BAM-Bericht no. 23 (Berlin: Bundesanstalt für Materialprüfung 1973; L. Burchardt, Wissenschaftspolitik im Wilhelminischen Deutschland, vol. 1 in the series Studient zur Naturwissenschaft, Technik und Wirtschaft im 19. Jahrhundert, W. Treue, ed. (Göttingen: Vandenhoeck & Ruprecht, 1975); U. Troitzsch, "Technisches Schulwesen, Wissenschaftsorganisation und Wissenschaftspolitik in Deutschland (1850-1914)," Technikgeschichte 42 (1975): 35-43.

52. K. H. Manegold, "Eine Ecole Polytechnique in Berlin. Über die im preussischen Kultusministerium in den Jahren 1820 bis 1850 erorterten Pläne zur Gründung einer höheren mathematisch-naturwissenschaftlichen Lehranstalt," Technikgeschichte 32 (1965): 182-96.

53. Manegold, Universität, technische Hochschule und Industrie, pp. 33, 128 ff., 147 ff. Riedler's report to the minister is printed in Verhandlungen des Vereins zur Beförderung des Gewerbefleisses in Preussen 72 (1893): 381 ff. He also gave a report to the Association of German Engineers that is printed in Zeitschrift des Vereins deutscher Ingenieure 38 (1894): 405 ff., 507 ff., 608 ff., 629 ff.

54. J. Kocka, "Von der Manufaktur zur Fabrik. Technik und Werkstattverhältnisse bei Siemens 1847-1872," in K. Hausen and R. Rürup, eds., Technikgeschichte, vol. 81 in Neue Wissenschaftliche Bibliothek (Köln: Kiepenheuer & Witsch, 1975), p. 281; W. Fischer, "Ein Jahrhundert der Landtechnik. Die Geschichte des Hauses Heinrich Lanz 1858-1958" (unpublished MS).

55. Pfetsch, Zur Entwicklung der Wissenschaftspolitik in Deutschland 1750-1914, pp. 63-68. There is, apart from some

disturbing printing errors, an element of arbitrariness in his assessment of what is related to economic purposes. His main criterion is expenditure that can be used directly by the economy, but this is difficult to decide. For instance, Pfetsch puts all expenditure for technische Hochschulen in the category of "economy-related," but not (partly because it was not possible to separate them from the general university budgets) the corresponding expenditure for science at the universities.

56. Ibid., p. 329, Table 6.

57. Ibid., p. 158, Table 3.

58. W. Treue, "Die Bedeutung der chemischen Wissenschaft für die chemische Industrie 1770-1870," Technikgeschichte 33 (1966): 25-51, esp. 50; Beer, "Coal-Tar-Dye Manufacture," p. 131, footnote 16.

59. Burchardt, op cit., pp. 98 ff.; G. Wendel, Die Kaiser-Wilhelm-Gesellschaft 1911-1914, zur Anatomie einer imperialistischen Forschungsgesellschaft, vol. 4 in the series Studien zur Geschichte der Akademie der Wissenschaften der DDR (Berlin: Akademie Verlag, 1975), p. 212.

60. Pfetsch, Zur Entwicklung der Wissenschaftspolitik, pp. 109-27; K. H. Manegold, "Zur Emanzipation der Technik im 19. Jahrhundert," in K. H. Manegold, ed., Wissenschaft, Wirtschaft und Technik, Studien zur Geschichte (Festschrift für Wilhelm Treue) (München: Bruckmann, 1969), pp. 379-402.

61. J. Kocka, "Siemens und der unaufhaltsame Aufstieg der AEG," Tradition 17 (1972): 125-42. For the different business strategies of these giants in the field also see Kocka, "Expansion-Integration-Diversifikation. Wachstumsstrategien industrieller Gossunternehmen in Deutschland vor 1914," in H. Winkel, ed., Vom Kleingewerbe zur Grossindustrie. Quantitativ-regionale und politisch-rechtliche Aspekte der Wirtschafts- und Gesellschaftsstruktur im 19. Jahrhundert (Berlin: Duncker & Humblot, 1975), pp. 203-26. For the role of scientists and engineers in the Siemens firm, see Kocka, Unternehmensverwaltung und Angestelltenschaft am Beispiel Siemens 1847-1914, no. 11 in the series Industrielle Welt Schrifteureihe des Arbeitskreises für moderne Sozialgeschichte (Stuttgart: Klett, 1969).

62. Manegold, Universität, technische Hochschule und Industrie, pp. 120 ff.; and Wissenschaft, Wirtschaft und Technik, p. 398.

63. A. Heggen, Erfindungsschutz und Industrialisierung in Preussen 1793-1877, vol. 5 of the series Studien zur Naturwissenschaft, Technik und Wirtschaft im Neunzehnten Jahrhundert, W. Treue, ed. (Göttingen: Vandenhoeck & Ruprecht, 1976); U. Troitzsch, "Die Auswirkungen der preussischen Patentbestimmungen auf die Eisenindustrie in den 50er und 60er Jahren des 19. Jahrhunderts," Tradition 17 (1972): 292-313.

64. K. H. Manegold, "Der Wiener Patentschutzkongress von 1873. Seine Stellung und Bedeutung in der Geschichte des deutschen Patentwesens im 19. Jahrhundert," Technikgeschichte 38 (1971): 161; Heggen, op. cit., p. 40.

65. Calculated from Heggen, op. cit., p. 78, Table 2.

66. Calculated from Hoffmann, Das Wachstum der deutschen Wirtschaft, pp. 266 ff.

67. Ibid., pp. 196 ff. There is no estimation of the capital stock or investments in different industries in Hoffmann's opus. He gives only estimates of capital formation in all industries, as compared with agriculture, railways, and so on.

68. Ibid., p. 63.

69. See Troitzsch, "Die Auswirkung der preussischen Patentbestimmungen"; E. Schmauderer, "Der Einfluss der Chemie auf die Entwicklung des Patentwesens in der zweiten Hälfte des 19. Jahrhunderts," Tradition 16 (1971): 157 ff.

70. A survey of the methods and the underlying theoretical assumptions is H. Walter, Der technische Fortschritt in der neueren ökonomischen Theorie (Berlin: Buncker & Humblot, 1969).

71. D. André, Indikatoren des technischen Fortschritts, pp. 101-15.

72. The leadership of railways in the 1840s to 1860s or 1870s is now well established. See Fremdling, op. cit.; R. Spree, Die Wachstumszyklen der deutschen Wirtschaft von 1840s bis 1880—mit einem konjunkturstatistischen Anhang (Berlin: Duncker & Humblot, 1976). The metal industries as a leading sector have not yet been subjected to similar studies.

73. Hoffmann, Das Wachstum der deutschen Wirtschaft, pp. 266, 389 ff.

74. For more detailed data, see W. Fischer, "Bergbau, Handwerk und Industrie 1850-1914," in Aubin and Zorn, op. cit., II, pp. 527-62.

75. Hoffmann, op. cit., pp. 266 ff.

76. Calculated from ibid.

77. W. Zierold, Forschungsförderung in drei Epochen. Deutsche Forschungsgemeinschaft-Geschichte-Arbeitsweise-Kommentar (Wiesbaden: Steiner, 1968); T. Nipperdey and L. Schmugge, 50 Jahre Forschungsförderung in Deutschland. Ein Abriss der Geschichte der deutschen Forschungsgemeinschaft 1920-1970 (Berlin: 1970).

78. S. Richter, Forschungsförderung in Deutschland 1920-1936, Technikgeschichte in Einzeldarstellungen no. 23 (Düsseldorf: VDI-Verlag, 1972), pp. 11, 37.

79. Science Indicators 1974. Report of the National Science Board (Washington, D.C.: U.S. Government Printing Office, 1975), pp. 162 ff.

80. I take this phrase from T. P. Hughes, "Technological Momentum in History: Hydrogenation in Germany 1898-1933," Past and Present 44 (1966): 106-33.

81. Richter, op. cit., p. 58.

82. C. Weiner, "The Refugees and American Physics in the Thirties," Perspectives in American History 2 (1968): 217.

83. K. H. Ludwig, Technik und Ingenieure im Dritten Reich (Düsseldorf: Droste, 1974), esp. ch. 6: "Naturwissenschaftlich-technische Forschung—Stiefkind des Systems," pp. 210-70.

84. Science Indicators 1974, p. 165.

85. T. Ihlau and L. Rall, Die Messung des technischen Fortschritts (Tübingen: J. C. B. Mohr, 1970), p. 168; M. M. Opp, Die räumliche Diffusion des technischen Fortschritts in einer wachsenden Wirtschaft (Baden-Baden: Nomos Verlagsgesellschaft, 1974), p. 31.

86. Bundesminister für Forschung und Technologie, ed., Fünfter Forschungsbericht der Bundesregierung (Bonn: Federal Government, 1975), p. 63.

87. Stifterverband für die Deutsche Wissenschaft, Forschung und Entwicklung in der Wirtschaft 1971, pp. 99, 101.

4

THE ROLE OF SCIENCE
AND TECHNOLOGY IN THE
NATIONAL DEVELOPMENT OF
THE UNITED STATES

Nathan Rosenberg

BACKGROUND

When the Americans declared their political independence
from England in 1776, their country was a colony in the economic as
well as the political sense of the term. Indeed, its distinctive features
would conform closely to those used to characterize the colonial condition in the years since World War II. Its population was overwhelmingly rural and agricultural. This was scattered along the eastern
seaboard, with slightly more than 200,000 living west of the Allegheny
Mountains. Only 5 percent of the American population (totaling about
3.9 million) lived in cities in 1790, and the cities were extremely
small by modern standards.

More than 75 percent of the labor force in 1800 was engaged in
farming activities or in other employment directly concerned with
primary product extraction, such as fishing and mining. Agriculture
alone employed 1.4 million out of a total labor force of 1.9 million,
with a much smaller number engaged in mining and fishing. [1] Manufacturing employment in 1800 probably accounted for less than 3 percent of the labor force. These were primarily small-scale craftsmen
producing basic articles: shoemakers, hatmakers, chandlers, and
coopers. Important activities were milling operations, especially
flour mills and sawmills, and, of course, the indispensable blacksmith. These activities were widely distributed geographically,
usually catering to small local markets.

When surrounded by the sophisticated manufacturing technology
of the late twentieth century, it is easy to forget how intimately
dependent earlier technologies were upon variations in environmental
factors, such as weather. In the colonial period, for example, it

was not only agriculture whose fortunes were closely linked to the vagaries of weather. "Many mills were shut down in dry weather, and a drought affected the price of provisions in the Philadelphia market as much by the scarcity of flour it occasioned as by its influence upon the grain crop. During a severe winter ice was equally effective in putting a stop to grinding."[2]

With respect to the country's role in international trade, its position was classically "colonial." That is, it had been throughout its colonial history, and still was in its early years as an independent nation, an exporter of primary products, or of primary products that had been subjected to modest processing operations, and an importer of finished manufactured products from the "mother country."[*] It became a substantial importer of British capital only after independence.

The composition of American exports from the earliest years had been shaped by the abundance of natural resources, on the one hand, and by the comparative scarcity of natural-resource-intensive products in certain foreign markets, on the other. Thus America's immense forest resources led to the immediate development of forest-based industries oriented to export markets. Lumber formed an important part of the earliest cargoes shipped from the Plymouth and Virginia colonies. It is recorded that, as early as 1671, New Hampshire, which had a tiny population, was shipping 20,000 tons of boards and staves to foreign markets, in addition to ten shiploads of masts.[3] America's great pine forests quickly became sources for the export of ship timber, pitch, tar, and turpentine. The white oak was readily shaped into barrel staves for sugar shippers and wine merchants, and also supplied containers for a wide assortment of colonial exports: fish, salt meats, flour, biscuits, rum and molasses, pitch, tar, turpentine, and whale oil. The forest provided the basis for four major colonial industries: lumbering, shipbuilding, naval stores, and potash making.[4] It should be recalled that the iron industry, which assumed increasing importance in the later colonial period, was also a forest-based industry, inasmuch as the prevailing iron

---

[*]"In 1770 the colonies' exports of finished manufactures were, at most, less than 1 percent of total exports. The main manufactured items they did export were spermaceti and tallow candles and soap, all closely related to their rich natural resources and dependent mainly on the existence of the fishing industry rather than on any particular advantage in manufacturing." Robert E. Lipsey, "Foreign Trade," in Lance D. Davis et al., eds. American Economic Growth (New York: Harper and Row, 1972), p. 568.

technology required charcoal as a fuel. Indeed, America's great abundance of wood for charcoal was a major part of the explanation of the fact that, on the eve of the Revolution, it was producing about 15 percent of the world's total output of iron.[5]

By the end of the colonial period, the impact of agricultural progress and its successful organization in the production of export surpluses for several overseas markets—not only Great Britain but also southern Europe and the Wine Islands, the West Indies, and Africa (the last market received primarily rum)—is evident.

Tobacco, bread and flour, rice, dried fish, and indigo accounted for more than 60 percent of the total value of American exports over the period. Other significant items were deerskin, naval stores, whale oil, bar iron and pig iron, potash, Indian corn, spermaceti candles, beef and pork, livestock, and wood products. The predominance of agricultural and primary products is apparent; indeed, there is little else.[6]

EARLY INCREASES IN PRODUCTIVITY

Our main task in the following pages will be to examine the roles played by technology and science in bringing about a transformation of this isolated colonial economy into the world's largest and most powerful industrial colossus. Much remains to be learned about the long-term growth experience of the American economy, especially for the period before 1840, a year that provided the first reasonably reliable economic census. Nevertheless, the broadest contours of the development can be discerned, at least with respect to the growth in per capita output, our most direct measure of improvement in material well-being. According to Robert Gallman's estimates, output per capita grew between 1840 and 1960 at an annual rate of slightly more than 1.5 percent—something like a sixfold increase in output per capita. Linking this experience with the pre-1840 period must be much more conjectural, in view of the absence of reliable data. Nevertheless, there is a consensus that eighteenth-century growth levels were a good deal more modest than those that prevailed after 1840. Gallman's estimate for the rate of growth of per capita output for the long period from 1710 to 1840 is between 0.3 and 0.5 percent per year.[7]

In moving from these data on the long-term growth performance of the American economy to a consideration of the role played by science and technology, it is necessary to make some basic distinctions. The aggregate growth of the American economy can be thought of as involving two components. On the one hand, the total output of the economy has grown because the volume of inputs—land, labor, and capital—has grown. Thus the long-term history of the American economy has had as a central feature the growth in population and the size of the labor force, high rates of capital formation, and, until about the last decade of the nineteenth century, an expansion of the available natural-resource inputs as population and the transportation network moved westward and made vast new tracts of land and mineral resources available for economic exploitation. The growing supply of valuable economic inputs would naturally lead us to expect a growth in the total output of the economy. What is not so obvious is why it should lead to a growth in output per person. Indeed, it is possible to visualize an extensive growth of the economy that simply continues to replicate existing conditions, with no improvement in output per person.

In fact, of course, a distinctive feature of American growth has been an increase in aggregate output at rates that have substantially exceeded the growth in the supply of inputs. Such long-term observations inevitably raise serious conceptual and methodological problems, but the overriding and unshakable fact is that both total output and per capita output have grown at rates well beyond what could be accounted for by the associated growth of conventionally measured inputs. It is this central phenomenon of the growth in the productivity of resources, and not just their growth in volume, that serves to focus our attention upon the role of technological change in America's long-term growth experience.[8]

What has been the role of technological advance in generating the long-term economic growth of the American economy? There is a strong intuitive as well as theoretical case for believing that it has been the primary factor. Moreover, one can compile massive amounts of evidence in documenting the productivity-increasing effects of individual innovations and innumerable improvements. In addition, as Moses Abramovitz and others have shown, the historically observed growth in inputs alone can plausibly account for only a rather small fraction of the observed growth in output. And yet, it is also apparent that there have been many sources of productivity growth other than those emanating from the realm of technology.*

---

*It may also be argued that the contribution of technological innovation to the growth of output and ultimate consumer well-being is understated in our conventional national income and national product accounts. This is so because these measures do not adequately reflect the results of technological changes that take the form of new and superior products or product improvement. Furthermore, product innovation (as opposed to cost-reducing innovations) has played a

Per capita output may grow for many reasons. The proportion of the population entering the labor force may rise, perhaps because of favorable changes in age composition or a decline in fertility that allows women to participate more readily. Improvements in the organization and flow of work resulting from time-and-motion studies or innovations in factory design or the redesigning of general-purpose machinery to suit the more highly specialized needs of individual classes of users may bring drastic increases in productivity. Purely managerial improvements may result in the assignment of workers to tasks for which they are more suited, or may bring about improvements in work scheduling or reductions in inventory requirements. More effective remuneration schemes may raise the incentive of workers to work harder, or the reduction in hours worked may lead to increases in output per man-hour. Indeed, each of these factors, as well as others, has played an important role in American development.

----

particularly important role in American economic growth since the 1870s. As Simon Kuznets has argued, much technological innovation has been associated with the rise of new industries producing new products. This has not been a random or adventitious association. Rather, high aggregative growth rates in an industrial economy are a reflection of a continuous shift in product and industry mix. According to Kuznets, all rapidly growing industries eventually experience retardation in growth as the cost-reducing impact of technological innovation in each industry eventually approaches exhaustion. A continuation of rapid growth therefore requires the development of new products. In view of the typically low long-term income and price elasticity of demand for old final consumer goods, further cost-reducing innovations in those industries will have a relatively small aggregative impact. In Kuznets' view, a sustained high rate of growth depends upon a continuous emergence of new inventions and innovations that provide the bases for new industries with high rates of growth that compensate for the inevitable slowing down in the rate of invention and innovation; it also depends upon the economic effects of both, which retard the rates of growth of the older industries. A high rate of overall growth in an economy is thus necessarily accompanied by considerable shifting in relative importance among industries, as the old decline and the new increase in relative weight. Simon Kuznets, Six Lectures on Economic Growth (Glencoe, Ill.: Free Press, 1959), p. 33. For a more detailed presentation, see Simon Kuznets, Secular Movements in Production and Prices (Boston: Houghton Mifflin, 1930).

Considerations such as these are highly germane to an examination of the economic performance of the American economy during its colonial period and perhaps for some time afterward. For, although growth during the colonial period was not rapid by later standards, growth nevertheless did occur that, by the end of the colonial period, was sufficiently great to have raised living standards of the colonists to levels at least comparable with, if not equal to, those prevailing in England. What role did improvements in technology play in raising living standards over the century and a half of colonial status?

The prevailing view of both older and more recent scholarship is that such improvements were not of great importance during the colonial period. Certainly in agriculture technological improvement appears to have been insignificant.[9] In fact, the rather quaint notion of "Yankee ingenuity" notwithstanding, while the Americans borrowed technology freely and extensively from Europe, there was little genuinely inventive activity in the long colonial period.

What, then, accounted for the slow but steady increase in colonial productivity? Apparently an undramatic but cumulatively powerful combination of factors centering upon improvements in the organization of industries and the more effective functioning of markets and related institutions. The changes collectively, gradually transformed the colonial economy from one in which subsistence (or, at best, localized) activities predominated to one characterized by a reasonably advanced degree of specialized activities integrated into well-developed commercial markets. The benefits included not only the well-known gains from an increasing division of labor of the sort celebrated in the opening pages of Adam Smith's Wealth of Nations but also the benefits of a growing geographic specialization that required for their realization the organization of trade on an increasingly interregional basis. The rise in the proportion of productive activities directly oriented toward markets was a major factor in the growth of productivity. Such orientation allowed the colonial economy to concentrate upon those specific activities for which it was best endowed by its abundance of natural resources, and permitted it to acquire other goods through purchase.*

---

*Guy Callender long ago emphasized the decisive role of the market as a necessary condition for the prosperity of newly settled regions. The mere presence of a certain number of industrious people in a country having abundant fertile soil, forests, mines, and fisheries by no means ensures a rapid development of those resources

These developments were closely tied to improvements in shipping and distribution, where major advances in productivity, and consequent cost reductions, had occurred in the century preceding independence. Improvements in this sector, in turn, were crucial to breaking the pattern of agricultural self-sufficiency or localized bartering or trade. The speed with which the colonial economy was integrated into the larger world trading community, with all the benefits flowing from such integration, was accelerated by productivity improvements in shipping and distribution. The growing population of the colonies and their wide geographic distribution provided increasing opportunities for mutually beneficial trade and specialization, but a lowered cost of trading was an essential precondition for such benefits.

Although there are substantial economies of scale in shipping (labor costs per ton-mile decline as ship size increases), it was difficult to employ large ships in the colonial market without encountering substantial increases in underutilization. This was because colonial markets were small and widely scattered. Moreover, the

---

and a consequent large production of wealth. The economic advantages possessed by such people consist simply in the ability to produce food and raw materials with a small outlay of labor. Before they can utilize these advantages, certain favoring conditions must be present. They must be able to exchange these commodities for the commodities they need; must have a market. Guy Callender, "The Early Transportation and Banking Enterprises of the States in Relation to the Growth of Corporations," as reprinted in Joseph T. Lambie and Richard V. Clemence, eds., Economic Change in America (Harrisburg, Pa.: Stackpole, 1954), p. 526. Such markets need not, of course, be foreign markets. In fact, one of Callender's central concerns is with the transportation improvements that linked the agricultural Midwest with the markets of the urbanized eastern seaboard and the South, which, because of its specialization in cotton culture, became a substantial food-deficit region. The linkage with these markets in turn depended upon an improved transportation system, which accounted, in Callender's analysis, for the willingness of state governments to provide immense financial support for transportation projects. Indeed, according to Callender, the opening of the West after the War of 1812 marked the shifting of the center of interest in American economic development, and the beginning of what has been the chief object of its economic activity ever since, the application of capital to the settlement of the interior and the development of natural resources. Ibid., p. 525.

pervasiveness of piracy and privateering in Atlantic and Caribbean waters meant that merchant vessels had to carry substantial armaments, and such armaments increased crew requirements. The increasingly effective control over, and eventual elimination of, piracy and privateering in the eighteenth century played a major role in the progressive decline in shipping costs. It made possible a reduction in armaments and crew requirements and permitted the adoption of some variant of the "flyboat," a highly efficient Dutch innovation in merchant shipping design introduced around 1595.

> Sharply different from other contemporary ships, the flyboat was a specialized merchant vessel with a design favourable to the carriage of bulk commodities. Its bottom was nearly flat; it was exceptionally long compared to its width; it was lightly built because armament, gun platforms, and reinforced planking were eliminated; and it was known for the simplicity of its rig. In contrast, English and colonial vessels were heavily built, gunned, and manned to meet the dual purposes of trade and defense. They cost more per ton to build because of heavy construction and armaments, and they used more men per ton because of more complex sail patterns, for manning armaments, and for additional protection in the event of attack.[10]

A major cost component of the commercial sector was the length of the voyage. Here, although we observe no significant increase in the average speed of ocean-going vessels in the Atlantic trade, we do observe a sharp decline in the average length of voyage due to improvements in the organization of colonial markets.[11]

Thus, the growing security of trade, the increasing size of markets, and the more effective organization of markets all contributed to the long-term growth of the colonial economy.

EARLY MANUFACTURING

The persistence of some of the main features of the colonial economy into the decades after the achievement of independence from England suggests that these features had not simply been rooted in British mercantilist regulations but, rather, reflected the operation of more pervasive economic forces. American exports, for example, continued to consist primarily of resource-intensive products from the land, the forest, and the sea. Even as late as 1850, finished manufactured goods constituted less than 10 percent of American exports.

Within the category of resource-intensive products, those from the agricultural sector assumed a role of increasing importance in American exports until the Civil War. The movement of American population into the Midwest and the construction of a transportation network (canals, railroads, and steamboats) made possible the commercial exploitation of immense quantities of fertile land that provided the resource basis for food exports. At the same time the expansion of cotton cultivation in the South was given an immense forward thrust by Eli Whitney's invention of the cotton gin in 1793, which provided an easy mechanical alternative for the highly labor-intensive manual removal of the seeds from the cotton after the cotton had been picked and eliminated a serious constraint upon the westward spread of cotton cultivation. It was now economically feasible to grow the short-staple upland cotton, a fact that released cotton from its seacoast confinement and permitted it to be grown in Alabama, Mississippi, Louisiana, Tennessee, Arkansas, and Florida. Cotton exports accounted for more than half of the total value of exports over the period 1815-60. [12]

The reduction of British protection for its domestic agriculture reinforced the effects of America's comparative advantage in agriculture during the middle of the nineteenth century. In the period 1820-60, agricultural products accounted for more than 90 percent of the value of American exports. Although agricultural exports declined in importance in the post-Civil War period, they continued, in most years, to account for more than three-quarters of all American exports until the 1890s. Subsequent to that period, a persistent secular decline finally set in. [13] Finished manufactured goods increased as a proportion of American exports in the twentieth century, rising from about a third of the total in the second decade of the century to about 60 percent of all exports after 1940. (See Table 4.1.)

Even within the category of finished manufactured goods, there have been significant compositional shifts that have reinforced the movement away from the resource-intensive end of the spectrum.

> The rising share of finished manufactures in U.S. exports tells only a small part of the story of changing U.S. comparative advantage, because manufactured goods are an enormously varied collection of products. At one extreme are simple transformations of agricultural or mineral products, the main value of which consists of the raw material itself, such as . . . candles and spermaceti wax . . . which were closely tied to the fishing industry. At the other end of the scale are complex machinery and scientific equipment in which the cost of the raw material is insignificant. American manufactured-goods exports have steadily shifted toward the latter group. [14]

TABLE 4.1

Share of Finished Manufactures in U.S. Exports
and Imports, 1770 to 1964-68
(percent)

Period	Exports	Imports
1964-1968	60.2	45.1
1954-1963	59.4	30.7
1944-1953	59.9	18.6
1939-1948	62.3	17.7
1929-1938	45.6	22.2
1919-1928	37.7	18.8
1909-1918	33.3	17.7
1904-1913	27.0	21.4
1899-1908	23.3	22.4
1889-1898	16.1	24.8
1879-1888	14.1	28.3
1869-1878	15.9	34.6
1859-1868	15.7	43.5
1850-1858	12.6	47.2
1850	12.6	54.6
1840	9.8	44.9
1830	8.5	57.1
1820-1821	5.8	56.4
1770	0.8	n.a.

n.a. = not available.

Source: Lance Davis et al., American
Economic Growth (New York: Harper and Row,
1972), p. 568.

All this reflects a long-term transformation of the American economy. In the early colonial period it possessed an abundance of natural resources but very little capital and limited supplies of labor. Since the nation's independence, high rates of population growth and capital formation, including investment in education and training,* have gradually but systematically shifted the comparative advantage away from resource-intensive goods and toward more capital- and skill-intensive production. Indeed, it is widely suggested that America's present comparative advantage is no longer even in the output of the capital-intensive, mass-production industries but, rather

> . . . in research-intensive products characterized by constant innovations, the production of which requires heavy expenditures on research and development and extensive employment of highly educated members of the labor force. Several studies have shown that the U.S. share of international trade tends to be high in such industries as aircraft, machinery, and drugs, in which scientists and engineers form a large proportion of those employed and research and development expenditures are high relative to sales.[15]

---

*It is essential to note that the American commitment to the importance of education has roots deep in the country's history and that the (nonslave) population was already one of the best educated in the world, according to certain crude measures, early in the nineteenth century. According to Albert Fishlow: "The fact of the matter is that from the very beginning, the inhabitants of the new world attached considerable importance to education. Mulhall's conjectural 1830 enrollment rates place the United States second only to Germany, and well beyond France and the United Kingdom; an adjustment to reflect education of whites alone, as an allowance for the institutional barrier slavery imposed upon education of Negroes, would place America in the first rank. The Compendium of the Seventh Census reflected this same leadership in an international comparison made 20 years later. Literacy rates tell the same tale. It should come as no surprise that even earlier in the nineteenth century the United States probably was the most literate and education conscious country in the world." Albert Fishlow, "The American Common School Revival: Fact or Fancy?" in Henry Rosovsky, ed., Industrialization in Two Systems (New York: Wiley, 1966), pp. 48-49.

The changing composition of American imports further supports our characterization of historic shifts in comparative advantage. Over half of American imports generally consisted of finished manufactured goods until about the middle of the nineteenth century. The domestic textile industry began its rapid growth under the artificial conditions of the Napoleonic Wars and the War of 1812, when the importation of British products was severély curtailed and, for a while, completely cut off. In the postwar period the fledgling cotton industry entered a period of remarkably rapid growth, and a protective tariff placed restrictions upon the importation of competing foreign products. A similar process of import substitution occurred later in iron and steel, also supported by import restrictions in the form of a tariff. The role of tariff policy has long been a subject of intense debate. In general the predominant view of modern scholarship is that although cotton textiles and iron and steel were supported by tariff protection, such protection was not of decisive overall importance in shaping their long-term growth.[16]

The share of finished manufactures in American imports declined in the second half of the nineteenth century and remained at 20 percent or lower through the first half of the twentieth century. Then, departing from a long-term trend of over a century, they rose abruptly in the years after World War II.

Although imports constituted á large share of several important industries in 1869, these shares fell drastically in subsequent decades. At the same time the share of imports in domestic consumption rose for the nonmanufacturing sectors: agriculture, fisheries, and mining. It is interesting to note that the one large increase in the share of a manufacturing industry between 1909 and 1947 occurred in the resource-intensive category of paper products.

THE ROLE OF EUROPEAN TECHNOLOGY

The discussion so far has suggested that the changing factor endowment of the American economy from colonial days onward shaped and restructured the nation's economic opportunities in decisive ways. The evidence provided by America's gradually changing role as a participant in international trade (in addition to such epiphenomena as wars, harvest failures, and changes in government policies) faithfully mirrors the underlying resource endowment of a growing and maturing economy. But this mode of description is excessively static in failing to make explicit the role of a dynamic technology in adapting to and in exploiting the conditions just described. It is to such considerations that we now turn.

The American colonies, and later the United States, must be understood as an extension of European (primarily British) culture in the North American wilderness. The earliest technologies employed by the European settlers were those they had acquired prior to their emigration. The mechanism of the transfer was the knowledge and the technical skills incorporated in each settler, plus the simple tools, utensils, and implements brought by the settlers. Throughout the entire colonial period, it should be remembered, the European technology was of a preindustrial nature. The skills transferred were of two basic kinds: those of the artisan and handicraft trades, such as the carpenter, mason, and blacksmith, on the one hand, and the food-growing skills of the husbandman, on the other.

In the early years the European patterns of specialization usually did not persist in the colonies. The small size of the early settlements did not encourage the emergence of full-time craft specializations; and even in later years such patterns emerged only in the larger cities. It was much more common in the New World than in the Old World for a person to engage in a diverse range of activities rather than in a single one. Moreover, the farmer, carpenter, or blacksmith would more commonly fashion his own tools rather than acquire them through purchase; and the individual household typically provided for a much wider range of its own needs—especially clothing, simple household requisites, and food preparation—than was the case in Britain.

In the more complex and capital-intensive industries, such as metallurgy, there was a greater and more prolonged reliance upon the importation of British equipment. In almost all cases, however, the dependence upon skilled workers in effecting the transfer remained critical. Even with the later emergence of industrial technologies, the skilled worker or artisan remained the vital carrier from the more advanced to the less advanced society. The most celebrated instance, of course, was that of Samuel Slater, who, because of British constraints upon the export of machines, carried the technological know-how of the newly industrialized British textile industry indelibly etched into his memory and transplanted it to Rhode Island, where he established the first American textile mill.

The United States was a beneficiary of those extensive technological innovations in Great Britain that we now call the Industrial Revolution. The American experience with industrialization did not necessarily involve an inventive process but, more commonly, the transfer of technologies that had been developed elsewhere. This transfer process was not an easy one; in fact, it required a great deal of skill, mechanical competence, and technical knowledge. Technologies are always embedded in social matrices involving differences in skills, managerial competence, and organizational forms in addition to the unique aspects of each geographic environment. As a result, the successful transfer of technology was much more than the mere transportation of a piece of hardware from one geographic location to another. It involved frequently complex questions of selection, adaptation, and modification to make it possible for a technology to function effectively in an environment different from the one in which it originated. Apparently minor differences in resources often required major alterations in technology; such was the case, for example, in metallurgy and other activities involving chemical processes. [17]

Although the early dependence upon European technology, therefore, was very great, it was not total. Since the differences in resource endowment and environment generated problems and presented opportunities outside the range of European experience, European solutions and techniques were sometimes impossible and more often highly inefficient. For example, whereas forest resources were increasingly scarce and costly in Britain, they were almost embarrassingly abundant in America. Indeed, wood and its valuable by-products, such as potash and pearlash, were often quite literally the waste products of the land-clearing procedures preceding agricultural settlement. As late as 1810, Tench Coxe could claim that the potash and pearlash derived from clearing the trees of new farmland for cultivation would "nearly compensate the settler" for the expense thus incurred, at least where the land was "convenient for boat navigation."[18] Whereas throughout the eighteenth century (and earlier) the high price of timber as a fuel and as a building material provided a powerful incentive in Britain to innovate in iron production (including the substitution of mineral fuels for wood or charcoal), Americans were intensely exploring the possibilities of a wood-intensive technology. There existed no pool of relevant skills and techniques in the Old World upon which the New World could draw without adaptation.

Thus, much of what distinguished the American experience was attributable to the fact that, when the nation commenced industrialization in the first half of the nineteenth century, it did so with the pool of British experience upon which to draw, but also from a distinctly more favorable resource position. The direction of technological innovation was a consequence of this circumstance. Much of it was specifically geared to the intensive exploitation of natural resources that existed in considerable abundance relative to capital and labor. In spite of America's late industrial start compared with Britain, it quickly established a worldwide leadership in the design, production, and use of woodworking machinery. These included a whole range of machines for sawing, planing, mortising, tenoning, shaping, and boring, in addition to an entire armory of woodworking

TABLE 4.2

Lumber Consumption in the United States and United Kingdom, 1799–1869

| | United States | | United Kingdom | |
Year	Consumption in Board Feet (thousand)	Per Capita Consumption	Consumption in Board Feet (thousand)	Per Capita Consumption
1799	300,000	58	102,703	10
1809	400,000	57	121,916	10
1819	550,000	59	244,745	17
1829	850,000	67	319,306	20
1839	1,604,000	98	430,267	23
1849	5,392,000	239	1,024,565	50
1859	8,029,000	259	1,796,596	79
1869	12,755,543	328	2,419,390	95

Source: Nathan Rosenberg, "America's Rise to Woodworking Leadership," in Brooke Hindle, ed., America's Wooden Age (Tarrytown, N.Y.: Sleepy Hollow Restorations, 1975), p. 56.

machines for more specialized purposes.[19] It was characteristic of these machines that they were wasteful of wood. Given the relative factor scarcities in the United States, however, such machines, which essentially substituted abundant and cheap wood for scarce and expensive labor, were admirably adapted to American needs.[20] American lumber consumption per capita was several times as great as the corresponding figure for the United Kingdom, and rose very sharply in the early years of industrial development.* (See Table 4.2.)

If American manufacturing industries are ranked on a value-added basis, the lumber industry component of wood use alone ranks as the second most important industry in the United States in 1860, close behind cotton textiles. On a similar basis, lumbering was the largest single manufacturing industry in the South and West in 1860.[21] By the time of the Civil War, a good deal of this initial resource advantage had been dissipated. The falling price of iron that was associated with the rapid expansion of the iron industry, combined with the rising price of wood, served to place the American economy back upon a somewhat more conventional "European" track. A distinct shift can be observed from cordwood to coal as fuel, and from wood to iron as a construction material.

American technology was from the earliest times particularly devoted to innovations that assisted in the utilization of wood or that reduced the cost of complementary inputs. The very first patent issued for a mechanical invention provided protection for an improved sawmill. This was in 1646.[22] Sawmills, although not invented in America, were widely used there long before their adoption in England.[23] American inventiveness early focused upon nailmaking machinery, and 23 patents for such machinery had been granted by the Patent Office before 1800.[24] The resulting dramatic reduction in the price of nails in the first half of the nineteenth century was an important element in the reduction in the relative cost of wood products.

---

*In addition, the early American cheapness of wood had led to a common practice of designing domestic fireplaces so that they could accommodate large logs, an arrangement that wasted wood fuel but economized on the labor-intensive activities of sawing or chopping wood. Stoves, of course, utilized wood supplies more efficiently for heating purposes, but were more expensive and raised the labor cost of preparing the wood. Eventually, of course, stoves replaced large fireplaces as wood prices rose substantially.

America's abundance of forest resources led to all sorts of adaptations that involved substituting natural-resource inputs for other, scarcer factors of production. The American builder relied upon wood, a most convenient and tractable construction material, in uses where his European counterpart employed stone or iron. In the construction of houses, Americans developed a distinctively new technique in the 1830s. This was the balloon-frame design, which eliminated all the heavy members of the traditional New England house (or barn). Its lightness and simplicity—it was essentially nailed together with light 2-inch by 4-inch studs—sharply reduced the total labor requirements of house construction and made it possible to substitute relatively unskilled labor for the skilled carpenter.[25] Indeed, European visitors often noted that Americans employed wood in unaccustomed ways and places, not only in building large bridges and aqueducts but also in more improbable uses, such as the framing of steam engines, canal locks, and pavements. They even, astonishingly enough, built roads (the famous plank roads) out of wood.

The increase in the relative price of forest products and the cheapening of coal and iron in midnineteenth-century America signaled a shift away from the use of wood both as a fuel and as a building material.[26] In 1850 mineral fuels supplied less than 10 percent, whereas wood supplied more than 90 percent, of all fuel-based energy. In the second half of the nineteenth century, however, the changes in the relative costs of fuel sources, as well as technological changes favoring the use of mineral resources in the manufacture of iron and steel and the production of steam power, brought about a shift to coal. Some of these changes occurred very rapidly. For example, although almost all the energy needs of the railroads were supplied by cordwood at the outbreak of the Civil War, 20 years later the railroads were using 20 times as much coal as wood, and were consuming more than 25 percent of the country's bituminous coal output. In 1899 steam power generated by coal accounted for over 80 percent of primary power capacity in manufacturing. By the early twentieth century, coal had largely displaced wood among material sources of energy. In subsequent decades coal declined in importance—a decline in which the increased use of diesel engines in transportation and the loss of household markets played major roles—and liquid and gaseous fuels increased in importance. In contrast with the prolonged dominance of coal in other industrialized countries, the supremacy of coal as an energy source in the United States was relatively short.

The drastic reduction in the reliance upon increasingly expensive wood as a fuel source had a direct counterpart with respect to the use of wood as an industrial raw material. The price of timber products quadrupled between 1870 and 1950. This increase triggered substitution of other inputs, including minerals, and appears to have induced significant technological changes that limited the utilization of timber. Whereas consumption per capita of mineral products increased almost ten times between 1870 and 1954, consumption of timber products per capita rose to a peak in the first decade of the twentieth century and then declined to almost half its 1900 level by 1954. Even in absolute terms the consumption of timber products was no greater in the mid-1950s than it had been in 1900. Timber products, which accounted for 4 percent of GNP in 1870, accounted for a mere .69 percent in 1954.[27]

This change was brought about through a broad spectrum of responses. Iron and steel were substituted for wood across a range of investment goods in the nineteenth century, as early as the pre-Civil War period. Machinery, ships, and bridges, which were made of wood in 1800, were made of iron or steel in 1900. In construction, by far the largest consumer of lumber, there has been an increasing reliance upon traditional masonry and other mineral building materials and upon aluminum. More recently, science-based technological change has produced plastics and fiberglass materials that have served as substitutes for wood (as well as for other natural materials, such as leather) over a wide range of uses. New materials, such as plastics and aluminum foil, and older ones, such as glass, have replaced forest products as a packaging material. Further technological changes economize upon wood requirements without the substitution of competitive materials, or substitute cheaper woods for more expensive ones—for example, plywood, fiberboard, and wood veneers.

Other technological changes, such as the self-powered chain saw, the tractor, and the truck, have reduced the cost of extracting and transporting the timber from its forest stands. Still other technological changes have, in effect, significantly increased the size of the forest resource base by making possible the utilization of low-grade materials that previously had gone unused. Until the 1920s the wood-pulp industry utilized only the spruce and fir trees grown in the northern portions of the country. Improvements in sulfate pulping technology during the 1920s made possible the exploitation of faster-growing southern pine. By the mid-1950s the South accounted for over half of the country's wood-pulping capacity. The rising cost of timber in the second half of the nineteenth century led the railroads to search for new techniques that would prolong the life of railroad ties. The eventual replacement of untreated railroad ties with those impregnated with creosote was estimated roughly to double the expected life of a tie, from 14 to 28 years, thereby saving labor to replace ties and permitting the use of otherwise inferior types of wood.[28]

AGRICULTURAL TECHNOLOGY

The main features of the long-term transformation in American agriculture since the nation's independence also bear the imprint of a shift from relative abundance to greater relative scarcity of natural resources.* No situation comparable to American abundance had existed in any of the other societies of Western Europe from which settlers had come to America. In agriculture this took the form of a high land-to-labor ratio, a feature that left a deep impression upon the direction of technological innovation in America. A major thrust of agricultural innovation under these circumstances, and one that became particularly conspicuous around the middle of the nineteenth century, were those designed to increase the acreage that could be cultivated by a single farmer. This was achieved by a process of mechanizing agricultural operations that differed from mechanization in industry in one fundamental respect. Mechanization in agriculture in the nineteenth century involved no extensive reliance upon the new power sources that were playing such a central role in industry. Although there had been a great deal of interest and discussion, and a good bit of actual experimentation, the steam plow was never widely successful; and the role of steam power in agriculture was confined to the postharvest operations of threshing, milling, and ginning. The mechanization of field operations in agriculture relied entirely upon animal power, a situation that persisted until the large-scale introduction of the tractor in the 1920s.

There was, however, a shift in the animal population as a result of mechanization. Horses were substituted for oxen on a large scale, since the latter did not move fast enough to serve as efficient sources of tractive power for the new machinery. It was the horse that supplied the power for the great mechanical innovations in nineteenth-century agriculture: the steel plow; the cultivator, which replaced the hand-operated hoe in the corn and cotton fields; and the reaper, which, by its swift acceleration of harvesting operations, greatly increased the acreage that could feasibly be cultivated by a single farmer. The reaper removed what had earlier been a basic constraint upon grain cultivation: the seasonal variations in labor requirements, which reached a sharp peak during the harvest season.

---

*The seeming circumlocation "greater relative scarcity" is deliberately employed because, although the relative scarcity of natural resources in America is not as great as in most other industrial economies, it is much greater than it was at earlier periods in our own history.

Later, output per worker was further augmented by binders and threshing machines and, eventually, by combine-harvesters. The introduction of barbed-wire fencing in the West, an area with few natural materials for fencing purposes, provided cheap fencing that could be put in place with relatively small amounts of labor.[29] The westward movement itself had a significant initial labor-saving effect on American agriculture, since it involved a movement from largely forested land to nonforested land, and since the clearing of the latter involved a much smaller labor cost per acre than did the clearing of the former.[30]

In the South, as we have already seen, the introduction of the cotton gin provided a substitute for the highly labor-intensive activity of manually removing the seeds from the cotton, and in so doing it made possible the rapid westward spread of cotton cultivation. The introduction of the mechanical cotton picker in the 1920s was so labor-saving that it alone was responsible for a vast outmigration of black labor from the states of the Confederacy, a fact closely connected with the growth of urban ghettos in later years.[31] It is worth noting that the introduction of the mechanical cotton picker also had a powerful impact upon the relocation of cotton cultivation. Since the mechanical cotton picker works much more effectively on a flat terrain than on a hilly one, this innovation (together with the increased use of liquid nitrogenous fertilizer) played a further important role in shifting the cultivation of the crop from its traditional region and into the broad, flat, previously barren expanses of western Texas, Arizona, and California.

Thus, the major technical innovations in nineteenth-century American agriculture (as well as some important twentieth-century ones) consisted primarily of labor-saving and land-using mechanical devices, often extremely simple, that drew extensively upon the cumulating pool of technical skills and knowledge in the growing industrial sector. Such machinery had the primary effect of raising agricultural output per worker, but not the productivity of land, the abundant nineteenth-century input. Twentieth-century agriculture, in which land has become an increasingly scarce input and capital a more abundant one, presents a more complex picture. The process of mechanization of agriculture has continued, especially in conjunction with the tractor and associated changes, and a range of new possibilities offered by electrification. However, a particularly interesting side effect of the tractor, which was widely adopted during the interwar years, was that it was an immensely important land-saving innovation. It has been estimated that by substituting gasoline for hay and oats (the basic "fuel" of the old technology), the tractor freed some 90 million acres of land for other agricultural uses. More generally, however, it may be said that while the main thrust of mechanization in agriculture was, and continues to be, labor-saving, the growth of the stock of biological knowledge in recent decades has imparted a pronounced land-saving bias to technological change.

The period after 1940 marks, quite simply, the most dramatic improvement in the history of American agriculture; the continuing thrust of mechanization merged with another new and powerful source of innovations: the life sciences, which, together with new industrial inputs (especially of a chemical-based sort) produced opportunities for vast increases in output per acre. From 1940 on, output per acre of land, which had been roughly stable at least as far back as 1880, exhibits a new growth pattern.[32] Output per acre (as well as output per man-hour) for the major crops—wheat, corn, and cotton—moved upward at rates for which there was no earlier precedent. Mechanical improvements continued to play an important role, especially as the rising demand for labor in nonfarm activities during and after World War II raised labor costs and provided a continuing inducement to mechanization. Machinery was made more versatile, size and design were changed to accommodate specific geographic conditions or crop requirements, a growing family of implements was made for the tractor, and the price of machine power itself declined with the introduction of such items as higher-compression engines. Furthermore, changing patterns of regional specialization tended to increase output per acre by concentrating production in areas better suited to particular crops. Expansion of irrigation facilities was important in some areas.

Nevertheless, the dominant factor was the successful exploitation of the slowly developing body of knowledge of the life sciences, knowledge dealing with the fundamental biological processes of life and growth, especially genetics and biochemistry. The application of such knowledge to the work of the plant breeder permitted the development of new and far more productive seed varieties, the most spectacular instance of which was the introduction of hybrid corn in the late 1930s. Such knowledge also made possible the development of a wide range of new plant seeds with specific characteristics. Plant breeding has raised productivity in many ways: by increasing the size and improving the quality of the plant, by developing plants that are resistant to specific diseases or require a shorter growing season, and by producing new plant strains that are highly responsive to fertilizer inputs.*

---

*"Biological innovations of the yield-increasing type involve the development of crop varieties which can respond to higher levels of

TABLE 4.3

Commercial Fertilizers: Quantities Used and Average
Primary Plant Nutrient Content, 1940-66
(thousand tons)

| Year | Quantity | Percent | | |
		Nitrogen	Phosphoric Oxide	Potash
1940	8,556	4.9	10.7	5.1
1950	20,345	6.1	10.4	6.8
1955	21,404	9.0	10.5	8.8
1960	24,374	12.4	10.9	8.9
1965	33,071	16.1	11.8	9.7
1966*	35,731	16.9	12.1	10.1

*Preliminary figure.
Source: Statistical Abstract of the U.S., 1968.

The responsiveness to fertilizer has been crucial because a
significant feature of American agriculture since World War II has
been the growing utilization of chemical inputs. The inputs have
included new herbicides, insecticides, and, most important, the
synthetic nitrogenous fertilizers. Fertilizer inputs into agriculture
increased more than four times between 1940 and the mid-1960s.
(See Table 4.3.) The increase was induced by a sharp decline in
fertilizer prices relative to product prices and the prices of other
inputs.[33] (The motivation to raise output per acre had been further
strengthened by government programs that centered on the use of
acreage restrictions.) Underlying the relative decline in fertilizer

fertilization. . . . In the United States the introduction of hybrid corn
(and other high-yielding crop varieties) is closely associated with the
growth of fertilizer consumption. A major factor in the development,
introduction, and adoption of hybrid corn, and other new crop varieties, was greater responsiveness to the higher-analysis commercial
fertilizers which were becoming available at continuously lower real
prices." Yjiro Hayami and Vernon Ruttan, Agricultural Development
(Baltimore: Johns Hopkins Press, 1971), p. 121.

prices was a range of technological innovations not only in chemical engineering but also in power production. The latter was important because energy is a major component in the cost of synthetic fertilizer. Lower fertilizer prices, it must be emphasized, would have yielded only modest benefits in the absence of the new fertilizer-responsive crop varieties.

The rising levels of income in the twentieth century led to major changes in the composition of agricultural output that we cannot pursue here. One point, however, is very relevant. Increases in consumer incomes brought a shift in farm output toward those products for which there was a high income elasticity of demand, particularly beef and dairy products. Associated with the resultant increasing importance of livestock was a growing recourse to less land-using methods of feeding them, the growing use of concentrated feeds and a decreasing reliance upon the range.

The result of the developments just described is that American agriculture has become a massive purchaser of industrialized inputs. The American farmer today depends heavily in his productive operations upon the purchase of inputs from a chain of industrial suppliers for commercial fertilizers, feed for livestock, seeds and insecticides, as well as machinery and fuel. We may summarize this trend in the following way. In 1870 the typical American farm provided most of its own inputs. According to one estimate, the value of intermediate products supplied by the nonfarm sector amounted in that year to less than 9 percent of the value of gross farm product. By 1900 such purchases of intermediate inputs still amounted to only about 13 percent of gross farm product. In recent years, however, such nonfarm inputs have come to exceed 60 percent of the value of gross farm product. In effect, the growing output of American agriculture has been achieved by technological advances embodied in the inputs that the farmer purchases from the machinery, chemical, feed-processing, and related industrial sectors. It follows from these observations that the small proportion of the total labor force remaining in agriculture is therefore an extremely misleading index of the economy's total resource commitment to the production of agricultural products.

One further, and related, observation needs to be made concerning the impact of technological change upon the agricultural sector. A major thrust of technological change in the twentieth century has been the introduction of products whose effect has been to satisfy some important need through activities in the nonagricultural sectors of the economy, thus reinforcing the relative decline of the agricultural sector. The replacement of the horse by the automobile and tractor is a classic example, but there have been many others. The introduction of synthetic fibers as substitutes for cotton and wool, of chemical for natural fertilizers, of petroleum for wood as a fuel, of plastics for leather and wood, of synthetic for natural rubber—all of these innovations have meant a shift away from the products of the agricultural sector (including, in some cases, imported as well as domestic agricultural products).

MANUFACTURING TECHNOLOGY

We consider some of the central features of technological development in the manufacturing sector of the American economy in the course of the nineteenth and twentieth centuries. We earlier emphasized America's role as a borrower of foreign, primarily British, technology. Let us recall some of the most distinctive features of Britain's industrial technology as it began to emerge in the last third of the eighteenth century. It was, to begin with, a machine-using technology, one that substituted machine-controlled operations for the handicraftsman who directed the movements of a tool through his own perceptions and skills. Second, the technology of the Industrial Revolution involved the utilization of vastly increased quantities of energy inputs per worker, as epitomized by the steam engine. Third, and closely related, there was a shift to the reliance upon more abundant forms of raw material inputs, in large measure a substitution of inorganic for organic ones. In Britain coal displaced wood as a fuel, and iron (and later steel) became the quintessential industrial raw material and building material. The substitution of coal for wood in the manufacture of iron was a critical technological breakthrough that ranks in importance with the development of the steam engine. The introduction of coke into the blast furnace, together with Henry Cort's reverberatory furnace and rolling mill (introduced in the 1780s) made possible dramatic increases in the output of iron. The rapid expansion in the range of uses to which iron was put in the nineteenth century rested upon these technical achievements.

Americans, it should be noted, were highly discriminating in their borrowing patterns and highly selective in the uses to which imported technologies were put. The steam engine was rapidly transferred to the United States but, in contrast with Britain, its early uses were primarily in transportation. Although the demand for power was great and the capacity to produce steam engines was certainly well established by the 1830s, an excellent and cheaper substitute in the form of waterpower was provided along the fall line by the many fast-flowing rivers and streams of New England, where early nineteenth-century American industry was heavily concentrated. The growing reliance upon steam power was closely connected with the westward movement of population and industry into geographic regions with fewer sources of water power and where, as a result, the economic significance of steam power was greater.

American practice and design with respect to power sources, moreover, reflected the nation's unique resource conditions in crucial ways. Where waterpower was used, it was typically exploited by the construction of inefficient forms of waterwheels that involved a smaller initial capital expenditure than the efficient ones.[34] The choice was a substitution of abundant resources for scarcer capital. Similarly, although the high-pressure steam engine had been developed simultaneously by Richard Trevithick in Great Britain and Oliver Evans in the United States in the opening years of the nineteenth century, British practice strongly favored the low-pressure stationary engine, while Americans overwhelmingly adopted the high-pressure stationary engine.[35] Although high-pressure steam engines were wasteful in their utilization of fuel, they were cheaper to construct, which made them quite appropriate to the environment of the United States.[36]

The situation changed after 1840 as the limited number of attractive water sites, even in New England, was exhausted and demand for power continued to grow. The decisive influence of geography and the natural-resource environment on the selection of technology is indicated by the fact that, as late as 1869, by which time steam had just surpassed water as a source of power (on a national basis), less than 30 percent of the power employed in New England manufacturing establishments was derived from steam.[37] New England industry long remained heavily concentrated in communities that offered superior access to waterpower: Lowell, Lawrence, Hadley Falls, Holyoke, Chicopee, and Springfield, Massachusetts; Waterbury, Connecticut; Manchester, New Hampshire.

Although the availability of a good substitute slowed the pace at which the stationary steam engine was introduced in the United States, the situation was vastly different in the application of steam power to transportation. So long as population was confined to a relatively narrow strip of land east of the Appalachians, the Atlantic Ocean and bays, sounds, and tidal rivers offered a reasonably adequite basis for the cheap movement of goods. But the exploitation of the vast resources of the trans-Appalachian west was dependent upon a technology that would free commerce from the prohibitively high cost of land transport and the upstream haulage of goods. It is perhaps not too much to say that the major economic consequence of the acquisition of the steam engine in the United States during the pre-Civil War period lay in its application to new forms of transport—the steamboat and later the railroad—which (together with the canal) provided a network for the cheap movement of goods, particularly bulky agricultural products. The contrast between the relatively slow adoption of the stationary steam engine and the rapid exploitation of the steam engine for transport purposes is highly instructive. It is an excellent demonstration of the manner in which social needs, as expressed in the marketplace and as they influence business profit expectations, shaped and directed the pattern of innovative activity.

The extent to which the early American exploitation of steam was dominated by the opportunities for the development of new and cheaper transportation forms may be indicated quite simply. According to a government report on steam engines, almost 60 percent of all power generated by steam in 1838 was accounted for by steamboat engines.[38]

The situation with respect to the transfer of the metallurgical innovations of Britain's Industrial Revolution was rather more complex. The reasons have to do with the intricate ways in which metallurgical processes are enmeshed with qualitative aspects of the resource environment. The need to adapt technological innovation to specific qualitative dimensions of available natural resources is one of the most persistent themes in the history of metallurgy. Indeed, in iron and steel, the most important of the metallurgical industries, such adaptation has often amounted to nothing less than entirely new major technological innovations.

Qualitative, locational characteristics lie at the heart of the protracted delay in America's adoption of one of the most significant innovations in the history of metallurgy: the substitution of a mineral fuel (coke) for charcoal in the blast furnace and the consequent transformation of blast-furnace operations. Although this shift, as we have seen, had been essentially completed in Great Britain by 1800, as late as 1840 almost 100 percent of all pig iron produced in the United States was still made with charcoal. As late as 1860 only 13 percent of American pig iron was being smelted with the "modern" fuel, coke.[39] As Peter Temin has shown, Americans failed to adopt the new British blast-furnace technology because bituminous coal was required for coking purposes. And, although America possessed vast deposits of such coal, these deposits suffered from either of two fundamental limitations in the early nineteenth century: (1) they contained substantial amounts of sulfur, the presence of which resulted in poor-quality pig iron, or (2) they were located west of the Allegheny Mountains, far from the country's population centers.

America did possess some very rich coal deposits in eastern Pennsylvania. But they were of anthracite, a coal containing neither gas nor sulfur. Although the absence of sulfur meant that the coal could be used to produce high-quality pig iron, the absence of gas made ignition much more difficult, and useless for British blast-furnace technology. The introduction of anthracite into the blast furnace was made possible only with the development of the hot blast in the 1830s. Although the technique was responsible for substantial fuel economies, its great importance in America was that it permitted the exploitation of a resource that was both very abundant and readily accessible to America's main population centers.* As a result, by 1856, only 17 years after the first successful use of anthracite in the blast furnace, anthracite furnaces accounted for fully half of American pig iron output.[40]

The eventual shift to a coke-smelting technology came only after the Civil War. The breakthrough was linked to the subtle interplay between the specific resource requirements of a particular technology and the historical sequence in which the available resource base of the country was uncovered. As a result of the westward movement of population and the intensive exploration of the trans-Appalachian West, the high-quality coking coal of the Connellsville region of western Pennsylvania was discovered. The physical structure of

---

*In Great Britain, on the other hand, the hot blast had a very different regional impact, reflecting each region's endowment of coal. Since the main feature of the hot blast was the fuel economy it made possible, it was adopted most rapidly in regions where coal supplies were poor (Scotland) and most slowly where coal supplies were good (Wales). Charles Hyde has pointed out that "the hot blast clearly revolutionized the Scottish iron industry. It was adopted much more slowly by the rest of the British iron industry and there its impact is less clear. There were great regional variations in the potential cost savings from the hot blast and these variations explain the varying speed at which the hot blast was adopted by the iron-masters in the other major ironmaking districts. The drastic reduction in fuel consumption brought about by the hot blast accounted for virtually the entire cost saving associated with the new technique. The potential fuel savings from the hot blast was a function of the carbon content of the coal used at the furnace. In regions such as Scotland with coal with a relatively low carbon content, more coal needed to be coked to yield a given tonnage of pure carbon. The higher temperatures realized with the hot blast allowed the ironmasters to use raw coal and completely avoid coking. The lower the carbon content of the coal available to a region, the higher the probability of achieving significant cost reductions with the hot blast. . . . One would have expected the hot blast to have spread most rapidly into the regions with relatively poor coal resources. This was in fact the pattern of adoption." Charles Hyde, "Technological Change in the British Iron Industry, 1700-1860" (Ph.D. diss., University of Wisconsin, 1971), pp. 183-84.

this coal and the absence of sulfurous components made it possible
to produce high-quality pig iron. These fortunate qualities, together
with the development of a low-cost transportation network and further
technical developments favorable to coke, assured the eventual domination of that fuel in the blast furnace and, as a result, a sharply
reduced price of pig iron. The dominance of anthracite in 1860, when
it accounted for more than half of U.S. pig-iron production, therefore
proved to be short-lived. The proportion of pig iron smelted by coke
was less than 10 percent in the late 1850s. By 1870 it rose to 31 percent, by 1880 to 45 percent, by 1890 to 69 percent, and to more than
90 percent in the first years of the twentieth century.[41]

The delicate linkage of technical innovation in the iron and steel
industry to specific qualities of a heterogeneous resource base became
even more conspicuous in the innovations at the refining stage, beginning with the development of the Bessemer process in the 1850s.
Since most of these developments took place in Europe, it will be
necessary to refer to them briefly, the more so since they were
eventually responsible for placing metallurgy upon a scientific basis.

The original Bessemer process could be employed only when
certain chemical conditions were precisely fulfilled. The method
required phosphorus-free iron. The fact that Bessemer's methods
could refine only materials that fell within certain narrow limits of
chemical analysis had, of course, important economic consequences,
imparting a strong comparative advantage to those regions possessing
the nonphosphoric ores. Britain's (acid) Bessemer process grew
rapidly after exploitation of the large deposits of nonphosphoric
hematite ores located in the Cumberland-Furness area, supplemented
by imports of nonphosphoric Spanish ores from the Bilbao region.
Germany and France had only very limited deposits appropriate for
the Bessemer technique, and Belgium had none. The Bessemer process was useless for the exploitation of Europe's massive deposits of
high-phosphorus ore in Lorraine and Sweden.

The difficulties encountered with the Bessemer method had
other far-reaching consequences. British ironmakers who had purchased lease rights to the Bessemer process quickly found, to their
considerable dismay, that they could not employ it successfully. It
so happened that Bessemer had conducted his experiments with
Swedish charcoal iron, the purest form of pig iron available to him.
The resulting determination to establish the cause of the failure of
the technique with certain British ores led to a prolonged, systematic
study of the chemical processes involved in iron and steel production.
In a very real sense, the transition to modern metallurgical science
may be said to have begun in this immediate post-Bessemer period.
It rested upon an inquiry into the basic structure and composition of

metals that had been initiated by difficulties experienced in the production of iron and steel.

The ability to conduct this inquiry was immensely advanced by technological innovation in other spheres. Most important was Henry Clifton Sorby's development, in 1863, of a technique for examining metals under a microscope, by the use of reflected light, thereby opening the door to an understanding of the microstructure of steel.

> From steel, the technique spread to show the behavior of microcrystals of the nonferrous metals during casting, working and annealing. By 1900 it had been proved that most of the age-old facts of metal behavior (which had first been simply attributed to the nature of the metals and had later been partially explained in terms of composition) could best be related to the shape, size, relative distribution and inter-relationships of distinguishable micro-constituents. [42]

Another fundamental improvement in experimental technique, first announced in 1912, was the development of X-ray diffraction and its application to the study of solids, because

> . . . it at once gave a measurable physical meaning to structure on an atomic scale, and made this as real as the larger-scale structures that had been revealed by Sorby's microscopic methods half a century earlier. It was a physicist's method par excellence, and a fundamental one, which served to relate much of the unconnected data of the chemist and the metallurgist. [43]

The Thomas-Gilchrist technique, introduced in 1879 after a long search for methods that permitted the exploitation of phosphoric ores, drastically altered comparative advantages in favor of Continental steel producers. Their introduction of a "basic" lining in place of an "acid" one vastly expanded the range of ores that could be utilized in modern steelmaking, thus allowing the intensive exploitation of Europe's great phosphoric ore deposits.* The Thomas-Gilchrist

---

*Whereas the acid Bessemer process required low phosphorus content, the basic Bessemer process required a high phosphorus content, more than 1.5 percent phosphorus. Although this technique was well-suited to German ores, it was not as well-suited to Britain's phosphoric ores. Britain's Cleveland ores proved to have insufficient

technique thus made possible a great expansion of steel production in Germany, France, and Belgium after 1880, an expansion involving both the basic Bessemer and the basic open-hearth methods. What appears as a rather insignificant and humdrum technological event, the "mere" substitution of a new material for an old one in the lining, was in fact an event of immense economic and geopolitical significance.[44]

The open-hearth method could exploit a much wider range of inputs than the Bessemer processes could. In particular it could utilize ore with almost any proportion of phosphorus content. In the United States it could use a much wider portion of the available spectrum of the immense Lake Superior iron ore deposits.[45] Moreover, the process could accept a high proportion of scrap as a material input, a consideration of great and increasing importance in places with ready access to such supplies.

The gradual exhaustion of the richest iron ore supplies in the United States and elsewhere in the twentieth century shifted the economic payoff away from the earlier concern over phosphorus content and toward the development of methods that would make possible the exploitation of low-grade iron ore. The result has been the growth of a highly sophisticated technology focusing upon the use of poor-quality inputs. Ores with a low iron content are now subjected to beneficiation, an upgrading of their iron content before they are introduced into the blast furnace. Waste materials such as clay, gravel, and sand are removed and the ores are crushed and washed, so that the material entering the blast furnace is cleaner and more uniform in quality. The implications of such techniques have been great because they have made possible the utilization of huge resource supplies that would formerly have been ignored. One of the most significant examples of this is the present exploitation around Lake Superior of huge reserves of taconite, a low-grade iron ore.

These developments in iron and steel are by no means unique. Although discussions of the impact of new technologies usually concentrate upon resulting improvements in productivity, it is essential to note that the main technological innovations in the iron and steel industry in the second half of the nineteenth century also had the immensely important effect of substantially widening the range of usable natural-resource inputs. New techniques elsewhere have, in

---

phosphorus for the basic process. See Peter Temin, Iron and Steel in Nineteenth-Century America (Cambridge, Mass.: MIT Press, 1964), ch. 6, for a good discussion of the Bessemer and post-Bessemer innovations in steelmaking.

effect, augmented America's "dwindling" supply of other minerals in parallel ways. The flotation process, originally applied to the exploitation of low-grade porphyry copper ores, has been applied to a wider range of ores, both of lower mineral content and of more complex chemical forms. Techniques of selective flotation have played a major role in offsetting the decline in the quality of available resources, not only of copper but also of lead, zinc, and molybdenum.

For a long time, natural gas was treated as essentially a waste product and was burned off at the oil site or, in some cases, consumed nearby.

> The chief obstacle to better use was lack of a means of transporting gas to places of consumption. This had to wait upon the development of high-pressure pipe and of improved welding techniques for joining lengths of pipe into leak-proof pipelines. Even when this was accomplished, toward the end of the 1920s, it was unusual to find pipelines that extended for more than two or three hundred miles. The next improvement that spurred the industry was the development of heavy power equipment to lay pipe; by the mid-thirties natural gas was being transported economically over as long a distance as 1,000 miles.[46]

With the rapid growth of the long-distance pipeline system in the 1940s and 1950s, natural gas completed the transition from the status of a nuisance to one of the most important energy sources.

A major economic function of technological innovation, therefore, has been to offset the increasing scarcity of specific natural resources by progressively widening the range of exploitable resources. As a result, although particular resources of specified quality do inevitably become increasingly scarce, the threat of a generalized natural-resource constraint upon continued economic growth by no means follows. As the discussion of iron and steel suggests, it is not very illuminating to ask how long it will be before we "run out" of specific resource inputs, defined and estimated in physical units. That is simply not an interesting question, partly because there are seldom sharp discontinuities in nature. But, more important, by making it possible to exploit resources that could not be exploited before, technological change is, in economic terms if not in geological terms, making continuous additions to the resource base of the economy. What we are more normally confronted with are limited deposits of high-quality resources and then a gradually declining slope toward lower-grade resources, which typically exist in abundance. One of the greatest benefits of a dynamic technology has been to shift the economy from dependence upon scarce sources of materials to dependence upon more abundant sources, a shift that has been dramatized in the twentieth century by the extraction of nitrogen from the air and magnesium from seawater.[47]

The technological innovations in metallurgy and power generation eventually provided a cheap and abundant supply of metals and energy that formed the basis for industrialization throughout the economy. There is an additional aspect of those innovations, however, that must be emphasized. This is that they were interrelated, or mutually reinforcing, in a way that further contributed to the overall dynamism of the industrializing process. It was frequently the case that one innovation could not be extensively exploited without one of the others in the cluster, or that the introduction of one innovation enabled the others to operate far more effectively. Metallurgical improvements made it possible to build more efficient steam engines: cheaper, with greater precision of component parts, capacity to withstand higher pressures and temperatures, and so on. Steam engines were used to provide a hot stream of air inside the blast furnace. This hot blast improved the efficiency of the combustion process, lowered fuel costs, and thereby reduced the price of iron. In a real sense, then, cheaper metal meant cheaper power, and cheaper power meant even cheaper metal. Also, cheap iron made possible the building of the railroads. Railroads lowered the very great costs of bringing bulky iron ore and coal together. By thus lowering transport costs, railroads reduced the cost of making iron. Cheaper iron meant lower transportation costs, which meant even lower costs for producing iron. Thus, the end result of the spread of industrial technology was a vast increase in human productivity, but it is impossible to understand why this growth in productivity was so great without an appreciation of the interrelated and mutually reinforcing nature of the technological innovations. To a considerable extent, it was a system of interlocking changes that fed upon themselves.

EFFECT OF NEW INDUSTRIAL TECHNOLOGIES

The spread of industrial technology brought with it certain distinct effects. In comparison with the handicraft technology that it displaced, the new technology was relatively capital-intensive and created persistent pressures to enlarge the scale of the productive operation. Both the matter of capital intensity and the matter of scale had important consequences. For example, since the new technology had much greater capital requirements than the old, it meant that, at least for some time, a greater quantity of resources had to be committed to capital formation. An increased fraction of available resources had to be devoted to activities that made no immediate contribution to present consumption and living standards. The capital resources absorbed in the building of the American railway network were immense, and the limited volume of domestic resources was supplemented by substantial foreign borrowing, especially from Great Britain.

The growth in scale, as new technologies brought with them larger and larger optimum size plants, had far-reaching consequences It meant an unprecedented concentration of labor in large factories and in a relatively small number of places. The factory system, in turn, involved totally new ways of organizing the labor force. It meant new forms of work discipline. It meant new work rhythms, the tyranny of the clock and the need to synchronize the efforts of large numbers of people. It meant allowing the pace of work to be determined by the pace of an impersonal machine process, whose pace was in turn set by a capitalist employer anxious to achieve the most intensive use of his highly expensive capital equipment. He was driven further in the same direction by the intensity of competitive pressures and market forces in general. Along with this rising importance of employee status, in which the worker was becoming a "hand," the family unit became less and less relevant as a social unit for the organizing of productive activity.

Under earlier technologies, however organized, manufacturing costs had consisted mainly of labor and raw materials. Often the worker had labored in his own home, typically utilizing only a modest amount of equipment, perhaps no more than a few simple tools or perhaps a small hand loom. Most of the costs of production varied directly with the actual volume of output. By contrast, a distinctive characteristic of the new factory technology was that it necessarily involved a massive increase in fixed costs, which were incurred regardless of the volume of output, which existed even if output were to decline to nothing. The factory was, both literally and figuratively, an overhead cost.

The growing role of the fixed-cost component was of great significance for several reasons. First of all, the amount of control over financial resources that was required for entry into a factory industry was very high, and resulted in drastic changes in the nature of the producing organization itself. Financial requirements came to exceed what could usually be provided by an individual proprietorship or partnership. What was required was an organization that could pool the financial resources of many people. In this sense the creation of the modern business corporation was closely connected with the emergence of a new technology involving a large fixed-cost component.

These organizational changes may be clearly observed in the operation of the railroads, with their unprecedented capital requirements.

Because of the large size of the financial commitments involved, the new technology involved a greater risk to the business enterprise. The size of that risk is also linked to the growing predominance of the corporation in American life. From this point of view, the corporation may be looked upon as an institutional device for the pooling not only of financial resources but also of risk. The essential feature of the corporation, from the perspective of the prospective investor, is that of limited liability. Unlike, say, a partnership or single proprietorship, where personal risk is unlimited, the legal definition of a corporation strictly limits the possible financial liability of the individual investor.

The prefactory system, although it was less productive, was in certain respects much more flexible. If a businessman's market temporarily collapsed under the old system, he could stop production and incur little or no costs, then resume production when market conditions improved. (The unemployed, or partially employed, worker in effect bore the social costs of flexibility.) Under the factory system the businessman was saddled with huge fixed costs that he could not avoid by reducing his output. He was, in a very real sense, a captive of his financial commitments. Many of the special features of industrial capitalism flowed from this situation.

Finally, one of the obvious features of this system is the tremendous pressure it created for the expansion of output and the search for new markets. It follows directly from the fact that if a cost is truly fixed, then it becomes less and less of a burden as the volume of output and sales rises. That is, fixed costs per unit of output, or average fixed costs, decline steadily with the expansion of output. A firm with large overhead costs incurs tremendous per-unit costs at small volumes of output, and steadily declining per-unit costs as output expands. The resulting drive to maintain high volumes of output is therefore a consequence of the peculiar nature of industrial technology.

METALWORKING TECHNOLOGY

We have stressed so far the importance of technological innovations in power and metallurgy. It is time we paid explicit attention to a third category, the group of innovations that, more than any other, accounted for eventual industrial leadership in the United States—the innovations that made possible the precision shaping of metals and, therefore, the making of complex machinery and final products. These metalworking machines include machine tools that impart a shape to metal by the use of a cutting tool and the progressive cutting away of chips, as well as machines that shape metal by stamping, punching, forging, bending, and so on.[48]

Much of the industrialization process in nineteenth-century America was really the emergence and maturing of firms whose specialty was the making of machines. In 1800 machinery was frequently made by the eventual user, not by a well-defined group of specialized capital-goods producers. Textile firms, for example, produced textile machinery as an essential adjunct to their textile operations in the first half of the nineteenth century. Some of these shops, such as those attached to the Lowell Mills in Lowell, Massachusetts, and the Amoskeag Manufacturing Company in Manchester, New Hampshire, became so successful that they undertook to sell textile machinery to other firms. Their machine-making skills became so widely recognized that they eventually accepted orders for many sorts of other machinery, including steam engines, turbines, mill machinery, and machine tools. Indeed, with the arrival of the railroads in the 1830s, both the Lowell Machine Shop and the Amoskeag Manufacturing Company entered into the production of locomotives. Out of the internal requirements of the textile and railroad industries a new capital-goods sector emerged that specialized in making heavy, general-purpose machine tools such as lathes, planers, and boring machines.

A common pattern was for specialized firms to split off from the parent firm after their market had reached a certain size. Hence machine tools created to solve a set of problems in one industry were made available to create new opportunities in others. Using a common basic machine tool in related industries obviously enlarged its market and lowered its cost of production.

Similar emergent processes were taking place elsewhere in the economy. The technical requirements of the critically important arms industry were quite different from those of textiles and railroads. The making of firearms required the development of more specialized machinery, lighter and capable of producing at high speed in order to manufacture large quantities of small component metal parts. Beginning with the work of Eli Whitney, who turned to contracting with the government for the production of muskets after failing to make any money on his cotton gin (invented in 1793 and widely pirated), the firearms industry remained a critical center for learning, invention, and diffusion of new techniques. It is interesting to note, by the way, that the public and private sectors worked closely in a symbiotic relationship, visiting each other's plants, sharing new technical knowledge, and even occasionally "borrowing" each other's workmen. Much of the most important work in the development of mass-production technology either was undertaken on government contract, as in the cases of Eli Whitney and Simeon North, or was actually developed in government armories, especially those at Springfield, Massachusetts, and, to a lesser degree, at Harper's Ferry, Virginia. Out of the technical requirements for firearms production came an array of special-purpose machines specifically adapted for large-scale production of uniform metal parts.

The essential economic point is that the earlier production of firearms by handicraftsmen involved an extensive subdivision of labor, but contained an immensely costly stage when the component parts had to be fitted together. A firearm is a rather delicate and complex mechanism—at least it was by early nineteenth-century standards—and the separate parts had to be adjusted very carefully if the piece was to discharge effectively. Under a handicraft system, lacking the technical means to achieve a high degree of precision and standardization, it was necessary to employ an army of skilled workment to adjust the separate pieces into a well-functioning weapon. The craftsman's "precision instrument" was the file, and a large fraction of the cost of a firearm was expended in this laborious, painstaking use of the file to adjust the separate pieces to one another. (Henry Ford was later to assert his technical triumph in the technology of automobile production by pointing out that he employed no "fitters" in his plant.) Moreover, under battle conditions, when firearms were rendered useless as a result of minor damage, they had to await the ministrations of a skilled, and usually unavailable, armorer instead of a simple replacement part.

For these reasons it is not surprising that the American firearms industry played such a central role in developing an array of technological innovations upon which the large-scale production of uniform metal parts depended: drilling and filing jigs, fixtures, taps and gauges, and the systematic development of die-forging techniques. Its contributions in the first half of the nineteenth century included three machine techniques of wide-ranging importance: the stocking lathe, which replaced the highly labor-intensive hand-shaping of gun stocks, and eventually was employed over a wide range of uses involving the reproduction of irregularly shaped objects; the milling machine, an extraordinarily versatile device for producing uniform shapes out of metal that provided an effective substitute for very expensive hand-filing and hand-chiseling; and the turret lathe, which held a cluster of tools on a vertical axis and thus made it possible to perform a rapid sequence of operations on a piece of work without having to reset or remove the piece from the lathe.

By the 1850s the precision machine-tool method allowing the use of interchangeable parts for firearm manufacture had achieved such a state that it was attracting European attention. The "American system of manufacture" was the label applied to it. So impressive was it that an entire American musket factory—machines and workers—was purchased by Britain, which had experienced severe skilled labor difficulties in gun manufacture during the Crimean War.

Space limitations make it impossible to pursue the story further in detail. Suffice it to say that these inventions contributed to an expanding pool of metalworking machines that constituted the basic production technology for a growing number of industries: not only firearms but also clocks and watches, hardware, sewing machines, agricultural machinery, typewriters and office machinery, bicycles, and, at the beginning of the twentieth century, automobiles. In the twentieth century the techniques were further applied to an expanding array of increasingly complex consumer durable goods. A growing pool of technical skills was applied to an ever-widening circle of material goods. Techniques developed in response to the specialized requirements of one industrial subsector were usually found to have useful application elsewhere. But there is nothing surprising or coincidental about this, for at the heart of nineteenth-century industrialization, as we have seen, was the widespread application of metals of increasing cheapness to an expanding range of uses, in many cases involving totally new products as well as drastically altered old ones. Out of the historical process came an industrial technology devoted to the large-scale production of standardized products that involved the use of sequences of highly specialized machines producing components of sufficient precision that they could be merely assembled, and not fitted. Thus the assembly line and assembly plants have come to constitute a distinctive feature of twentieth-century technology.

The great diversity of final products should not distract attention from the essential fact that there is a common underlying body of techniques. The use of machine methods for the precision shaping of metals involves a limited number of operations and, therefore, of machine types: turning, boring, drilling, milling, planing, polishing, and others. Furthermore, the machines performing these operations have to deal with a similar collection of technical problems involving power transmission, control devices, feed mechanisms, friction reduction, and a range of problems concerning the behavior of metals, such as their capacity to withstand stresses and pressures and their heat resistance. It is because these processes and problems became common to the production of a widening range of commodities that we may speak of a common technology and shared technological solutions even where the nature and uses of final products were totally dissimilar.

THE ROLE OF SCIENCE

What has been the role of science in bringing about the transformation of the American economy since 1776? Although reference has been made several times to the contribution of science to some technological innovation, we have not so far addressed the question explicitly. In particular, what is the relationship between the growing stock of scientific knowledge and the technological improvements upon which we have concentrated? Obviously the answer to this question will depend in part upon how narrowly or comprehensively one defines science. Let us, for purposes of this discussion, think of science as a carefully articulated theoretical model that orders and systematizes the interrelationships among a large body of natural phenomena, and regard technology, as we have throughout this chapter, as a pool of useful knowledge concerning the production of material goods and services.

Our discussion must be conducted in broad strokes. Two propositions seem defensible at the outset. First of all, no single description can be applied to all scientific disciplines. It is therefore more appropriate to speak of the experiences of individual sciences rather than to speak of "science." Different disciplines have had different time patterns of development and each therefore has an individual history with respect to its relationship to those industrial groups where its knowledge has had useful applications. [49] Second, although there is considerable disagreement among scholars about the extent to which technology drew upon scientific knowledge in any specific period or industry, there is widespread agreement that the long-term trend, especially since the mideighteenth century, has been toward an increasing dependence upon scientific knowledge.

One of the most common pitfalls in trying to understand the past is that it is perilously easy to read the present back into it, or simply to assume that relationships that dominate the world today must have been similarly dominant at an earlier date. Specifically, because it is true today that technological developments in many sectors of the economy are dependent upon scientific knowledge—and often upon recent increments to the stock of scientific knowledge, as is true of penicillin and the transistor—it does not follow that this dependency characterized earlier periods in history.

In fact, throughout most of history, science and technology were not only totally separate activities pursuing quite separate goals, as they still do, but they were also practiced by different individuals from distinct and separate social classes and, most important, with little communication between them. Although there was increasing communication between these two realms in post-Renaissance Europe, it is still true that the major inventors of the British

Industrial Revolution were primarily technologists and engineers, usually quite innocent of the science of their day. Their work and inspiration came overwhelmingly from a concern with technical problems arising from the productive process. The assistance they received from contemporary science was usually minimal or nonexistent.

The limited contribution of science to technology before the middle of the nineteenth century has to be seen in the perspective provided by two basic considerations. First, with few exceptions, most scientific disciplines before the nineteenth century had not attained sufficient maturity for them to offer much valuable knowledge or guidance for technological processes. Second, most realms of productive activity had made great progress in traditional ways of accumulating useful technical information or methods that did not require scientific knowledge.

It goes without saying that technological progress would have been far more rapid in the presence of appropriate scientific knowledge. Trial and error, empirical search procedures were notoriously unreliable; and technological improvement was, as a result, fitful, spasmodic, and dependent to an unusual degree upon sheer chance. The absence, before the nineteenth century, of well-developed bodies of scientific knowledge and techniques placed sharp upper limits upon the extent to which technological progress could advance.

Nevertheless, one should not underestimate the extent to which improvement was possible prior to the availability of a serious scientific understanding of a subject. British urban mortality from infectious diseases had been sharply reduced by the provision of pure water, improved methods of waste disposal, and other sanitary reform measures before Louis Pasteur and others had identified the bacterial vectors of such diseases. To be sure, a much more effective assault upon a wide range of infectious diseases became possible after the emergence of modern microbiology in the second half of the nineteenth century, but considerable progress was possible prior to the availability of such scientific knowledge.

Until the last third of the nineteenth century, when modern metallurgical science began to emerge—a science based upon the systematic study of the structure and chemical composition of metals—metallurgy remained what it had always been, an activity dominated by a crude, trial and error empiricism, almost totally ignorant of the chemical transformations involved in the various production stages of iron and steel. Nevertheless, considerable technological progress continued to be made even though the supply of systematized knowledge was very limited or nonexistent. Indeed, Frederick W. Taylor, who developed high-speed steel in the Bethlehem Steel Works at the end of the nineteenth century, was not even trained in the metallurgy of the day. Rather, high-speed steel, an alloy that strikingly improved the capacity of a cutting tool to maintain its hardness at high temperatures, and thereby brought about immense productivity improvements throughout the machine tool industry, was the product of careful observation and painstaking experimentation by an efficiency expert, a man much better known for his contribution to scientific management. By contrast, the distinctively "modern" metal aluminum was from the beginning based directly upon scientific knowledge (of chemistry). "It could hardly be otherwise, since it was science and not empiricism that gave rise to electrolysis and broke the age-old dependence of metallurgy on carbon as fuel and reducing agent."[50] Charles Hall's innovations in the 1880s involved electrolytic techniques for producing aluminum from its oxide.

The fact of the matter is that many industries were able to make great technological strides without science, for the perfectly sufficient reason that technological progress did not require scientific knowledge. This was true of the central thrust of nineteenth-century American industrialization, which primarily involved the development of a machine technology. The invention of new machines or machine-made products—cotton gin, reaper, thresher, cultivator, typewriter, revolver, barbed wire, sewing machine, bicycle—involved the solution of problems requiring mechanical skill, ingenuity, and resourcefulness but not, typically, a recourse to scientific knowledge or elaborate experimental methods. Even in the twentieth century, when technology has learned to exploit scientific knowledge much more effectively, it is a mistake to take such dependence for granted. The realm of technology incorporates a considerable degree of self-reliance, in the sense that it exploits knowledge that has been produced within that realm—for instance, by engineers and other technologists—and not imported from the scientific world.

Indeed, it may be asserted that it is a major activity of the engineering profession to develop workable techniques that specifically bypass the need for scientific knowledge. Much of the most important work of engineers has been precisely of this sort. In aerodynamics and fluid mechanics, for example, engineers routinely produce information sufficient for a safe and workable solution to some technical problem or process even though no systematized theoretical model (or "scientific knowledge") is available. Technological progress on the steam turbine has continually proceeded in advance of rigorous scientific understanding. Similarly, over the years such important pieces of "hardware" as ship hulls and propellers, water turbines and airplane fuselages have been designed through use of purely empirical methods of optimization, well in advance of theoretical knowledge. Even an activity so obviously "science-based" as petroleum refining progressed until World War I without any substantial assistance from science. (The petrochemicals industry, by contrast, was heavily science-based from its inception.)

> Early refining techniques were empirical and crude; in
> general, knowledge of the chemistry of refining processes
> lagged behind practice. Petroleum is a complex mixture
> of hydrocarbons, and the chief problem of the late nine-
> teenth century was to separate the marketable kerosene
> fraction, widely used as an illuminating oil, from the rest
> of the crude. The methods were already at hand; frac-
> tional distillation techniques were well known and widely
> used in such industries as coal oil, alcohol, and tar. At
> a fairly early date, refiners did some "cracking," that
> is, using heat to split some of the larger and heavier
> hydrocarbon molecules into lighter ones in the kerosene
> range. Thus, when the French chemist Berthelot pub-
> lished in 1867 his basic researches into the action of heat
> on various hydrocarbons, he merely provided a basis for
> interpreting what was happening in practice in the oil
> industry. . . . Most improvements in the refining process
> were the work of practical refiners rather than trained
> chemists. The most serious chemical problem facing
> refiners before the end of the century, the sulfur content
> of crude from the Lima, Ohio, fields, was solved by
> Herman Frasch, a German-born technician with only
> minimal formal training in chemistry. Similarly, early
> by-products such as mineral lubricating oils or "Vaseline,"
> the first useful petroleum jelly, were developed by prac-
> tical men rather than scientists. [51]

The transition to a science-based technology can be seen clearly in the case of metallurgy. Improvements in experimental techniques that originated in metallurgy contributed to and joined with a widening stream of understanding that pertains not just to metals, but to all materials. Indeed, this intellectual merger is now coming to receive explicit terminological recognition in the increasing substitution in recent years of the term "materials science" for "metallurgy." This broader, fundamental advance in our understanding of the physical world derives from our insight into the atomic and molecular basis for the behavior of all matter, an understanding that dates from Dmitri Mendeleev's formulation of his periodic table of the elements in 1869.

The knowledge revolution in materials science since the 1870s is based upon a continuous deepening of our understanding of the gen-eral rules determining how atoms and molecules combine to form

progressively larger and more complex groups. Once these rules
are mastered, it becomes possible to manipulate materials, to alter
their characteristics, to maximize desirable properties, and even
to create entirely new and synthetic materials with desired combinations of properties. Our recently acquired knowledge of molecular
architecture makes it possible to create synthetic materials with
combinations of properties that have no counterparts in the natural
world. The burgeoning new industries exploiting these man-made
organic polymers—the vast range of plastics, synthetic fibers, packaging materials, synthetic rubber, lightweight thermal insulation,
water-repellent coating, and high-strength adhesives—are the direct,
legitimate offspring of this knowledge of high-polymer chemistry.

It is important to note that the basis for technical progress in
these industries is totally unlike that in the old metallurgical industries, where trial and error and intelligent empiricism led to considerable advances. No conceivable amount of experimentation, such
as once brought slow progress to metallurgy, could ever have generated modern synthetic polymers. The production of polymers absolutely required an adequate theory of molecular structures,[52] and a
full appreciation of the staggering complexity of these structures in
turn required a highly sophisticated collection of instruments: X-ray
diffraction equipment, the ultracentrifuge, the electron microscope,
the viscometer. Similarly, the remarkable advances in the electronics
industry after World War II—the breakthrough in semiconductor technology at Bell Laboratories in 1947-48 with the development of the
transistor, the replacement of the vacuum tube by the transistor, the
application of semiconductor devices to electronic data processing
and to an expanding field of military and commercial uses—were
dependent not only upon complex techniques of instrumentation but
also upon the development of quantum mechanics in the mid-1920s.
Quantum mechanics provided the essential theory that made it possible
to understand the determinants of electrical conductivity in terms of
the atomic structure of crystalline solids.[53]

Although the reliance upon science is incontestably much
greater than a century ago, it is important that the extent of this
dependence, great as it is today, not be exaggerated. Technological
improvement continues to draw upon a range of sources. Much technological progress consists of a cumulation of innumerable small
improvements that are, individually, invisible to all except specialists. It includes a flow of improvements in materials handling and
containerization. It involves the redesign of productive techniques
for greater convenience, or the modification of a technique to accommodate the specific requirements of individual, specialized users.
It involves new methods for reducing maintenance and repair costs,
as in modular design. Much technological improvement continues to

be generated, as it always has been, by the highly motivated exercise of human ingenuity by individuals who possess extensive technical knowledge of an industry but have no particular reliance upon science.

What is the present relation between the size of a firm and technological innovativeness? Although the view is widely held that giant corporations are the fountainhead of technical progress, empirical studies support no such simple conclusion. Technical innovativeness seems to depend less on the size of firms, at least beyond some minimum threshold size, than on other characteristics of an industry. One can find many industries dominated by large firms where the overall rate of innovation is low, and one can find many industries where the average size of firm is small, but where the rate of innovative activity is high. Even in industries dominated by large firms, the small firm and the lone inventor continue to be important sources of key technological improvements. In industries where large and small firms coexist, large firms do indeed make more innovations than small ones, but often not disproportionately more. It remains difficult to provide convincing empirical support for the view that large firms contribute disproportionately to technical advance. Here, as elsewhere, technological change seems to create conditions that defy easy and sweeping generalizations.

INSTITUTIONAL AND ORGANIZATIONAL CONTEXTS OF SCIENCE

The central thrust of this discussion is that science has been transformed from an unimportant source of technological development in the early history of American economic development to a vital and increasingly indispensable source of economic improvement since the late nineteenth century. To enlarge our understanding of this role it is necessary to look at American science in its changing institutional and organizational contexts. This raises questions somewhat different from the questions of intellectual innovation and priority that are usually the main concern of historians of science. Moreover, their primary interests lie at the pure or basic end of the spectrum of scientific research and not, typically, with the practical applications of such research. Thus, a major preoccupation of American science historiography has been with the reasons for the paucity, at least until the twentieth century, of first-rate, creative scientists whose work fully measured up to the best work in Europe. The most distinguished American claimants for international recognition in the nineteenth century were J. Willard Gibbs, for his theoretical work in thermodynamics and statistical mechanics, and Joseph Henry for his earlier, independent discovery of electromagnetic induction. But it would be difficult to add to the list of serious claimants on the basis of work done by Americans before 1900.

From the point of view of the development of the American economy, however, the number of first-class theoretical scientists and the reasons for their infrequent appearance are not really so important as the growth of forms of scientific knowledge or methodologies that have practical applications, and the capacity to realize these useful applications. Indeed, the evidence is strong that societies that had effectively institutionalized the capacity to make useful applications of scientific information, when such information became available, were not greatly handicapped by the absence of a first-rate domestic scientific establishment.

America's scientific establishment was a long time developing. During the colonial period, sciences were still relatively nontechnical in nature and could be pursued by men of modest formal training, often by self-trained men. Indeed, there was little science (or "natural philosophy," as it was then called) to be learned in American universities, since such institutions were devoted essentially to the classics and theology. [54] Although institutions were established that sounded national in scope—American Philosophical Society, American Academy of Arts and Sciences—they were, in fact, regional or even local. Such American science as existed through the eighteenth century was not so much the activity of an autonomous intellectual community as a remote outpost of the European scientific establishment. Colonists conducting research of potential scientific value sought recognition (typically from the Royal Society of London) and, most important, publication in Britain. Those who had the financial means pursued their education abroad.

The main thrust of American scientific research was long decisively shaped by forces that we have already discussed in a different context: the extraordinary resource abundance of the American continent. This abundance served to focus attention upon the description and classification of its natural environment for two very different reasons. First, the flora, fauna, and geology offered an enormous potential increment to the stock of European knowledge concerning the natural world. European scientists were intensely interested in accumulating such information. Descriptive accounts and all forms of reporting on the natural environment of the New World were eagerly sought and enthusiastically received in Europe. The comparative advantage of aspiring American scientists was, thus, emphatically in the "export" of primary products: relatively unprocessed, descriptive accounts or specimens of the natural environment of the New World. As one author expressed it:

> During the first century and a half of its existence, Amer-
> ican science was a branch of British science and depended
> on Europe for inspiration and ideas, for models of scien-
> tific achievement, for medical training, for avenues of
> publication, for books and instruments, and for museums
> and herbaria—in short, for everything except native talent
> and the raw data provided by nature. In science, as in
> trade, the American colonies were purveyors of raw mate-
> rials—astronomical observations, seeds, plants, shells,
> and fossils—which were worked into finished products in
> the scientific centers of the Old World. [55]

Second, from the point of view of the Americans, who had come into
possession of a continent of immense but uncataloged natural treas-
ures, their economic well-being was directly and intimately linked
to the rapid acquisition of much more detailed information on the
natural resources of the continent.

As a result, surveying, measurement, and exploration remained
a main concern, perhaps the main concern, from the earliest coastal
settlement until the post-Civil War period. The importance and
growth of maritime commerce placed a great economic value upon
an accurate survey and mapping of the coast and the provision of
accurate charts. The work of the Coast Survey was generously supported, at least according to the standards of the times, by the federal government. With the movement across the Mississippi after
1830, the army played an increasingly important role, often lending
engineers to the railroads for surveying purposes.* Although the
Corps of Engineers played a prominent role in the exploration of the
west, it eventually relinquished these functions to a permanent civilian
agency with the establishment of the U. S. Geological Survey in 1879.

No single cluster of institutional innovations was more important for the future useful application of scientific knowledge than those
in agriculture. The support of scientific agriculture that began with
the Morrill Act of 1862 was eventually incorporated into a network
of institutions at the federal, state, and county levels. President
Abraham Lincoln established the Department of Agriculture in 1862,
a department that in succeeding decades would greatly centralize

---

*For many years West Point was the only educational institution
in America engaged in the training of engineers. The first true engineering school, Rensselaer Polytechnic Institute, was founded in
1824. Almost without exception, the early American railroad engineers were graduates of West Point.

some research activities of general interest to farmers, in the Bureaus of Plant Industry, Entomology, Animal Industry, and Chemistry. At the same time, the Morrill Act donated public lands "to the several States and Territories which may provide colleges for the benefit of agriculture and the mechanic arts."

This was supplemented by the Hatch Act of 1887, which established federally supported agricultural experiment stations in each state. Little serious scientific research was conducted in the early years, largely because the appropriate life science disciplines were just beginning of concentrate on the kinds of knowledge—in biology, genetics, botany, entomology, and chemistry—that would revolutionize agricultural technology. When this knowledge was developed, it could be rapidly exploited through an institutional network in which research activities were conducted at several levels and in which a highly effective communications and information-delivery system had been established by the Department of Agriculture and its Agriculture Extension Service. Although the accomplishments of this system were relatively modest before World War I, it provided the framework within which the great agricultural breakthroughs eventually took place. [56]

Perhaps the most interesting feature of the American experience with respect to the organization of the innovation process as it applies to agriculture has been the diversity of institutional forms that have been combined into a highly successful whole. One authority has made the point as follows:

> Organization for research and development in the farm supply industries varies widely. At one extreme are such industries as hybrid seed corn, where a substantial share of research and development has been conducted at USDA and land-grant college experiment stations. At the other extreme is the farm equipment industry where public contributions have apparently been considerably smaller relative to the contributions by private industry. A third variation is represented by the fertilizer industry where the TVA has provided what is in effect an industry research institute for the fertilizer industry with a research program ranging from fundamental chemical and biological research to applied engineering and economics. [57]

By the 1830s the cumulation of knowledge in individual scientific disciplines led to the permanent displacement of the gentleman-amateur and to the professionalization of scientific work. Thus the institutionalization of science reflected both a growing expectation of the practical value of such knowledge and the increasingly technical and highly specialized nature of its contents. By the 1870s the evidence of maturing scientific disciplines is unmistakable: specialized scientific journals, specialized professional societies, and the rise of graduate institutions for the training of professional specialists. [58]

Gradually, toward the end of the nineteenth century, a different theme appeared, one that foreshadowed concerns that were to become increasingly important in the twentieth century. Its arrival was announced by one of the most influential essays in American history: Frederick Jackson Turner's announcement, at the 1893 meetings of the American Historical Association, of the closing of the American frontier. [59] Turner's "frontier hypothesis" takes the year 1890 as marking the closing of the American frontier—the end, in other words, of the extreme resource abundance that, as we have argued, characterized the earlier stages of American national development.

It is clear that, by the opening of the twentieth century, this growing concern had already received or was shortly to receive institutional as well as political expression. During the Progressive era there was abundant evidence of a growing preoccupation with the conservation of resources. The Geological Survey in the Department of Interior focused scientific research upon natural-resource problems. The Reclamation Service became a separate bureau in the Interior Department in 1907, and the Bureau of Mines was established in 1910. Toward the end of the nineteenth century one of the most conspicuous features of the American environment, to many observers, was the progressive elimination of forests; and by the turn of the century there was a powerful political movement directed toward the permanent preservation of forest lands, something that would have seemed inconceivable a century earlier, in the days of the Lewis and Clark expedition. The Division of Forestry in the Department of Agriculture was transformed into the Forest Service under the energetic leadership of Clifford Pinchot and emerged as an important scientific agency devoted to the preservation of forest resources. There also were other agencies with parallel concerns in the federal government: "In lesser degree, the Biological Survey, the Fish Commission, the Division of Agrostology, and the Office of Irrigation Investigations were conservation agencies."[60]

The twentieth century has witnessed a full maturing of the role of science in the technological development of the American economy. Fundamental to these events has been the coming of age of individual scientific disciplines. Chemistry has a long preindustrial history, but began its modern emergence with John Dalton's atomic theory in the early nineteenth century. Electricity, in spite of Benjamin Franklin's marvelous work, may fairly be described as a new nineteenth-century discipline dating from the early discoveries of Humphry Davy, Michael Faraday, and Joseph Henry, and from James Clerk Maxwell's predictions of the existence of electromagnetic waves in Electricity and Magnetism, published in 1873. The electricity-based industries that emerged in the last third of the nineteenth century may be regarded as the first industries to be based, from their inception, upon scientific knowledge and research. (Perhaps the telegraph, an invention of the 1840s building directly upon the recent discoveries of electromagnetic induction, should be regarded as the first totally science-based industry.) The biological sciences produced Darwinian evolution, cell biology, and genetics in the second half of the nineteenth century (although Gregor Mendel's monumental discoveries lay unrecognized until they were, in effect, rediscovered at the beginning of the twentieth century). Nuclear physics has been, of course, a distinctly twentieth-century science.

As these separate disciplines matured, they became institutionalized in various way during the twentieth century. Much of the basic research and experimentation has been centered in the university communities.* As the conditions of research became transformed and as the transition was made from little science to big science, the federal government came to play a dominant role as financial patron of the sciences, a role that became massive after World War II had demonstrated the decisive military significance of the research establishment in such fields as synthetic chemicals, aircraft, electronics, and nuclear power. By the years just after World War II a

---

*Some crude measure of America's movement to a position of worldwide leadership in the basic sciences since the beginning of the twentieth century may be gauged from the following information about the nationality of Nobel Prizes in science (including physics, chemistry, physiology or medicine).

Country	1901-30	1931-50	1951-66
United States	5	24	44
Belgium	1	1	0
Canada	1	0	0
France	14	2	4
Germany	26	12	7
Japan	0	1	1
Netherlands	7	1	1
Great Britain	16	13	18
U.S.S.R.	2	0	7
Other countries	22	17	0

Note: Information is from Encyclopedia Britannica (1967), XVI, pp. 549-51.

vast, permanent federal establishment had come into being, controlling huge sums of money and dispensing them through such agencies as the National Science Foundation and the National Institutes of Health. The public support of such research reflects an increasing awareness that the social returns to the production of new scientific knowledge often vastly exceed the private returns and, therefore, that a reliance upon private profit calculations and incentives alone would be far from socially optimal.

At the same time, however, purely private profit calculations have led to one of the uniquely twentieth-century institutional developments, the industrial research laboratory. As it became apparent that individual sciences were reaching the point where they offered information or techniques of great potential economic value, practitioners of those sciences were hired in growing numbers by private industry. Although there were isolated harbingers of such arrangements in the late nineteenth century, the new system emerged in the opening years of the twentieth century with the foundation of some of the most famous industrial research laboratories. General Electric established its laboratory in 1900, and other well-known firms, such as du Pont, American Telephone and Telegraph, and Eastman-Kodak followed within the next decade or so. In the economy as a whole, the number of such laboratories grew rapidly during the period of World War I, then moved into a period of truly exceptional growth in the 1920s. According to surveys by the National Research Council, there were almost 1,000 industrial research laboratories in the United States in 1927, more than 1,600 by 1931, and 4,834 in 1956. [61]

Most of this research has been at the applied, development end of the spectrum; little of it has been basic. However, such research on the part of private industry was made possible by the fact that basic research in many scientific areas had reached the point where industrial applications could be immediately perceived. As a result, scientific knowledge of chemistry, electricity, and biology became increasingly pervasive in the industrial economy of the twentieth century. The increasing employment of scientists by private industry, then, should not to taken as a measure of research being conducted anywhere near the forefront of scientific knowledge. Nevertheless, what was provided was extremely valuable. Much of the work of the industrial scientist involved such mundane activities as testing materials, various forms of quality control, and a more precise identification of the materials employed in the productive process. Such activities, while scientifically uninteresting and humdrum in the extreme, could be, and often were, of great economic value. Consider Andrew Carnegie's enthusiasm with the "findings" of his German chemist, a Dr. Fricke:

. . . great secrets did the doctor open up to us. Iron
stone from mines that had a high reputation was now
found to contain ten, fifteen, and even twenty per cent
less iron than it had been credited with. Mines that
hitherto had a poor reputation we found to be yielding
superior ore. The good was bad and the bad was good,
and everything was topsy-turvy. Nine-tenths of all the
uncertainties of pig-iron making were dispelled under
the burning sun of chemical knowledge. [62]

The industrial exploitation of specialized bodies of scientific
knowledge underlies the continuing long-term growth of the American
economy in an era of increasing resource scarcity. In spite of an
increasing concern over the limits of growth imposed by the growing
significance of natural-resource scarcities, there is some comfort,
and perhaps wisdom, to be found in the realization that we have so
far managed to evade these constraints upon growth, which have
worried industrial man ever since the gloomy prognostications of
Thomas Malthus and David Ricardo in the early nineteenth century.
The past is not necessarily prologue; and anyone who cannot find much
to be despondent over in the present state of the human predicament
is, to say the least, a confirmed Pollyanna. Nevertheless, American
history strongly suggests at least the possibility that we can continue
to underestimate our capacity for creative innovation and adaptation.
At the very least, it is worth reflecting upon that expenditure for
natural-resource inputs today constitutes a much smaller percentage
of our gross national product than it did when Frederick Jackson
Turner announced the closing of the American frontier in 1893.

NOTES

1. Stanley Legergott, Manpower in Economic Growth (New
York: McGraw-Hill, 1964), p. 510. It should be noted that well over
a quarter of the labor force in 1800 (530,000 out of a total of
1,900,000) were slaves.
2. Victor Clark, History of Manufactures in the United States
I (New York: McGraw-Hill, 1929), p. 180.
3. Ibid., p. 94.
4. Ibid., p. 73.
5. Kendall Birr, "Science in American Industry," ch. 2 in
David Van Tassel and Michael Hall, eds., Science and Society in the
United States (Homewood, Ill.: Dorsey, 1966), p. 38.

6. James Shepherd and Gary Walton, Shipping, Maritime Trade and the Economic Development of Colonial America (Cambridge: Cambridge University Press, 1972), ch. 6 and app. IV.

7. Robert Gallman, "The Pace and Pattern of American Economic Growth," ch. 2 in Lance Davis et al., American Economic Growth (New York: Harper and Row, 1972), pp. 21-22.

8. Moses Abramovitz, "Resource and Output Trends in the United States Since 1870," American Economic Review Papers and Proceedings 46 (May 1956): 5-23.

9. This is the conclusion of Shepherd and Walton, op. cit. The following discussion of the colonial economy draws heavily upon this valuable study. For two earlier, careful studies of agriculture that support the conclusion of insignificant technological improvements during the colonial period, see P. W. Bidwell and J. I. Falconer, History of Agriculture in the Northern United States, 1620-1860 (Washington, D. C.: Carnegie Institution of Washington, 1925); and Lewis C. Gray, History of Agriculture in the Southern United States to 1860, 2 vols. (Washington, D. C.: Carnegie Institution of Washington, 1933). For a useful description of colonial manufacturing technology, see Clark, op. cit., I, ch. 8.

10. Shepherd and Walton, op. cit., p. 81.

11. Ibid., pp. 77-80.

12. George Rogers Taylor, The Transportation Revolution (New York: Rinehart and Co., 1951), app. A, table 2.

13. Robert E. Lipsey, "Foreign Trade," ch. 14 in Davis et al., op. cit., pp. 566-69.

14. Ibid., p. 571.

15. Ibid.

16. For cotton, see Robert B. Zevin, "The Growth of Cotton Textile Production After 1815," ch. 10 in Robert Fogel and Stanley Engerman, eds., The Reinterpretation of American Economic History (New York: Harper and Row, 1971). For iron and steel, see Peter Temin, Iron and Steel in 19th Century America (Cambridge, Mass.: MIT Press, 1964).

17. For further discussion, see Nathan Rosenberg, "Selection and Adaptation in the Transfer of Technology: Steam and Iron in America, 1800-1870," in Perspectives on Technology (Cambridge: Cambridge University Press, 1976), pp. 173-88.

18. Tench Coxe, A Statement of the Arts and Manufactures of the United States of America for the Year 1810 (Philadelphia: 1814), p. xvii.

19. See Nathan Rosenberg, "America's Rise to Woodworking Leadership," in Brooke Hindle, ed., America's Wooden Age (Tarrytown, N. Y.: Sleepy Hollow Restorations, 1975), pp. 37-62.

20. Nathan Rosenberg, "Innovation Responses to Materials Shortages," American Economic Review Papers and Proceedings, 63 (May 1973): 111-18.

21. Eighth Census of U. S., Manufactures (Washington, D. C.: 1865), pp. 733-42. Value added by manufactures was $54,671,082 for cotton goods and $53,569,942 for lumber.

22. Clark, op. cit., I, p. 48.

23. Rosenberg, "America's Rise to Woodworking Leadership," p. 42.

24. Temin points out that, in colonial days, abandoned houses were often burned down in order to recover the nails. Temin, op. cit., p. 42.

25. Siegfried Giedion, Space, Time and Architecture (Cambridge: Harvard University Press, 1941), pp. 347-55. See also John Kouwenhoven, The Arts in Modern American Civilization (New York: Norton, 1967), pp. 49-52 (first published in 1948 under the title Made in America).

26. This and the next paragraph are taken, with minor alterations, from Rosenberg, "Innovation Responses to Materials Shortages," pp. 114-15.

27. J. Fisher and E. Boorstein, The Adequacy of Resources for Economic Growth in the United States, Study Paper no. 13, prepared in connection with the Study of Employment, Growth, and Price Levels, 86th Cong., Dec. 16, 1959, p. 36.

28. Sherry Olson, The Depletion Myth (Cambridge, Mass.: Harvard University Press, 1971), p. 132. This book includes an account of the railroads' response to rising timber prices.

29. See Earl W. Hayter, "Barbed Wire Fencing—A Prairie Invention," Agricultural History 13 (October 1939): 189-207.

30. See Martin Primack, "Land Clearing under Nineteenth-Century Techniques," Journal of American History 22 (December 1962): 484-97. It is often assumed that because land was very abundant in America, farmers could readily ignore deterioration of soil fertility. Where clearing costs were high, however, "abundant" land was not necessarily equivalent to "free" land.

31. For a study of the technological factors responsible for the push out of southern agriculture, see James H. Street, The New Revolution in the Cotton Economy (Chapel Hill: University of North Carolina Press, 1957), pt. III.

32. Yujiro Hayami and Vernon Ruttan, Agricultural Development (Baltimore: Johns Hopkins Press, 1971), fig. 6-2, p. 117; app. table C-2.

33. For an analysis of fertilizer price changes relative to the prices of farm products and other inputs, and of the contribution of

technological change to the relative decline in fertilizer prices, see G. S. Sahota, "The Sources of Measured Productivity Growth: U. S. Fertilizer Mineral Industries, 1936-1960," Review of Economics and Statistics 48 (May 1966): 193-204. See also Zvi Griliches, "The Demand for Fertilizer: An Economic Interpretation of a Technical Change," Journal of Farm Economics 60 (August 1958): 591-606.

34. Clark, op. cit., I, p. 406.

35. Peter Temin, "Steam and Waterpower in the Early Nineteenth Century," Journal of Economic History 26 (June 1966): 187-205. See also Carroll Pursell, Jr., Early Stationary Steam Engines in America (Washington, D. C.: Smithsonian Institution, 1969).

36. Louis Hunter, Steamboats on the Western Rivers (Cambridge, Mass.: Harvard University Press, 1949), pp. 129-33.

37. Allen H. Fenichel, "Growth and Diffusion of Power in Manufacturing, 1839-1919," in Dorothy Brady, ed., Output, Employment and Productivity in the United States After 1800 (New York: National Bureau of Economic Research, 1966), app. B, p. 456.

38. U. S. Congress, House, Report on the Steam Engines in the United States, H. Doc. no. 21, 25th Cong., 3d Sess. 1839), p. 10. The figures, mostly reported but partly estimated, were steamboats 57,019 horsepower, railroads 6,980 horsepower, others 36,319 horsepower, for a total of 100,318 horsepower.

39. Temin, Iron and Steel in Nineteenth-Century America, p. 82 and app. C, table C.3. This paragraph draws heavily upon Temin's excellent study.

40. Ibid., app. C, table C.3.

41. For an interesting treatment of the role played in early American industrialization by the opening up of the anthracite mines of eastern Pennsylvania, see Alfred D. Chandler, Jr., "Anthracite Coal and the Beginnings of the Industrial Revolution in the United States," Business History Review 46 (Summer 1972): 141-81.

42. C. S. Smith, "Materials and the Development of Civilization and Science," Science (May 14, 1965): 915. The same author has remarked elsewhere that ". . . the structure-property relationship has been the central theme of the last century of metallurgy. . . . While the atom and its substructures are undoubtedly at the bottom of it all, the properties of materials that are experienced and exploited by man are most directly a result of the level of structure represented by the molecule and the microcrystal." C. S. Smith, ed., The Sorby Centennial Symposium on the History of Metallurgy (New York: Gordon and Breach Science Publishers, 1965), p. xvii.

43. C. S. Smith, "Materials," Scientific American 217 (September 1967): 75.

44. The spectacular growth of German steel production in the 1880s and 1890s was directly based upon this new process. See J. C. Carr and W. Taplin, A History of the British Steel Industry (Cambridge, Mass. : Harvard University Press, 1962), p. 172.

45. Clark, op. cit. , III, p. 17.

46. Hans Landsberg and Sam Schurr, Energy in the United States (New York: Random House, 1968), p. 45.

47. Rosenberg, "Innovative Responses to Materials Shortages," discusses these points at greater length.

48. The following discussion of the machine tool sector draws freely upon Nathan Rosenberg, "Technological Change in the Machine Tool Industry, 1840-1910," Journal of Economic History 23 (December 1963): 414-43.

49. For further discussion see Nathan Rosenberg, "Science, Invention and Economic Growth," Economic Journal 84 (March 1974): 90-108.

50. C. S. Smith, "The Interaction of Science and Practice in the History of Metallurgy," Technology and Culture 2 (Fall 1961): 357-67.

51. Birr, op. cit. , pp. 60-61.

52. Much of the basic scientific work on the relationship between the structure and the properties of synthetic materials was done by Wallace H. Carothers while he worked for du Pont from 1928 until his death in 1937. See, for example, "Polymerization," ch. 4 in The Collected Papers of Wallace H. Carothers on High Polymeric Substances (New York: Interscience, 1940), where he discusses solubility properties, mechanical properties, and crystallinity as they relate to structure. This paper, written in 1931, together with Hermann Staudinger's finding in 1935 that polymerization involves three distinct steps—chain initiation, chain growth, and chain termination—provided the basic theoretical understanding that made it possible to produce plastics with predetermined characteristics.

53. See National Academy of Sciences-National Research Council, Physics: Survey and Outlook (Washington, D. C. : U. S. Government Printing Office, 1966), p. 69.

54. See the evidence on this subject compiled in Theodore Hornberger, Scientific Thought in the American College, 1638-1800 (Austin: University of Texas Press, 1945).

55. John Greene, "American Science Comes of Age, 1780-1820," Journal of American History (June 1968): 22.

56. See A. Hunter Dupree, Science in the Federal Government (Cambridge, Mass. : Harvard University Press, 1957), ch. 8, for a useful exposition of institutional developments between 1862 and 1916. For a masterly survey of long-term growth in American agriculture,

see the chapter by William N. Parker (ch. 11) in Davis et al., <u>American Economic Growth</u>, loc. cit.

57. Vernon Ruttan, "Research on the Economics of Technological Change in American Literature," <u>Journal of Farm Economics</u> 42 (November 1960): 735-54.

58. From 1818, Silliman's <u>American Journal of Science and Arts</u> had served as a kind of catchall for virtually the whole range of scientific disciplines. For chemistry, see Edward H. Beardsley, <u>The Rise of the American Chemistry Profession, 1850-1900</u> (Gainesville: University of Florida Press, 1964), ch. 4. See also the useful chronologies in Van Tassel and Hall, op. cit., pp. 303 ff. It is worth noting that in 1874 the AAAS divided itself into sections, in recognition of the increasingly specialized nature of scientific pursuits.

59. Frederick Jackson Turner, <u>The Frontier in American History</u> (New York: Holt, 1950). Turner's original essay bore the title "The Significance of the Frontier in American History."

60. Dupree, op. cit., p. 246.

61. National Academy of Sciences–National Research Council, <u>Industrial Research Laboratories of the United States</u> (10th ed.; Washington, D.C.: 1956). See also earlier editions.

62. Andrew Carnegie, <u>Autobiography</u> (Boston: Houghton Mifflin, 1920), p. 182.

5

THE ROLE OF SCIENCE AND TECHNOLOGY IN HUNGARY'S ECONOMIC DEVELOPMENT

Mária Csöndes
Lajos Szántó
Péter Vas-Zoltán

The ancestors of the Hungarians, a Ural-Altaic nomad ethnic group speaking a Finno-Ugric language, settled in what is now Hungary in the ninth century of the Christian era. In 1001, King Stephen I, founder of the state, received his crown from the pope. Feudalism in Hungary began at that time and remained the only form of social organization until the mid-1800s, after which time, although no longer absolute, it continued to exert a strong influence until the end of World War II. Hungary had native-born kings until the early fourteenth century, when it came, successively, under French, German, Austrian, and Polish rule. In the second half of the fifteenth century it again had a Hungarian ruler and briefly experienced the Renaissance. For most of the sixteenth and all of the seventeenth centuries, Hungary was a part of the Ottoman Empire. Most of Hungary was ceded to the Austrian House of Habsburg in 1699; and between 1867 and 1918 it belonged to the Austro-Hungarian monarchy.

Capitalism in Hungary began in the mid-1800s, with an interruption of some four months during the 1919 Republic of Councils, during which period the present national borders were drawn. Since 1948 the Hungarian people have been building a socialist republic.

During its 1,000 years of existence, the social structure of Hungary has changed three times, the ruling dynasties several times, the extent of its territory and population often. This was all part of the general development of Eastern Europe, which, compared with the western portion of the continent, has until recently been characterized by severe economic and cultural underdevelopment. Hungary's frequent role as a geopolitical buffer zone and battleground for such diverse armies as those of Genghis Khan and Napoleon accounts in part for some of the slow progress.

Some basic data on the present Hungarian People's Republic will, however, indicate what has been accomplished:

The area of the country is 93,032 square kilometers (35,921 square miles).

As of January 1, 1976, the population was 10,572,000. Population density is 114 per square kilometer. The average rate of population increase is 0.4 percent per annum. [1]

The number of workers employed is 5,085,500 (48.5 percent of the population). Agricultural workers account for 22.7 percent of the total work force. [2]

The national income in 1975 was 402 billion forints, of which 56.6 percent was generated by industry (including the construction industry) and 15 percent by agriculture and forestry. [3]

The net national product (net national income plus services, NNP) amounted to 400 billion forints at current prices in 1974, according to the United Nations (UN) national accounts statistics. Accordingly, the per capita NNP was $1,630 in terms of the exchange rate used in the UN statistics ($1.00 = 23.38 forints).

Hungary's population is only 0.3 percent of the world's total, yet its industrial output is 0.8 percent of that of the world. [4]

The socialist sector—state enterprises and cooperatives in industry and agriculture—accounts for 99.2 percent of the gross production.

Of the land under cultivation, 30.4 percent belongs to the state sector and 63.8 percent to the cooperatives. [5]

Hungary has an open economy, since the early 1970s more than 40 percent of the value produced in the country has been realized through foreign trade.

R&D employed 80,542 people (1.6 percent of the total number of employees) in 1974, including 33,784 researchers (21,758 full-time equivalents). R&D expenditures·in 1974 totaled 12.4 billion forints (3.3 percent of the national income) including development requiring no research. [6] Currently 130 research institutes, 55 higher educational institutions with 1,085 departments and 261 industrial and other research units (such as libraries, archives, hospitals, planning agencies, and service companies) are active. [7]

FROM THE BEGINNING OF THE HUNGARIAN NATION
TO THE DECLINE OF FEUDALISM (1848)

The Magyar tribes, perhaps the last of the nomads moving westward, settled in the plains region of the Carpathian Basin. They abandoned their nomadic life style and, in addition to traditional animal husbandry, they started to cultivate the soil. This latter skill was acquired from the Slavs already in the region and, to a greater extent, from the missionaries coming from the West, as is indicated by the foreign words, adopted at the time, for things connected with agriculture. Agricultural activities combined the necessary with the possible; wheat, for example, was grown and used for making bread. Weaving, leather working, and woodworking were all done at home. Wood was plentiful. Construction in the modern sense of the term was applied only to churches and forts, and was done mainly by Italian masters.

During the Turkish occupation most of these buildings were destroyed, central national leadership was scattered, and intellectual life became disorganized; but agricultural technology continued to improve. The Hungarian language contains some 300 Turkish words that are linked to agriculture and public life.

Science did exist, but its influence on the economy was negligible throughout the period. It had an isolated existence, marked by occasional outstanding individual achievements. A foremost characteristic was its dependency on developments abroad. The roots of science in Hungary, as in all other European countries, were in the church. In the twelfth and thirteenth centuries there were several hundred Hungarians studying at the University of Paris. Studying abroad was later a regular occurrence; and there were many Hungarians at the Universities of Bologna, Padua, and Cracow. The first Hungarian universities were established in the fourteenth century. It is noteworthy that in 1367, King Louis I of the Angevins established a university without a chair of theology at the town of Pécs. Eventually all of these institutions of learning closed; and the first that may be regarded as the ancestor of the modern university was established in 1635 at Nagyszombat, a town outside of the jurisdiction of the Turkish occupiers. In 1777 it moved to Buda.

By the eighteenth century it became obvious that, in addition to the universities that taught mainly theology, jurisprudence, and philology, there was a need for institutions dealing with technical and economic matters. In 1763 the Institute of Mining was established at Selmecbánya and became the first to teach mining engineering. It was not long-lived, but its fame spread; and on September 23, 1794, the French chemist Antoine de Fourcroy made a motion in the French National Convention to have an institution, later called the École Polytechnique, founded in France on the model of the Hungarian one.[8] The Collegium Oeconomicum at Szencz, established in 1763, was the first to teach economics. It lasted only seven years. In 1782 the Institutum Geometricum was founded at Buda, to focus on science and technology. The first European agricultural college, called Georgikon, was opened in 1797 at Keszthely.

In the eighteenth century some Hungarian scientists and scholars became world-famous. János Apáczai Csere (1625-59), who was influenced by René Descartes, had to emigrate because of his progressive views. His Magyar Encyclopaedia, popularizing Copernicus' system, was published at Utrecht in 1653. Sándor Kőrösi Csoma (1784-1842), commissioned by the British, prepared the first Tibetan-English dictionary during his stay in Tibet. János Segner (1704-77), a physicist, invented the Segner wheel, a basic instrument in hydrodynamics that was used to prove the law of action-reaction. Pál Kitaibel (1757-1817), who engaged in wide-ranging scientific activities, began research on Hungary's flora and conducted experiments in chemistry and meteorology. It was also in the eighteenth century that the idea of founding an academy of sciences was first proposed, although one was not established until 1825.

By the end of the eighteenth century, although it was rather random and limited to a few scientists, scientific research in Hungary had reached a level that may be called significant in terms of the overall situation of the country. Unfortunately we cannot assess its effect on production, either directly or indirectly. Most technology was found in the handicraft industries and in mining, transportation, construction, and agriculture. None of it was based on scientific research, and all was backward in comparison with the standard attained in Western Europe.

Economic development of the sort connected with scientific research and technology did not begin until the first half of the nineteenth century. The national independence movements following the French Revolution and culminating in the national independence wars of the midcentury, gave an impetus to reforms of the Hungarian language, research into Hungarian history, and the development of national literature. They also served as a catalyst for economic and technical progress. Agriculture remained the basis of the national economy, but production was gravely endangered each year by extensive floods. A vital prerequisite for the unfolding of commerce was the development of safe waterways and, later, of bridges and a network of roads. The Industrial Revolution reached Hungary in the first third of the nineteenth century. Nationwide flood- and river-control works were started, the construction of permanent bridges was begun, and the industrialization of agriculture and the building of the railway developed slowly prior to the 1848-49 revolution and war of independence. A rather large milling industry arose in this agricultural nation, in order to meet the needs of the Austrian Empire. The first industrial enterprises began to appear, and by the middle of the nineteenth century Hungary had some 400 handicraft workshops. [9]

Social interest in science was indicated by the formation of medical associations and the establishment of the Natural Science Society, the Hungarian Geological Society, and the Hungarian Academy of Sciences. However, this did not mean a change in the relationship between scientific research and production. Basic research included that conducted by János Bólyai (1802-60), former bodyguard and military engineer of the Habsburgs, who wórked out and published his theory on absolute (non-Euclidean) geometry at about the same time as Karl Friedrich Gauss and Nikolai Lobachevski, but independently of them. Another example is that of Ányos Jedlik (1800-95), inventor of the dynamo. Of the Hungarian scientific achievements of the era, only the work of Ignácz Semmelweis (1818-65) was placed directly at the service of society. He discovered the cure for peritonitis some 30 years ahead of the rest of Europe's bacteriological researchers.

DEVELOPMENT OF HUNGARIAN CAPITALISM WITHIN THE AUSTRO-HUNGARIAN MONARCHY

On March 15, 1848, a bourgeois democratic revolution for national independence broke out in Hungary. Its economic line favored capitalism, demanding an end to feudal tenure and the right to use land as collateral to obtain credit. This uprising was defeated in 1849 and Hungary suffered reprisals and, until 1867, totally lost its independence. In 1867 a compromise was reached between the Hungarians and the Habsburgs that permitted the operation of Hungarian ministries and Hungarian autonomy within the monarchy, except in foreign affairs, defense, and finance. The eighteen years of Austrian rule had paralyzed the nation's cultural and scientific life, but it had also created a framework for the role of the state in directing the development of capitalism. The period that followed the Compromise, as the accord of 1867 came to be called, was marked by an economic resurgence in which industry gained strength, the large feudal estates became capitalistic, and production became more modern in technique.

Between 1867 and 1873, 60 percent of the capital invested in Hungary came from abroad. Between 1873 and 1900 this proportion changed, with capital of domestic origin accounting for 55 percent; this trend peaked at 75 percent between 1900 and 1913. It is obvious, however, that the import of capital, mainly from Austria, played a decisive role in the initial period of Hungary's modern economic development. [10]

The development of heavy industry was mainly in relation to coal mining. Coal and iron-ore mining doubled during the period after the Compromise. [11] The first modern heavy industrial factory, the Ganz Works, which produced chilled-cast railway carriage wheels, became a stock company in 1868. It revolutionized and made Hungary's steel industry world-famous by introducing the manufacture of the roller frame. [12] Between 1863 and 1884 the total horsepower of steam engines in use in Hungary's industries increased from 8,100 to 63,900. The location of these engines illustrates the dominantly agrarian nature of Hungarian society: 60 percent were used in the food industry. [13]

We have already mentioned the sources of the capital used to develop Hungary's industries. The other essential, cheap labor, was provided by the poor agrarian masses leaving the big estates. By the end of the century, there was such a surplus of labor that some 3 million people, mostly poor peasants, emigrated, mainly to the United States.

The growth of industry in no way meant a growth in mass education. In 1880, 58.7 percent; in 1890, 49.4 percent; in 1900, 40.7 percent; and in 1910, 33.3 percent of the adult population was illiterate; in 1920 the figure dropped to 15.2 percent. In comparison with the developed West European countries, this was a lag of 10-15 years. [14] The Hungarian economist F. Jánossy's conclusion that the course of economic development is determined neither by capital investments nor by research, but primarily by the structure of specialized manpower, certainly holds true for that period. [15]

Although the development of science continued to be divorced from production activities, particularly in regard to basic research, there was a new development, in that certain research served immediate social ends, especially in industry and agriculture.

Although science can never become a total direct force in production, the tendency was nevertheless generally present in the period following the Industrial Revolution. Such a trend was apparent at the end of the last century in Hungary, but two specific factors prevented the widespread use of scientific achievements. One was the tendency for researchers to be erudite encyclopedists, whereas elsewhere in Europe there was a fairly well established division of labor in science. [16] The second factor was the Hungarian language, which proved to be as much of an obstacle to the spread of knowledge as Latin had been half a century earlier. One clear example of this was Endre Hőgyes (1847-1906), a professor of medicine, who published his studies on vestibulo-ocular reflexes in the late 1870s and early 1880s in Hungarian. When his main treatise was translated into German in 1911, under the impact of the fundamental discoveries by Robert Bárány (1876-1936) on caloric nystagmus (for which he won the Nobel Prize in 1914), they exerted little influence, in spite of many parallels between the two men's work. It was another 40-50 years until the fundamental novelty of Hőgyes' work could be appreciated.

By the end of the nineteenth and the first few decades of the twentieth centuries, internationally significant results were attained in mathematics, chemistry, and astronomy, with scientific schools being established in these fields. Loránd Eötvös (1848-1919), one of the great figures in experimental physics and physical chemistry around the turn of the century, demonstrated by unprecedented exact measurements the equivalence of the inert and the gravitational mass, which became a key issue of the general theory of relativity. By constructing the torsion balance, he laid the foundations of geophysical research in Hungary. His discovery was also significant in practical terms. His instruments have been used to explore the oil and ore layers and the stratification of the deep layers of the earth throughout the world.

The application of research-based technology in Hungary started in the last decades of the nineteenth century. The degree of technological development was remarkable for a country with relatively backward technology. A co-worker of Thomas Edison, Tivadar Puskás (1845-93), worked out the principle of the telephone exchange and presented his invention at the 1881 World's Fair in Paris. The carburetor, an important part of combustion engines, was invented by two Hungarian engineers, Donát Bánki (1859-1922) and János Csonka (1852-1939). Ottó Titusz Bláthy (1860-1939), Károly Zipernovszky (1853-1942), and Miksa Déri (1854-1938) invented the transformer. Kálmán Kandó (1869-1931) made possible the application of electric energy in railroad transport by developing the phase-converter electric locomotive.

In this period there seems to have come into being, for the first time in Hungary's history, a close relationship between economics, science, and technology that also kept pace with the international level. The country was experiencing an extremely rapid and highly concentrated industrialization. At the same time it became apparent that the most significant technological inventions surpassed the limitations not only of Hungary but also of the Austro-Hungarian Monarchy, since, owing to the shortage of capital, they could not be exploited for domestic use and, thus, could achieve practical application only abroad. This explains the extensive international contacts enjoyed by Hungarian science.

Scientific forums were established during the second half of the nineteenth century, the foremost being the Hungarian Academy of Sciences. Institutions were founded to conduct R&D in science, technology, and production. The first research institutions dealing with agriculture and the food industry, established in the last third of the century, included the Economic and Machine Research Station, the Agricultural-Chemical Research Station, the Seed-Grain Research

Station, the National Phylloxera Research Station, and the National Entomological Institute. [17] In addition to their research activities, these institutes participated in the organization of the economy and health protection, and in some respects even carried out duties as public authorities.

Although for the purposes of this study we are not exploring developments in the social sciences, we must mention the role of the economic sciences in development. Economic research as such was first published at the beginning of the nineteenth century, in Latin. It quickly entered the public realm through the works and debates of Count István Széchenyi (1791-1860), the founder of the Hungarian Academy of Sciences, and Lajos Kossuth (1802-94), the governor-president of the bourgeois democratic revolution. The abolition of serfdom, the capitalist industrialization, the modern system of credit and banking, and the recognition of the need to establish an infrastructure to meet the needs of the age were among their accomplishments.

In the second half of the nineteenth century, Hungarian scientists established international scientific contacts. Some of them joined international scientific organizations, and scientific congresses were held in Hungary. In addition to personal meetings or correspondence, the international contacts of Hungarian scientists were fostered by the Academy through the honorary memberships conferred upon famous foreign scholars and scientists who included the German philosopher Friedrich Schelling, the Czech historian Frantisek Palacky, the German mathematician Karl Friedrich Gauss, the German physicist-chemist Robert Bunsen, the British scientist Michael Faraday, the German linguist Jacob Grimm, the German geographer Alexander von Humboldt, the German chemist Justus Liebig, the German historian Theodor Mommsen, the British naturalist Charles Darwin, the German scientist Heinrich Helmholz, the German pathologist Rudolf Virchow, the French chemist Louis Pasteur, and the Russian chemist Dmitri Mendeleev.

The first representative of Marxism in Hungary was Leo Frankel (1844-96). Ervin Szabó (1877-1918) was the first to translate works by Marx and Engels into Hungarian, thereby making them accessible to the masses.

In other words, during the second half of the nineteenth century and in the period preceding the outbreak of World War I a trend to modernization was under way in Hungary, both economically and socially; its essence lay in the dismantling of feudalism, in the rise of capitalism, and in the economic, technological, and social manifestations of the Industrial Revolution. This development lagged behind that of Western Europe. Whereas the latter grew into industrial countries during the Industrial Revolution, Hungary failed to change fundamentally despite major alterations in its domestic economic and social structure. Its international relationships remained static, and the nation remained the agrarian supplier for the more highly industrialized parts of the Austro-Hungarian Monarchy. In addition to the relatively low level of industrialization, the traces of the earlier underdevelopment remained in every sphere of the economy, including agriculture. [18]

DEVELOPMENT BETWEEN THE TWO WORLD WARS

When World War I ended, the Austro-Hungarian Monarchy collapsed and the oppression of the masses sparked the bourgeois democratic revolution in the autumn of 1918, which was followed by a proletarian revolution in the spring of the following year. For 133 days Hungary was governed by the socialist Republic of Councils, which was overthrown by a combination of Entente intervention and by internal counterrevolutionary forces assisted by the former. A quarter-century of counterrevolution followed, and Hungary became the first fascist country in Europe. Allied to the Axis Powers, Hungary came under increasing German influence and economic domination, culminating in the Nazi-German occupation on March 19, 1944. In short, after ridding itself of the Monarchy, the nation experienced a period that was far more reactionary and only seemingly independent.

The new (and present) borders of Hungary were set after World War I. The total land area amounted to 32.7 percent of the former total, while the population was reduced to 42 percent of the former total. In the course of the border revisions, several industries lost their sources of raw material. In terms of prewar production levels, the iron and other metals industry had its capacity reduced to 50.7 percent, that of the machine industry to 82.2 percent, that of the building material industry to 59.7 percent, and that of the wood industry to 22.3 percent. The breakdown of the population by occupation also revealed changes, with those in agriculture dropping from 64.5 percent to 55.8 percent of the total, while those in mining and industry rose from 17.1 percent to 21.4 percent. The country seemed more industrialized and, having been freed of the Monarchy's almost self-sustaining closed economy, it was forced to become foreign-trade-oriented, without the benefit of any transitional period whatsoever. [19] In 1920 agriculture was responsible for 58 percent of the national income and industry for 28 percent. [20]

In the economic chaos that followed World War I, great inflation occurred (it was surpassed in 1946). The only development that took place was due to foreign loans, and it was thwarted by the 1931 worldwide economic crisis. Only increasing preparations for war during the 1930s brought relief, but that was negated by the outbreak of World War II.

The progress of science and technology encountered serious obstacles during the period, but certain results were nevertheless attained. The basic reason for this was that domestic science had received a good foundation during the preceding years and that training was adequate in terms of the requirements of the times. Development in technology was based partly on domestic resources, but most of it came from Germany.

One of the first science policy measures of the Horthy government was the ousting of progressive professors and teachers from the universities and other educational institutions. The second measure was the introduction of the numerus clausus in higher educational institutions, the first racist measure of fascism. The Horthy government also sought to regain—under the slogan of irredentism—those areas with Hungarian inhabitants that under the Paris Peace Treaty (1919) had been awarded to the allies of the victorious Entente powers. The irredentist policy sought to prove Hungary's "cultural superiority" over Romania, Czechoslovakia, and Yugoslavia, an ambition totally without a factual foundation.

All of these measures led to the emigration of many respected Hungarian scholars. The émigrés included the philosopher György Lukács (1885-1971); the economist Jenő (Eugene) Varga (1879-1964), who later was the director of the Institute of World Economics of the Soviet Academy of Sciences. Others became famous for their work done after they left. These included Thomas Balogh, Miklós Káldor, Theodor von Kármán, John von Neumann, György Pólya, Gábor Szegő, Leó Szilárd, Ede Teller, and Paul E. Wigner. Frigyes Verzár was still working in Hungary when he conducted research on vitamin chemistry. Albert Szent-Györgyi was also still in Hungary when he received the Nobel Prize in 1937 for his work on cell oxidation and the isolation of vitamin C. György Hevesi had left Hungary when he was awarded the Nobel Prize in 1943 for his work on the application of isotopes as indicators. György Békésy, who left Hungary after World War II, was awarded the Nobel Prize in 1961 for his research on acoustics and audiology. Though not directly relating to this field, here we must also make mention of people like Lászó Moholy-Nagy who taught at the Bauhaus of Weimar and later at that of Dessau and who was a co-worker of Walter Gropius. In addition to his activities as an architect and a painter, he was also the designer of the Parker pen. It was during that period that László József Biró began his career with several inventions while still residing in Hungary, but it was only abroad that he became famous as the inventor of the ball-point pen.

The domestic base of scientific research continued to be the university. Technological progress was achieved mainly in the research laboratories belonging to industrial plants. In iron metallurgy the major advance was the termination of the obsolete Bessemer steel process and the switch to electrosteel production. The significant upsurge in mining activities was due to the discovery of rich bauxite deposits in Hungary in the 1920s. Also important to the basic industries was the increased production of electrical energy. During the early 1920s per capita production of electrical energy was 35 kilowatts, an amount three times as great as during the period before 1914. By 1938 approximately 12,000 kilometers of power lines had been placed.

The development of the machine industry was in a class by itself. While the manufacture of machinery experienced a boom on a global scale, the Hungarian machine industry suffered a serious shortage of capital and several plants had to be closed. In 1938 the production of that branch was only at the pre-1914 level. There were a few attempts to start an automobile industry, but since only 200-300 cars could be produced per year during the second half of the 1920s, production was terminated. On the other hand, the manufacture of motorcycles became firmly established. [21]

Modern mass production based on modern technology and being on par with similar achievements elsewhere, was attained only in some branches of the electrotechnical-appliance industry. The Ganz factory was producing for the U.S. market, thanks to the electric meter developed by Ottó Bláthy in 1923. The United Incandescent Lamp Factory adopted the latest achievements of its U.S. counterpart and, beginning in 1931, started to manufacture light bulbs filled with krypton gas, the invention of Imre Bródy (1891-1944). This factory exported its products to 53 countries between the two world wars. It also manufactured radios and tubes; and by the end of the 1930s it had produced 40,000-50,000 radio sets, three-quarters of which were exported. It is indicative of the speed with which the electrotechnical industry was advancing that in 1938 it accounted for 30 percent of the production of the machine industry. [22]

Also between the world wars, the important work of Géza Zemplén (1883-1956) in organic chemistry contributed greatly to the extension of pharmaceutical research and production. This is exemplified by the great range of pharmaceutical and hormone compounds produced by the Chinoin and Richter Pharmaceutical Works. The preparations for war favored chemistry. In 1935 the Hungarian Fertilizer Factory began to manufacture heavy chemicals that had not been produced previously.

Remaining significant were the researches in machine construction by the Weiss Manfred Factory; the high-level motor research at the Csonka Machine Factory and the optical instruments of the Gamma Factory became widely known.

Certain other developments had no effect on the economy, although they definitely were part of the mainstream of technological progress. These include György Jendrassik's gas turbine, Oszkár Asbóth's helicopter, and Albert Fonó's work on jet propulsion. [23]

Thirty scientific institutions were operating, but in the present sense of the term only about half of them could be called research institutions. Aside from the institutes for medical psychology, sociography, and public health, most were connected with agriculture.

Between the world wars it was obvious that the government was somewhat more methodical in establishing institutions. For instance, a minister speaking at the 1926 Natural Science Congress said:

> Small countries have to be careful in organizing their institutions due to the limited funds available. We Hungarians can consider the establishment of separate research facilities only in those areas in which, due to geographical or climatic reasons or other unique factors certain prerequisites are present which only we possess.
> . . . [24]

Similar purposefulness may be seen in the new university curriculum of 1923, which provided for a more intensive and more differentiated course, with greater specialization of the departments. [25]

The various branches of science that received the most support from the Horthy regime were those that justified the system politically. Thus the greatest restrictions were placed on the social sciences. In the natural sciences, the school of mathematics continued to develop, and Hungary's medical research not only preserved its reputation but also managed to enhance it further.

Although outstanding scientists and scholars were working at the universities, the Hungarian Academy of Sciences was characterized by a bleak backwardness. Only seven out of its 24 leaders were scientists, the rest being church dignitaries, aristocrats, and so-called experts on cultural policy who were loyal to the Horthy regime. For a long time the president of the Academy was a Habsburg archduke. Even so, the foreigners elected to honorary membership included such eminent scientists as C. V. Raman of India, Max Planck of Germany, and Niels Bohr of Denmark, all three of whom won the Nobel Prize in physics.

Resting on a comparatively narrow social base, Hungarian science developed very little during the counterrevolutionary regime. The direct input of science to the economy was small. Technological development was only partly based on domestic effort, most of it coming from abroad, particularly from Germany. A career in science could offer a livelihood to only a few. The counterrevolution and the prevalent racism compelled many outstanding people to emigrate. By the final phase of the war, production was disorganized, most of the relatively modest base of scientific research and technological development was destroyed, and all research had ceased.

DEVELOPMENT UNDER SOCIALISM AFTER WORLD WAR II

Owing to the ravages of the war and to the fact that the retreating fascist army had carried away a great deal of what was valuable in Hungary, resuming production took a long time. All of the major Danube and Tisza bridges, and many of the rest, had been blown up; 40 percent of the railroads were destroyed; 1,260 of the some 3,000 locomotives had been taken away, and 60 percent of those left were damaged; 3,600 factories and 90 percent of the companies suffered damage; in Budapest 13,500 individual apartments had been almost totally demolished, 18,700 had become uninhabitable, and 48,000 had been seriously damaged, adding up to 30 percent of all the homes in the capital. [26] Altogether, some 40 percent of the national wealth had been destroyed. [27]

Except for the bourgeois democratic revolution and the 133 days of the Republic of Councils, Hungary had no democratic traditions; and the liberalization that occurred in the last third of the nineteenth century did not solve the problems of bourgeois democratic development. The postliberation anti-fascist coalition government had to institutionalize the bourgeois achievements: repealing fascist legislation, implementing a land reform to abolish the feudal remnants, extending suffrage, separating the state and church, electing local organs of the state in democratic fashion, and so on.

Political democratization went together with the large-scale reconstruction program, primarily involving the railroads, roads, bridges, and homes, and with starting the factories once again. The economic chaos that prevailed after the war made it necessary for the state to take a strong hand in the running of the economy. By the spring of 1946, 75 percent of the factories' incomes came from the state and 90 percent of the production of the mines and of the iron and machine industries was based on state orders. [28] Using only internal resources, in August 1946 Hungary achieved the stabilization of its currency. With a strict deflationary monetary policy and a maximally controlled credit policy, combined with central price-fixing, the government brought about a system of state control over the economy in which the traditional market system, based on supply and demand, was severely curtailed. [29] Between December 1945 and

December 1949, the mines, banks, and factories were gradually nationalized.

Understandably, it was years before the government could concern itself with the modernization of the sciences and of higher education. The Hungarian Working People's Party, which resulted from the merger of the Hungarian Communist party and of the Social Democratic party in 1948, was the first to raise the matter in its program, discussing expectations of science under socialism: ". . . scientific research . . . must be freed from its dependence on capital and it must be placed at the service of the people. This is the only way to guarantee the freedom of research and creativity."[30]

A concrete science program, however, was preceded by several preparatory measures. The democratization of the educational system made it necessary to raise the compulsory school age to 16. (In principle, it had been 12 before the liberation, but in practice it was only ten.) With state support, a scholarship system and a network of colleges, secondary, and higher educational institutions were opened to young people from working-class or peasant families. Further education was made accessible in the form of evening and correspondence courses for adults who had been subject to prewar restrictions. The training of professional manpower was extended. According to the demands of socialist industrialization, the high-level training of professional manpower, primarily that of engineers, was differentiated into two types.

In 1948 the government established the Hungarian Council of Science, which was to develop and modernize science in Hungary and to reorganize the Academy of Sciences in accordance with modern requirements. Essentially it is since 1949 that Hungarian science has taken a socialist direction, including the new type of institutional relations between science and the economy. This development was justified not only by the relative backwardness and the need to make up for the losses caused by the war, but also by the modern requirements of the scientific and technological revolution occurring throughout the world. These determinants being given, there were three sources of support for the development of Hungarian science. One was the increasing state support and guidance. The second was the aligning of individual initiatives with the main overall objectives. The third was assistance from abroad, chiefly through important scientific traditions being given an opportunity for further development and the initiatives of individual scientists and experts facilitating the adoption and domestic growth of modern technology.

Of great importance was the economic, cultural, and scientific assistance given by the Soviet Union in setting up modern industrial plants (including a ball-bearings factory and an integrated iron metallurgical works) and in acquiring the necessary scientific and technological knowledge. The Council for Mutual Economic Assistance, established in 1949, created the conditions for compensation-free exchange of industrially valuable scientific and technological knowledge among the member countries.

The stabilization of Hungary's currency in 1946 made the realization of a planned economy feasible. On August 1, 1947, the First Three-Year National Economic Plan was initiated. Its objective was the restoration of the nation's economy by attaining, within three years, the production level of 1938 (the last year before the war, yet strongly exhibiting the effects of the war boom).

The First Three-Year Plan significantly increased production and ordered investments. In the period following, despite further successes in the effort to modernize agriculture and to industrialize the country, serious distortions occurred, such as the stagnation of the living standard. Indeed, there were even some signs of slippage. The economic policy called for a forced pace in industrialization, and the trend of development was inconsistent with the country's possibilities. The autarkic attitude wasted much valuable energy. Industrialization chiefly meant the disproportionate development of the heavy industrial sector, with 90 percent of industrial investments going to heavy industry in the period 1950-55. As late as 1961-65 it was 81 percent. [31]

This economic policy, while in principle accepting the motivating role of financial interest, failed in practical terms to make the productive units interested in raising the level of their technology. The differences in the level of production technology also made the adaptation of new scientific and technological advances rather difficult. The effect of R&D on the economy was primarily felt in the manufacturing processes and production technologies, in exploring important basic resources and energy sources, in soil improvement, and in the development of agrotechnology. The most outstanding of the achievements of this period was the establishment of research institutes. The first coherent institutional system of science-technology-economy came into being because industrial ministries were also running research institutes.

The new political line and the accompanying consolidation after the 1956 counterrevolution produced an economic policy that favored a sound production structure and the reorganization of agriculture on a modern, large-scale basis. There was a decrease in centralization, prominence was given to creative initiative, and there was an opportunity for research to be conducted on a contract basis between research institutes and enterprises. In the interest of expanding the relationship between research and production, new economic and interest-eliciting regulators were introduced.

As a result of these measures, the structure of the national economy is significantly different from its prewar state. In 1974 national income had increased fivefold, the share of agriculture in production had decreased from 58 percent to 16 percent, and industry (including construction) rose from 25 percent to 55 percent. The two most dynamic industrial branches, chemicals and machines, grew by 23 times and 10 times, respectively. [32]

Besides the large investments of capital, a significant role was played in this development by the great increase in the number of employed workers, the solution of the unemployment problem by the late 1950s, and the incentive for greater technological development provided by the great manpower demand. In addition, scientific research and technological development were of great importance in economic growth. According to some studies, about 30 percent of the value of a new product represents R&D and education.

A decisive factor in the growth of science and technology was the influence of the state on research and development, a consequence of the very nature of a socialist society and the demands made upon it. From the middle of the nineteenth century data are available on the science budget of Hungary. (No such data are available concerning the enterprises, since the figures were regarded a part of "business secrets.") From the 1860s to 1938 the state expenditure in any one year did not exceed 1 percent of the same annual expenditures in the 1960s. [33]

The principle applied to R&D expenditures is that "although the relationship and interaction between science, technology and production is complex, it is clear that in the interest of boosting production, technological growth should surpass that of production and that science should advance at a more rapid rate than technology." [34] Since the late 1950s the growth of the input into R&D has surpassed that of the national income. As a percentage of the latter, it has increased from 1. 6 percent to 3. 3 percent. In the fields of technical and agrarian sciences we use 75 percent of the expenditures, those being most closely related to production.

It is important to note that in the R&D base, industry's share of the total expenditures is outstanding: 40 percent of all workers, 45 percent of researchers, 45 percent of the allocations, and 56 percent of the research areas directly serve the industrial branches of the economy. [35]

There are two basic characteristics of Hungary's research structure. First, the proportion of research conducted at institutes is high, while that conducted at university and higher educational institutions is low. Second, the proportion of basic research is generally high. There are reasons for this. Before World War II, the universities were the only site of basic research. The network of research institutes, 90 percent of which were established after 1945, attracted significant university resources from both the theoretical and the experimental sectors. Furthermore, while formerly the universities carried out mainly basic research, now they are also dealing with development and applied research. Because of the changes occurring in higher educational institutions, the overwhelming majority of material and intellectual resources are devoted to educational concerns. Thirty-six percent of the total research manpower is now working in universities; the rest is concentrated in research institutes and research units of business enterprises.

The comparatively high rate of basic research follows from several factors. Technically, part can be explained by the fact that in Hungarian research statistics, all social sciences are considered to belong to basic research. But the emphasis on basic research is consistent with the necessity to build a theoretical basis for the new technology that is used more and more as the country participates in the scientific and technological revolution. Scientific workers engaged in basic research also train professional manpower for both research and university staffs.

There has been a growing attitude in the 1970s that basic research, and primarily natural-science research, should be oriented toward social needs. Referring to some of the results of this kind of research will show the extent to which the results of basic research can affect the product structure and technology of a branch of industry. For instance, the Institute of Isotopes of the Hungarian Academy of Sciences supplies the country with isotope products to be distributed within the country and abroad. The institute has developed a complex instrument family, and every year some 100 measuring stations using them are set up, mainly for the automation of industrial processes.

Basic nuclear research started in the 1950s at the Central Research Institute for Physics of the Hungarian Academy of Sciences. This required the development of appropriate instruments that served as a basis for the family of nuclear devices produced by the Gamma Works. The institute also laid the groundwork for development of a multichannel analyzer that later was taken over by the Factory of Electronic Measuring Instruments. In cooperation with other industrial companies, the institute elaborated several important technological applications of the laser. It has achieved good results in laying the foundations of computation techniques and in disseminating them in Hungary.

The Agricultural Research Station at Martonvásár improved hybrid corn as early as the 1950s. In recent years, 75 percent of the corn, 50 percent of the wheat, and 40 percent of the barley cropland of the country have been sown with varieties and hybrids bred or maintained by the institute. Since the early 1970s new autumn wheat varieties have been certified every year as a result of the research done at the institute. These species are of excellent quality and provide 10 percent more grain than the best species of foreign origin sown in Hungary.

The work of the Research Institute for Technical Physics of the Hungarian Academy of Sciences contributed to the scientific foundation of the metallurgy industry. Besides developing semiconductor devices and tungsten processing methods necessary for light bulb production, it has elaborated important measuring methods for the physical qualification of materials and devices.

The Computer and Automation Institute of the Hungarian Academy of Sciences was among the first to train computer experts in the country. It has had a pioneering role in the automation of the machine industry and in introducing the numerical control system. The institute was also the first to use computers for planning purposes. For five years, it provided the program for the information system used to control inputs, output, and warehouse stock at the Duna Steel Works. The number of computers also has increased remarkably. Aside from pocket calculators, the number of computers trebled between 1970 and 1975, and at present there are about 38 computers per one million inhabitants.

In 1974, 430 patents were issued in Hungary, 636 foreign patents were licensed, and 3,212 innovations were accepted. [36] Of these, 91 percent originated within the framework of the technical sciences and technology.

The progressive scientific schools, especially those in the fields of basic research, developed continuously and extensively. New research trends have appeared in mathematics, including modern algebra, approximation theory, graph theory, probability theory, and mathematical statistics. Progress in physics has been characterized by the extension of research in theoretical physics and by the creation of experimental physics. By establishing a wide base for experimental research, Hungarian physics could participate in international-level research. This promoted the introduction of sophisticated technological processes and methods in precision engineering, measurement techniques, computation techniques, and nuclear instrumentation. In cooperation with the United Nuclear Research Institute in Dubna, USSR, many Hungarian researchers have conducted research in experimental physics. In astronomy, Hungarian scientists have taken part in the examination of variable stars, in their photoelectric detection, in tracing satellites, and in data processing.

In chemistry, research in analytical chemistry, and particularly in instrumental analysis, has attained a high level. Analytic chemistry has made possible the progress of gravimetry, which, in turn, has enabled scientists to gain a deeper insight into the nature of reactions through analysis according to weight, particularly by way of derivatographic examination. [37] Outstanding accomplishments in the rapidly developing field of physical chemistry are the catalysis, polymerization-kinetic, and material structure investigations and the research on nuclear chemistry and reaction mechanisms. The investigation of large molecular structures and the progress of colloid chemistry, which has yielded results applicable in production, have also become significant. In organic chemistry, research on carbohydrates, carotinoids, and compounds of pharmaceutical value has been extended. New trends have developed in work with alkaloids, polypeptides, and stereochemistry.

The geological sciences have played an important role in the exploration of Hungary's physical features, in the scientific grounding of the exploration for mineral resources (bauxite, nonferrous metals), and in the utilization of water reserves. In biology the experimental disciplines have advanced, and significant research is being carried out in systematics concentrating on ecology. Noteworthy results have been obtained in elucidating the neuronal interconnections in the cerebellum and in mapping those areas of the brain that regulate hormone release from the hypophysis. Muscle research has been a fruitful concern of Hungarian biophysics and biochemistry. A school of enzymology has won international recognition for its work on the structure-function relationship of enzymes.

In the agricultural sciences, which have always played an important role in the country's R&D work, the nature of research changed after 1975, mainly due to the large-scale reorganization of agriculture. In addition to the traditional fields of pedology, veterinary science, and plant breeding, new, intensive research was conducted in agrochemistry, animal husbandry, irrigation, business organization, and horticulture. The mapping of the country's soil composition and soil erosion was an achievement of pedology and agrochemistry. Important results were achieved in the theoretical foundation and practical elaboration of methods for improving sandy and saliferous soils in Hungary. In plant breeding, the production of highly fertile corn hybrids and the spread of their use in the country was the most significant achievement. In animal husbandry the major result was the increase of animal products, achieved by the breeding of swine and cattle hybrids for meat and milk. Large-scale poultry raising has been introduced. A widely recognized result of veterinary research is the method of immunization against hog cholera and the elaboration of methods of combating and preventing fowl diseases.

Owing to its pre-World War II dependence on German industry, the development of the technical sciences was uneven in Hungary. During socialist development this unevenness had to be eliminated and new needs had to be met. Technological research aimed at the exploration of the country's natural resources came to the fore: analytic investigation of mines, the technology of preparing and enriching ores. In metallurgical technology, the mechanism of large-scale plastic deformation has been worked out and the preparation of nonferrous metals from the scarce ore resources has been made possible. In the field of energy research, heat-energy and high-voltage electrotechnical researches have yielded important results. The utilization of poor-quality Hungarian coal for energy has made further research into heating technology imperative. One major development in this area is the Heller-Forgó air-cooled heat-exchange system, based on heat condensation, which uses a ribbed radiator. Conservation of heat and electric energy has been put on a scientific basis, and problems connected with the electrification of major railroads and the further development of Kandó's electric locomotive have been solved. The mechanization of agriculture also has progressed. It has become possible to use plastic machine components and modern machine-tool series.

Automation research is quite new in Hungary. In addition to theoretical work, it seeks to automate the production processes of certain branches of industry. Good results have been achieved in developing computer systems, with accompanying equipment and software, for the automation of planning in the machine industry.

Research has made it possible to develop tungsten-filament neon tubing. The production of electronic and light tubes of high durability has begun. Electroacoustical research, a new field in Hungary, has produced such items as electric billboards and amplifiers.

In accord with the social transformation, theoretical medical research—linked to research on prevention, protection, and curing—is to serve therapeutic purposes. Owing to the advances in medicine since the end of World War II, such infectious diseases as tuberculosis have been overcome. The average age of males rose from 48.7 years in 1930–31 to 66.9 in 1972, and that of females from 51.8 to 72.6.[38] The death rate per 1,000 births, of children under one year of age, decreased from 92.5 in 1948–49 to 33.8 in 1973.[39]

This uneven development is composed of three different phases. First, until 1957 emphasis was on reconstruction, the laying of foundations, and quantitative development. Immediately after the war only two or three research institutes were functioning, and the emphasis on quantitative growth was for the most part justified because the institutional base had to be created from practically nothing. That emphasis also bore the marks of the distortions of an autarkical attitude. Second, from 1957 to 1968 there was a more modest but deliberate quantitative development marked by more intensive methods, better planning, and a modern science policy. Third, progress since 1968 may be characterized by an essentially moderate quantitative development, by the advance of technology, by nationwide, comprehensive objectives, and by modern scientific apparatus.

Although with certain unevenness, the phase changes of science policy have tended to follow the phase changes.in the economic policy since the late 1950s. This situation has not, despite all good will, served the best interests of the development of the relationship between science and production.

Since the 1960s the Hungarian institutional system of research has become capable of participating in the scientific and technological revolution, and science and technology have become socially important. The research system is now able to follow international developments, to further the adoption of results achieved abroad, and to yield new results that meet international standards in mathematics, experimental physics, biology, and musicology. The staff of research units had risen to 80,000 in 1975, roughly as many employees as in metal mass-production or building-materials industry.

In the present stage of economic development, the intensive factors of growth have become most important. There is no excess of manpower in Hungary, and in the foreseeable future an increase in production can be achieved only by increasing productivity. The nation, poor in raw materials, must process what it has into structural materials of higher use value. As a result of the country's energy shortage, the consumption of energy must be carefully planned. In its program declaration, the Congress of the Hungarian Socialist Workers' Party, held in March 1975, took the following position on the present role of science:

> Scientific research work should be put, consistently and in a planned manner, at the service of the socialist society. Special attention is to be paid to the development of those high-priority branches of science which are most important from the aspect of performing our sociopolitical, economic and cultural tasks. We are supporting basic research which is indispensable for the development of science. We shall emphasize to an ever greater extent the advantages of the international division of scientific work, above all, in the scientific cooperation with the socialist countries. . . .[40]

The law on the "Fifth Five-Year Plan of the Hungarian National Economy 1976-1980," passed by the Parliament on December 18, 1975, provides that the effectiveness of scientific research and the practical application of the results of R&D should be increased. Three percent of the national income that can be used within the country can be spent on R&D.

> The objectives of scientific research and development should be selected so that the intellectual and material resources might be concentrated to a larger extent than before on tasks promoting the effectiveness of social production, primarily of technological development. Within this, the proportion of the development tasks needs to be increased. [41]

In Hungary today, relations and interactions between science, technology, and the economy conform to a system of objectives. Proper attention is given to socioeconomic planning, with reliance on the results of science and technology. At the same time, the introduction of scientific and technological research results into practice is still too time-consuming, the system of control and management is not effective enough, and not all the possibilities of the international division of labor in research are duly exploited. Essentially, it is still true that "the pace of economic development in the final analysis is not so much dependent on the development activity itself but on how fast the results spread."[42]

In Hungary the necessarily extensive development process for a long time pushed the intensive elements into the background; in science and technology more attention was paid to the quantitative development of R&D than to the utilization of its results. The level of technology achieved in various branches of production is still uneven, and in some places obsolete technologies still prevail. This makes it difficult to assimilate new scientific and technological results. The existing incentives of production and also some special characteristics of the stable socialist economy do not seem to give adequate encouragement to the assimilation of newly developed technologies. The Fifth Five-Year Plan points out that the possibilities for the international division of scientific and technological labor are underutilized. [43]

EXPANSION OF INTERNATIONAL SCIENTIFIC AND TECHNICAL CONTACTS AFTER WORLD WAR II

The oldest traditional form of international scientific contacts, the institution of getting foreigners to become honorary members of the Academy, has been developed further. Hungary's international technological and scientific contacts are no longer simply a matter

of honorary memberships, but are characterized by stable, institutionalized bilateral and multilateral agreements. After 1945 Hungary
had to go a long way before reaching the desired level of participation
in the international division of labor. The first government-level
technological and scientific cooperation agreements were concluded
with the socialist countries, followed by nongovernmental (such as
Academy-level) bilateral cooperation agreements. Relations with
nonsocialist countries were for a long time blocked by the Cold War,
the notion of the Iron Curtain, and by the "Hungarian question" that
arose in the forums of international organizations. Hungary now has
technological and scientific agreements at governmental and institutional levels with most of the developed capitalist countries. Similar
relations have been established with several developing countries as
well. Official contacts are revealed by travel statistics: The yearly
number of journeys abroad by persons traveling in nonprivate capacities rose from 44,400 in 1960 to 128,300 in 1968 and to 194,400 in
1974. The 1974 figure corresponds to 3.8 percent of the employed
population. However, it must be noted that a certain number of persons made several trips within a year.

Foreign visitors to Hungary in the same capacity numbered
42,800 in 1960, 197,400 in 1968, and 304,000 in 1974. Of the yearly
nonprivate travels abroad, 8,625 in 1968 and 18,834 in 1974 were
for scientific purposes. [44] The 1974 figure corresponds to roughly
one-third of the total number of Hungarian scientific workers, but
of course some persons made several trips.

The process of establishing multilateral contacts took longer
than that for the stabilization of bilateral relations. Prior to World
War II, Hungarian scientists had participated in scientific organizations as individual members. Several progressive international
organizations, such as the International Union of Academies, decided
to exclude the Horthy regime. After World War II, major quantitative
and qualitative changes took place in the international scientific
organizations. The most significant international organizations, such
as UNESCO and ICSU, envisaged a wide range of international programs instead of the former individual memberships. Corporate
membership, ensuring better continuity, has become important,
thereby multiplying the possibilities of cooperation. Before the war
some 80 to 100 Hungarian scientists and scholars were interested in
international scientific organizations; now the corporate Hungarian
membership in some 200 international scientific organizations permits the involvement of several thousand Hungarian scientists and
experts.

Hungary's role in international scientific organizations was
consolidated in the 1960s; and by about 1970, although lagging considerably behind the Western countries, it was considered to have

attained an acceptable position. Bilateral and multilateral cooperation with the socialist countries is of fundamental importance, especially with the members of the Council for Mutual Economic Assistance. This cooperation is based on the principle of internationalist assistance, as well as on mutual advantage. At the same time, Hungary is very much interested in scientific relations with the developed capitalist countries on the basis of mutual interests. It has been demonstrated that since the 1960s Hungary has reached the stage of scientific development at which it can assist developing countries in creating their own national scientific base, in developing their progressive traditions, and in freeing themselves from the limitations of monoculture.

In quantitative terms, Hungary's international scientific relations are relatively satisfactory. In respect to quality, however, there are still many possibilities for growth. First of all, in the spirit of the final document of the Helsinki Conference of 1975, the country's relations with the West should be directed toward joint research projects instead of the existing exchange of experts, the initial possibilities for which are already present.

THE SOCIAL ORGANIZATION OF SCIENCE IN HUNGARY

Figure 5.1 shows the institutional network for science policy, the organizing and planning bodies, and the structure of science in Hungary. It should be read from left to right, in order to comprehend the institutional groups and their various roles in shaping and directing science policy, as well as in planning research work and the popularization of science.

From the top-level state agency down to the smallest research unit the relationship is continuous. There is feedback in both directions. The supreme organ of state power in the Hungarian People's Republic is Parliament, which acts on general measures, approves the state budget, and enacts laws concerning science and technology. The chief executive power of the state is the government, which is the supreme body directing science policy. Figure 5.1 indicates all the ministries under the government, as well as organs of national authority that have an active role in guiding scientific efforts. Ministries, from which an unbroken line extends to research institutes and university departments, in addition to their controlling activity, perform operational duties that involve directing scientific activity. These include the Central Planning Office, the Ministry of Education, the Ministry of Finance, and the Ministry of Labor.

This brief outline reveals two general traits of the Hungarian organization of science. First is the polycentric nature of guidance:

FIGURE 5.1 Organization of Science in Hungary

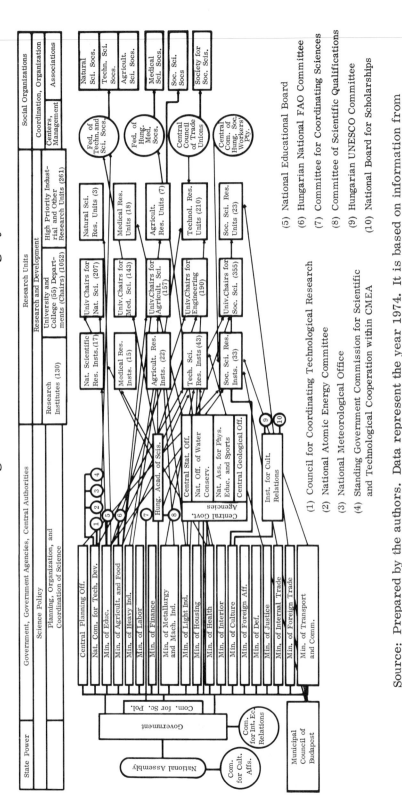

(1) Council for Coordinating Technological Research
(2) National Atomic Energy Committee
(3) National Meteorological Office
(4) Standing Government Commission for Scientific and Technological Cooperation within CMEA

(5) National Educational Board
(6) Hungarian National FAO Committee
(7) Committee for Coordinating Sciences
(8) Committee of Scientific Qualifications
(9) Hungarian UNESCO Commission
(10) National Board for Scholarships

Source: Prepared by the authors. Data represent the year 1974. It is based on information from "Tudományos kutatás 1974," Scientific research (Budapest: KSH, 1976).

the government marks out the main directions and coordinates nation-wide tasks, but most other duties are the responsibility of more than 20 national bodies. At present there is no intention to transfer the control of research to a single-center system. Second, the social character of the promotion of science is clear in all areas; state and social guidance is closely connected at all levels, and the creative participation and initiative of society are fully utilized.

The government's guiding of scientific activity is also corporative in character. One of the deputy prime ministers is responsible for science policy and, since 1969, the Committee on Science Policy has been working under his guidance. The Committee on Science Policy was set up in the interest of uniform state control of scientific research, for the coordination of the country's research effort, for determining the main direction of R&D, for ensuring the financial and material conditions of research work, and for controlling the effectiveness of expenditures.

Another deputy prime minister, who is responsible for economic policy, guides the activities of the Committee on International Economic Relations, which controls participation in the international division of scientific work and, primarily, Hungarian economic cooperation within the Committee on Mutual Economic Assistance. In this context it also acts as the coordinating agency of scientific and technological cooperation.

An organization of ministry rank, the National Committee on Technological Development, is the advisory, coordinating, and, in some areas, the controlling government organ handling all problems of technological development affecting the national economy. Its basic tasks are to elaborate scientifically sound technological and economic studies that meet the demands of modern technology and the national economy, to coordinate the activities of the national-level ministries and agencies in solving problems of technological development affecting the national economy, and to direct technological research affecting several branches of industry or a whole sector of the national economy. The experts involved may form and express their views to the best of their knowledge and scientific conviction, independent of the standpoint of their supervisory agency.

As the supreme scientific body of the Hungarian People's Republic, the Hungarian Academy of Sciences plays a leading role in the guidance of scientific work. From its foundation in 1825 until the end of World War II, it was a learned society acting as a social association. It was reorganized in 1949. According to Decree no. 41 of 1969, which defines its present status, statutes, and tasks, the Academy participates in the nationwide guidance of scientific research and, through the work of its institutions and members, provides for the promotion of science and controls the research conducted at its institutes. It acts partly as a scientific body and partly as the central agency managing the work of its institutions. Its activities are supervised by the government, and its expenditures are a separate item in the state budget.

One of the main characteristics of the socialist-type Academy is that apart from its participation in sciences through its members, it also runs a wide network of research institutes to carry out its own scientific tasks. Besides its own network, however, the Academy also directs and controls some of the research units active at the universities. In this sense the Academy can be compared not so much with its Western counterparts as with high-level scientific agencies having research institutes of their own, such as the Centre National de la Recherche Scientifique in France and the Max-Planck-Institut in the Federal Republic of Germany. In late 1974 some 140 research units belonged to the research base of the Academy. Of these 38 were research institutes, and the rest were research units within the departments of higher educational or other institutions. All were supported by the Academy.

SOCIALIST SCIENCE POLICY

Socialist science policy forms part of the general policy of the state, which sets its main objectives. Science policy in a socialist society is subordinate to the general interests of society. Taking into account the resources of the given society, its task is, on the one hand, to determine objectives that can be achieved by means of science and are necessary to social, economic, and cultural development, and, on the other, to create the possibilities for the attainment of these objectives. To achieve this end, science policy forms its own operational mechanism, determines the objectives and directions of scientific research, and ensures the proportional development of the individual sectors of research. It decides on the proportion of the national income that is to be spent on research, and generally the rate of increase in R&D expenditures surpasses that of the national income. (This occurred from the late 1960s to the late 1970s, then, for economic reasons, temporarily slowed, remaining at 3 percent of the national income.)

Science policy is also concerned with the location of research institutes. In Hungary, 20 percent of the country's population is concentrated in the capital; thus, in order to achieve balance, it is imperative to establish research bases elsewhere in the country. Moreover, science policy extends to the social and legal protection of research work and researchers, and to setting up appropriate organizations and institutions. Finally, it determines the content and degree of participation of the international division of scientific work. Of great importance is the institutional relationship between science and practice. On the whole, Hungarian science policy is characterized by thorough planning.

Generally it is not necessary that the science policy of a state be formulated in a law or government decree. It may be laid down in a series of more or less coherent measures, which would constitute a de facto science policy. Hungary, however, has a de jure science policy that was summarized in the science policy guidelines of the Central Committee of the Hungarian Socialist Workers' Party in 1969. At the same time Hungarian science policy does not want to (nor is it able to) regulate every field, every objective, and every element of scientific activity. Thus science policy, taken in a strict sense, covers only the essential elements; it also takes into account the intrinsic laws of development in different fields of science. Research has two simultaneous objectives: to meet concrete social, economic, and cultural demands and to produce the new knowledge necessary for further development. Depending on the country's intellectual and economic capacity, science policy can offer a large scope to basic research aimed at exploring the still-unknown laws of nature and society. Naturally it does not regard the whole of scientific effort as a social function yielding immediate benefit.

To realize the defined tasks and objectives of research work, science policy employs intermediary institutions. Through such institutions the ideas of science policy may become a material force. The most important elements of these intermediary elements are the system of management and the planning of research and science organization.

Economic planning in Hungary started in 1948, and the planning of scientific research a few years later. Under the present system of research planning, the long-term national plan (20 years) which deals with the main goals and the major structural changes, is primary. The medium-term plans of the state-budgeted research institutes, the lengths of which coincide with the Fifth Five-Year Plan (1976-80), implement tasks derived from the long-term plan. They also contain programs based on proposals of the research units themselves or on orders from other organizations. In collaboration with his research staff, the director of the research institute has a high degree of independence in formulating its plans. One important guideline in research planning is to spend about two-thirds of the material and intellectual resources on the realization of high-priority programs, thus ensuring that the decisive part of the research effort will serve social objectives. About one-third of the research capacity is spent on new tasks not included in the national research plan.

Planning in the recent past has suited the needs of research work more closely; but it is still predominantly thematic, being occupied mainly with the general goals without ensuring the means to realize them. In some cases the necessary foresight is absent, the approach being pure empirical. The adjustment of research to the national economy is not always perfect. These negative phenomena are explained by the facts that there is not yet sufficient experience in research planning and that the method of research planning is not keeping pace with the changes resulting from development of the economic policy. It goes without saying that the application of indirect economic and direct administrative regulators is not always successful. Some of the regulators are new, and therefore their effects cannot be judged. The tendency for research to be more closely intertwined with the economy has not yet reached a desired level. In some cases it runs counter to the main tendency and leads to wasted efforts.

The basic principle of socialist science policy, freedom of scientific research, is formulated in the 1969 Resolution of the Central Committee of the Hungarian Socialist Workers' party as follows:

> The properly interpreted freedom of science is an essential condition of the cultivation of science. There are no scientific problems, that are connected with our social practice and the development of our society, the research and analysis of which would not be—from the ideological aspect—in the interest of the people building socialism. Of course, there are no reactionary forces in nature, but conclusions drawn from the facts might be incorrect, even harmful. Scientific work cannot be forced to come to prescribed conclusions. [45]

At the same time, society can rightfully expect its scientists to be committed to social progress. Scientists should realize the demands of society, take into account the country's capabilities, and, with this knowledge, freely present their scientific convictions and independently choose their working methods.

CONCLUSIONS

Before the Industrial Revolution, science produced certain results and technology made its appearance in certain fields, but they were not connected with economic development. Technology made its effect felt only very slowly. Connections between science and technology and the economy started to develop only in the nineteenth century, but this process was uneven and it was not a decisive force in social development; results affected only certain fields.

The historical development of Hungary up to the end of World War II, and especially from 1848 to 1945, left the nation with a heritage of socioeconomic backwardness and, in spite of all these historical burdens, a very respectable intellectual tradition in relation to science. After the war, under altered social circumstances, the nation built upon these intellectual traditions and established the modern institutional relationship among science, technology, and the economy.

During the building of socialism, there evolved the economic and political conditions under which the interconnection of science and technology, and its positive effect on economic growth, have come into full display. If the development since the end of World War II is characterized only by the fact that, compared with 1938, agricultural production quadrupled and industrial output septupled, even these data show an extraordinary development. These economic results were not achieved through spectacular scientific discoveries and inventions; rather, they came about through the "everyday" work of science and technology, the introduction of foreign technology, and the efforts made to reach world standards.

Hungary's science policy—in accordance with the uniform government policy and with its various subdivisions (economic, foreign, and so on)—is promoting the domestic realization of social objectives through the international division of labor on the basis of mutual interest and also through cooperation with other countries.

By enhancing planning, financial interestedness, and the modernization of organizational frameworks, the relationship between research and planning may be further improved. There may be some unexplored factors in the social process of research and production that are essential for the improvement of the system of control and organization. There is also the hypothesis that in Hungary science is more developed than production, and therefore industry cannot fully utilize the results of research.

Participation in the international division of scientific and technological work is a necessity for a country like Hungary, which can be classed as small and relatively developed. As such, its internal market is narrow and does not influence the world market; therefore Hungary must adjust itself to international economic life. There are some special circumstances that require Hungary to join in the international division of labor:

1. Hungary has a considerable area under cultivation and has favorable climatic conditions. It is thus able to ensure its own food supply and can produce for export.

2. Hungary is poor in energy resources and in structural materials for industry, and is generally in need of raw materials and mineral resources.

3. Modern branches of industry (such as electronics, telecommunications, electric automation), which now characterize the development tendencies of the world, can develop in Hungary only through participation in the international division of scientific and technological labor.

4. Although a relatively adequate higher educational and scientific base has developed in the country, Hungary still cannot be self-sufficient in R&D, higher education, and scientific training. At the same time, however, its system of scientific institutions and its intellectual and material capacity enable it to participate in international cooperation.

NOTES

1. Magyar statisztikai zsebkönyv, 1976 (Hungarian statistical pocketbook) (Budapest: Statisztikai Kiadó, 1976), p. 35.

2. Ibid., p. 139.

3. Ibid., p. 55.

4. A Magyar Tudományos Akadémia másfél évszázada 1825-1975 (150 years of the Hungarian Academy of Sciences) (Budapest: Akadémiai Kiadó, 1975), p. 502.

5. Magyar statisztikai zsebkönyv, 1976, p. 95.

6. Tudományos kutatás, 1974 (Scientific research) (Budapest: KSH, 1976), p. 11.

7. Ibid., pp. 12-13.

8. Z. Magyary, A magyar tudománypolitika alapvetése (The foundation of Hungary's science policy) (Budapest: Tudományos Társaságok és Intézmények Országos Szövetsége, 1927), p. 236.

9. T. I. Berend and G. Ránki, A magyar gazdaság száz éve (A hundred years of Hungary's economy) (Budapest: Kossuth Kiadó and Közgazdasági és Jogi Könyvkiadó, 1972), p. 9.

10. Ibid., p. 77.

11. Ibid., p. 54.

12. Ibid., p. 55.

13. Ibid.

14. Magyary, op. cit., p. 23.

15. F. Jánossy, A gazdasági fejlődes trendvonala és a helyreállitási periódusok (The trend of economic development and the periods of adjustment) (Budapest: Közgazdasági és Jogi Köngvkiadó, 1966), pp. 115, 245.

16. Magyary, op. cit., p. 63.

17. Ibid., p. 439.

18. Berend and Ránki, op. cit., pp. 100-01.

19. Ibid., pp. 107-09.

20. Ibid., p. 112.

21. Ibid., pp. 160-62.

22. Ibid., p. 163.

23. Ibid., p. 167.

24. Magyary, op. cit., p. 370.

25. Ibid., p. 273.

26. Budapest székesfőváros statisztikai zsebkönyve, 1941 (Statistical pocketbook of Budapest) (Budapest: 1941), p. 48.

27. Berend and Ránki, op. cit., pp. 221-24.

28. Ibid., p. 229.

29. Ibid., p. 231.

30. Társadalmi szemle (Social review), April-May 1948, p. 262.

31. Berend and Ránki, op. cit., p. 271.

32. "Harmine év gasdesági fejlődésének adatai" (The data of thirty years of economic development), Gasdaság, 9, no. 1 (1975): 125; M. Timár, Gazdaságpolitika Magyarországon 1967-1973 (Hungary's economic policy) (Budapest: Közgazdasági és Jogi Könyvkiadó, 1973), p. 190.

33. P. Vas-Zoltán, "A nemzetközi tudományos szervezetek gasdasági hatékonysága" (The economic efficiency of international scientific organizations) (C. Sc. diss., Budapest, 1966), p. 40.

34. A. N. Kosygin, "A Close Relationship Between Science and Life," Izvestia, June 15, 1961.

35. Tudományos kutatás, 1974, pp. 18-19.

36. Ibid., p. 27.

37. A Magyar Tudományos Akadémia másfél évszázada 1825-1975, p. 478.

38. Magyar statisztikai zsebkönyv, 1976, p. 44.

39. Magyar statisztikai zsebkönyv, 1975 (Budapest: Statisztikai Kiadó, 1975), p. 75.

40. MSZMP XI kongresszusának jegyzokönyve (Minutes of the Eleventh Congress of the Hungarian Socialist Workers' Party) (Budapest: Kossuth Könyvkiadó, 1975).

41. "A Magyar Népkoztársaság V. Ötéves Terve 1976-1980" (Fifth Five-Year Plan of the Hungarian People's Republic) December 18, 1975, secs. 6.

42. Jánossy, op. cit., p. 140.

43. "A Magyar Népköztársaság V. Ötéves Terve 1976-1980" (Fifth Five-Year Plan of the Hungarian People's Republic) December 18, 1975, secs. 41-42.

44. Tájékoztató a kutátas—fejlesztés 1968 évi fontosabb statisztikai adatairól (Information on the more important data of the 1968 research and development effort) (Budapest: MTA Tudományszervezési Csoport, 1969), pp. 91, 130, 154; Tudományos kutatás, 1974, pp. 94, 122, 158.

45. MSZMP KB Tudománypolitikai irányelnei (Resolution of the CC of the Hungarian Socialist Workers' Party on the Science Policy Principles). (Budapest: Kossuth Könyvkiadó, 1969), p. 39.

6

SCIENCE AND TECHNOLOGY IN MODERN JAPANESE DEVELOPMENT

Shigeru Nakayama

In those countries that have most recently attained the modern stage of economic development, science and technology have usually been introduced by the public sector and transferred to the private sector when they have become productive. The Japanese experience in the late nineteenth century was typical and one of the earliest examples of this process. In the process of transferring a technology to the private sector, there are critical points at which it must be determined whether imported science and technology will become established in the private sector and flourish in an indigenous, self-perpetuating form, or whether their continued importation into the public sector will be necessary.

The next question is who was instrumental in initiating and transplanting science and technology and who supported this effort. I will discuss this in terms of the following tentative model: First, there must have been a group of native people who could become professional scientists and engineers, and who could act as leaders in the mass education of the people and in material construction. Next, the level of basic education must have been substantially raised from that of a nondeveloped economy for a successful transfer to occur; a modern state needs scientific manpower at all levels of society. Finally, in this process social mobility begins and class differences tend to diminish. Modern science and technology is then diffused and rooted among the populace, and its traditions become perpetuated.

The achievement of these three steps may be the major factor differentiating the countries that are to become technological from those that must experience an ever-increasing technological gap.

We must also consider that the degree of success in the transfer depends to a great extent on the kind of technology chosen. To find the conditions for successful transplantation, we shall examine several selected technologies of entirely different types, some imported and some autochthonous.

EARLY EFFORTS, 1868-85

The Utilitarian Image of Science

It is a common belief among historians of Western science that in premodern times science and technology were distinct activities with different social origins. In spite of the effort made by the Encyclopedists to liquidate this social interface, the dual structure of science and technology was still maintained, even in the nineteenth century, by socially separated groups. This was exemplified by such institutional separations as that between the German university and the polytechnical college. However, there was no particular reason for the midnineteenth-century Japanese to distinguish between science and technology when facing the impact of the modern West. To the Japanese it appeared that modern science and modern technology grew in a single Western tradition. It was not the science-versus-technology dichotomy but, rather, the traditional-versus-Western dichotomy with which the Japanese were seriously concerned.*

While science in nineteenth-century Europe was still in the main a cultural activity rather than a practical means of achieving economic growth, as is well illustrated by the issue of the theory of evolution, the Japanese image of science in the late nineteenth century was quite modern. It was exclusively utilitarian and pragmatic, planned for national interests if not purely for profit-making, specialized and compartmentalized. Emphasis was on physical and applied science rather than on biological, and hence the style was closest, for that period, to contemporary scientific technology.

The Institutionalization of Science

After the Meiji Restoration of 1868, the Japanese response to Western science was dramatically transformed. In the preceding

---

*Until the 1880s the Japanese language did not distinguish clearly between "science" and "technology." The separation between the concepts became real only toward the close of the century, when autonomous scientific communities were formed at the university-faculty level.

Tokugawa period, Western science was initiated and advocated mainly by scholars in the private sector. Only in the last period of Tokugawa rule were official training institutes for Western naval technology and related sciences opened at Nagasaki and elsewhere. Being hard-pressed by the urgent defense needs of the country, the administration lacked the foresight to build institutions providing modern, systematic education in science and technology. During the first two or three years (1868-70) of the new Meiji government, there was still an influential group that wanted a faithful restoration of the ancient imperial system of government; but the modernist leadership soon followed the policy of Westernization. Scientific education was completely institutionalized. It became firmly programmed in such a way that an institution was first created, and then European and American scientific and technical specialists were invited to meet selected Japanese students within the institution. Outstanding graduates were sent abroad for further study. [1]

Guidelines for Westernization

In the draft rules for sending students to study abroad, prepared in 1870, the following subjects and preferred countries of study are listed:[2]

Britain: machinery, geology and mining, steelmaking, architecture, shipbuilding, cattle farming, commerce, poor-relief
France: zoology and botany, astronomy, mathematics, physics, chemistry, architecture, law, international relations, promotion of public welfare
Germany: physics, astronomy, geology and mineralogy, chemistry, zoology and botany, medicine, pharmacology, educational system, political science, economics
Netherlands: irrigation, architecture, shipbuilding, political science, economics, poor-relief
United States: industrial laws, agriculture, cattle farming, mining, communications, commercial law.

In view of the history of nineteenth-century science, the above assessment was for the most part correct and objective. Presumably the policies for science and technology, including the above recommendation, were drafted mainly on the suggestion of G. F. Verbeck and other foreign advisers to the government.

Employment of Foreign Scientists and Engineers

About 1870, a policy was adopted for the employment of westerners to guide teaching in schools and governmental enterprises.
French engineers were employed in the army, in mines, and in dock-
yards; Germans taught medicine and basic sciences in the schools;
and a number of Americans were agricultural specialists for the
Commission for the Colonization of Hokkaido (the island on which
American large-scale cultivation was attempted).

The British were numerous in many fields. Generally, in
engineering they made the greatest contributions. Nearly 100, for
example, were employed in the construction of railways. At the
Imperial College of Engineering, which existed from 1873 to 1886,
the teaching staff was mostly British and was directed by a Scottish
engineer, Henry Dyer.

The number of government employees reached its peak in 1874.
The total of the employees' salaries reached its peak in the same
year, its ratio to the total national expenditure being about 2 percent.
Apparently the government wanted to reduce the number and cost of
foreign employees, who were paid salaries ten times greater than
those of the Japanese, and to replace them with natives who had
received scientific training in the West. The Ministry of Education
allocated funds for this purpose, but the amount was still far less
than the foreigners' salaries.

Japanese Students Abroad

The initial program of sending students abroad at government
expense was not well organized, and in 1873 it was abolished. In
1875, in its place, overseas scholarships were instituted by the
Ministry of Education. The Ministry of Technology sponsored a similar program beginning in 1880. Since they had already received basic
training from foreign teachers, science students abroad concentrated
on research, and technology students used their time to observe the
actual practice in the cities.[3] When they returned, professorial (or
equivalent) posts awaited them. This first generation of Japanese
university professors displaced the foreign instructors.

Until about 1877 education in Japan put more emphasis on
language study than on one's chosen specialty. There was, therefore,
a tendency among the first students who went abroad to choose the
country on the basis of the language they had studied rather than on
the basis of a particular subject. After 1881, however, the Japanese
government began to copy the German administrative system and
political institutions; accordingly, many students turned their

attention to Germany. It was the heyday of German science—students from England, France, and the United States were going there for advanced study—and the Japanese students must have been impressed by the German scientific leadership. Out of 26 Japanese doctorates of science conferred by 1891, 13 went to men with study and research experience in Germany (five had studied in the United States and three in Britain). [4]

Emphasis on Physical Science and Specialization

Japanese studying Western science during the late Tokugawa period were impressed by the Western process of inquiry into physical laws, rather than by the aggregate of facts and objects of nature. They came to feel keenly that although there was no great gap between East and West as far as the classificatory knowledge of natural history was concerned, Chinese and Japanese cultures seriously lacked the belief in the underlying regularity of nature and the "investigation of its principles": natural philosophy. * Along with this view of Western science, the first primary school curriculum prescribed by the Ministry of Education was not oriented to natural history and biology, as was the case in American primary education, but to physics. [5]

During the 1870s and 1880s the relative position of science and technology in the Japanese educational curriculum, from elementary school to university level, was much higher than in any other nation. For instance, mathematics and science occupied about one-third of the school curriculum in the lower grades (first four years) and two-thirds in the upper grades of the eight-year elementary education. However, the shortage of qualified teachers makes it somewhat questionable to what extent these ideal plans were put into practice. At the university level the emphasis on science and technology was evident in the high percentage of graduates in scientific disciplines from Tokyo University (85 percent in the 1880s, compared with 40 percent in the 1920s.

---

*This intellectual tradition started with B. Miura and T. Shizuki and extended until the time of Y. Fukuzawa. Its exponents appropriated the neo-Confucian concept of kyuri to translate "natural philosophy" (in Dutch, natuurkunde), but later, especially in the Meiji period, this term was specified to mean physics.

The Official and Planned Character of Science

At the frontiers of newly forming disciplines in nineteenth-century Europe, the scientist was free to select for his research any problem that interested him; and the scientific communities were formed by individual scientists drawn together by a common interest. The voluntary research activities of scientific or professional societies usually preceded their inclusion in the university curriculum. In Meiji Japan (and generally in the case of the artificial transplantation of a foreign discipline under state sponsorship) this process was reversed. The government first created institutions for training scientific personnel; only then did university graduates in each discipline form scientific societies, not purely for academic purposes but mainly with common interests in their new and still very weak scientific careers. Thus the scientific community in Japan had a "planned character"; it was planned for the purpose of catching up with the Western standard of science as quickly as possible.

In the nineteenth century, however, it was uncommon, even among advanced nations, to find established precedents or formulas for a national science policy. Thus, the Meiji government had to find its own way by trial and error.

On the practical level, it urgently needed qualified teachers and engineers to build a modern state. The new government built schools and factories, trained scientists and engineers in these institutions, and sent them off to their posts. Rather than having every scientist follow his own research interest, priority was given to certain basic tasks, the accomplishment of which was necessary for the operation of a modern state: geographical and geological surveys, weights and measures, meteorological observation, sanitation, printing, installation of telegraph and telephone lines, military works, railways, and surveys of natural resources. All these enterprises were carried out by the Ministry of Technology, the Ministry of Interior, the Ministry of Finance, the Commission for the Colonization of Hokkaido, the army, the navy, and other governmental agencies under the supervision of such engineers. To conduct such nationwide projects, these agencies had to have their own short training programs to provide field assistants for foreign employees. These agencies did not depend exclusively on the Ministry of Education, which was responsible for regular long-term educational programs.

We may label these activities as "public science" initiated by the government. This step in science for public service was the indispensable prerequisite for the industrialization carried out by the next generation. In addition, the government entered into private entrepreneurship, constructing and managing pilot plants, guiding and subsidizing new kinds of industries. The Ministry of Technology and the Commission for the Colonization of Hokkaido were two major institutional innovations. They carried out experimental programs and were centers of westernization and modernization.

Their enterprises were from the outset, however, exposed to financial risk. Many of the Ministry of Technology's projects eventually proved to be too far ahead of their times, for they tended to introduce the technology of an industrialized society into a preindustrial environment. For instance, the railway construction enterprise was economically unsuccessful at the time, and paid off commercially only after 1885, in the next phase of industrialization.

Thus, Y. Fukuzawa, Japan's foremost exponent of industrial revolution, concluded that "we should not blame them too much for their financial failure. After all, it was a costly tuition fee for the Japanese to learn civilization."[6]

Lack of Research

This planned character of Japanese science has one notable defect. The institutionalization of science and technology at the governmental initiative was certainly an efficient tool for transplanting and introducing foreign science and technology, but it was less effective for fostering original creative activity. Institutions were divided into specialized disciplines, each scientist and engineer assuming his specialized role, a situation permitting little cross-fertilization of ideas.

In fact, in comparison with traditional Confucian learning, the Japanese in the early Meiji period found the strongest point of Western science to be its specialization. This was a particular nineteenth-century aspect of science, not found earlier. The term coined by the Japanese in the 1880s as a translation of the word "science" was kagaku, which originally meant "study of one of the hundred departments" and was the equivalent of the German term Fachwissenschaft.

Thus, each scientist or engineer was concerned with absorbing the foremost achievements of the Western world, within a narrowly prescribed disciplinary barrier, rather than with cooperating with his Japanese colleagues in different fields.

The other side of science policy, research policy, was tactfully avoided. Research policy in any systematic form was the product of the mobilization of science during World War I. The Meiji government gave practically no attention to scientific research. One point should be noted, however. Unlike the European monarchies, Japan had no great interest in founding a national academy of science as a status symbol. The practically oriented Japanese totally subordinated pure research to manpower policy; thus Tokyo University enjoyed the position of the top educational institution as well as the highest academic prestige in the country. *

For the first generation of scientists in the early Meiji period, it was more important to establish institutions than to pursue piece-meal individual research topics. There were some internationally notable contributions made by early Japanese scientists, such as S. Kitasato's works on microbiology, but these usually originated during research apprenticeship in the West. In Japan research work-ers could not find colleagues for discussion, unless they trained their own students in the same research tradition. Thus they became ex-clusively involved in administration and education, eventually leaving their research activities completely aside.

The first generation of scientists concentrated on local science, the application of modern scientific methods to the analysis of local flora and fauna and earthquake and geological observations. Such disciplines as zoology, botany, and geology remained local sciences at least until World War II. Even in such method-oriented disciplines as physics and chemistry, the early scientists' concern was with geo-physical observations and the chemical analysis of local products. Until 1885 more than half of the research articles published by the faculty members of the College of Science at Tokyo University were on local science. [7]

The Samurai Spirit in Science and Technology

Modern scientific and technological professions were the arti-ficial creation of the new, Western-oriented government. The main practitioners of these new professions were former samurai, who were warriors by tradition but who, during the Tokugawa period, became primarily administrative bureaucrats. In the past they had received hereditary family stipends in exchange for their loyalty to the shogunate or local feudatories. They were the class long accus-tomed to thinking in terms of public affairs and to holding public office.

In the 1870s efforts were made by the Meiji government to curtail the inherited family stipends of the samurai class. While farmers, artisans, and merchants could continue their inherited

*The Japan Academy, founded in 1879, was a gathering of obsolete scholars of general Western learning and was often ridiculed as being a home for the retired.

vocations, the samurai lost their traditional source of revenue. The modern government needed less than 10 percent of the samurai population to meet its personnel needs, so the rest were forced to find an entirely new means of living. Since samurai could not compete with other classes in traditional business, they were invited to engage in such new business projects as the agricultural exploitation of Hokkaido, silk manufacturing, and textile industries. This was especially true after 1876 when the former privileges of the samurai were completely abolished. Science and technology was one of the new fields into which jobless samurai were attracted. It is reported that almost all of the early graduates of the engineering college were samurai. [8]

As late as in 1890, the percentage of samurai-born graduates from each school of the Imperial University was as follows: engineering majors, 85. 7 percent; science majors, 80. 0 percent; literature majors, 75. 0 percent; law majors, 68. 3 percent; agriculture majors, 55. 9 percent; medicine majors, 40. 8 percent. [9] The science and technology group had the highest rate of samurai-born graduates and the medical group the lowest. We may assume that graduates of the medical and agricultural schools were, respectively, mainly sons of medical practitioners and farmers, whereas many samurai, deprived of status and occupation, found their most promising career possibilities in the new professions of science and engineering. As the science teachers and promoters of the new Westernization policy and as technocrats building a modern state, both purely governmental enterprises of top priority, the former samurai were able to use their aptitude for administration to achieve a class vocation.

Apart from ideological considerations, and perhaps more important, the sons of the impoverished samurai class were attracted to government-sponsored careers in science and technology because tuition was free, grants were provided, and government service was obligatory on graduation. (For graduates of the Imperial College of Technology, seven years' service in government factories was required.) Thus, science and technology provided new careers for a substantial stratum of society.

Hence, modern Japanese scientific and technological professions were, at the beginning of their development, very "samurai-spirited. " Unlike the European pattern, in which science and technology attracted the rising middle class, Japanese scientific and technological professions in the last quarter of the nineteenth century were dominated by the samurai class, the top 5 percent of the total population.

REORGANIZATION AND TAKEOFF: 1886-1914

Reorganization of Institutions Around 1886

After two decades of intensive development programs, including the institutionalization of science and technology, Japan was prepared to become a modern industrial state. Around 1885 the Meiji government reassessed, reformulated, and reorganized many of its early policies. The most significant event was the dissolution of the Ministry of Technology in 1885. Many government enterprises were transferred to the private sector, mainly because both the government and its critics realized the inefficiency of government factories in profit-making industrial activities. Public-minded samurai engineers accordingly moved into the private sector.

In 1886, when Tokyo University was reorganized into the Imperial University by uniting the Imperial College of Engineering with the Ministry of Technology, most foreign teachers were replaced by native scientists who had just returned from studying abroad.

Until 1885 the scientific public services, such as surveying, had been conducted by transitional manpower who had received short, intensive instruction and had trained mainly on the job. After such services were reformulated, many transitional schools and offices were closed and, around the time of the founding of the Imperial University in 1886, the vocational engineering schools[*] were rearranged more systematically to provide engineers for Japanese industry, which at that time badly needed lower- and middle-level engineers.

The days of enthusiasm for Western science were over, and the government became more interested in having good-quality administrative bureaucrats, rather than entrepreneurial technologists, for the maintenance of regime. Law graduates of the Imperial University received governmental favor, and engineers were placed in positions subordinate to them.

The Language Problem

In the early days of the Meiji period, college teaching was still conducted, even by native Japanese teachers, in such foreign languages

---

*Major examples are Ashikaga Textile Training Center (founded in 1885); Kyoritsu Women's Vocational Training School, Tokyo Apprenticeship School of Commerce and Technology, and Kyoto Dyeing Training Center (founded in 1886); and Hachioji Textile and Dyeing Center and Kanazawa Engineering School (founded in 1887).

as English, German, and French, according to the country in which they were trained. By 1900 the Japanese language was dominant, though technical terms remained untranslated.

It was once seriously proposed by A. Mori, an exponent of modernization and Westernization, that the Japanese language should be wholly replaced by English in order to follow the international standard of knowledge as closely as possible, but his plan was never put into practice. If it had been enforced, it might have created a dual culture with an English speaking elite. It would have blocked or considerably delayed the diffusion of Western scientific culture among the populace, who would continue to speak Japanese.

In the nineteenth-century scientific community, English was not so predominant a language as it is now. At the Imperial University, while English prevailed in the School of Engineering, the Medical school adopted German as its academic language as early as 1869 and continued to use it until after World War II. Further, German was second only to English as the second language taught in schools that prepared students for the university. Japanese students were required to learn it as a scientific language.

The Dual Structure of Japanese Technology

Modern engineering and technology may be said to have two entirely different origins. One is the "community-centered" or "public-centered" engineering service practiced in the public sector, best exemplified by the military engineering taught and practiced at the French École Polytechnique. The other is "self-centered," profit-making, capitalistic engineering practiced in the private sector, as exemplified by the Watt-Boulton type of enterprises, power engineering, pharmaceutical technology, and so on.

The traditional Western dichotomy created in the Japanese technological world can be translated in socioeconomic terms into the private/public dichotomy. Private technology is not only "self-centered" and profit-making but also traditional and domestic, not based on science, and transmitted through apprenticeship in the nonsamurai sector, such as in sake brewing and in ceramics. Public science and technology are, on the other hand, not only "community-centered" but of Western origin, university-based (and hence science-based), and practiced mainly by the samurai elite.

Traditional craftsmanship, such as carpentry and fishery, remained outside the government system of modern science and technology. Established scientific and technological professionals were completely separated from traditional craftsmen and found employment in such government agencies as the office of the geological survey and in military arsenals.

I will now examine the actual state of this dual structure by using the list of patent applications. In 1885 the Patent Law was issued, and 100 patents were granted in the first year of its existence. Its purpose was to protect the honor and profits of native Japanese inventors against other Japanese, and had no international implications.

In the early Meiji period, Western industrial designs and inventions were not legally protected in Japan, and the Japanese were free to appropriate them. Westerners claimed that Japan should subscribe to the international patent regulations, since otherwise they could not give their technological advice freely, without the fear of unlicensed Japanese copying. But those who welcomed the enforcement of international regulations were, of course, the Westerners rather than the Japanese. Around 1885 a major political and diplomatic issue in negotiation with major Western powers was the adjustment of unequal treaties. The government intended to give patents to foreigners in exchange for Western denunciation of unequal treaty items,[10] and made a preliminary draft of a domestic patent system. The treaty amendment was postponed to 1899, when Japan subscribed to the International Industrial Property Regulations. Not until then did foreigners obtain Japanese patents.

Out of the first 100 patents issued in 1885, applications by former samurai accounted for only 17 percent.[11] This is in clear contrast with the fact that samurai made up a high percentage of the College of Engineering graduates. Many of the samurai applicants signed only as investors, rather than as inventors.

This contrast may well be explained by supposing that while elite graduate engineers were busy introducing and translating the technology imported from the West, inventive skill in the private sector was being demonstrated in local technological adaptation. (A collection of lives of Japanese investors to 1935 shows that only 7 percent were college graduates.) College graduates were engaged in public works, and thus it was perhaps not proper for them to apply for profit-making patent rights, whereas private-sector inventors were eager for the honor of official recognition, even if the invention was not put into profit-making practice. Only in the twentieth century, after Japan had subscribed to the International Industrial Property Regulations in 1899, were high-level engineers concerned with patent rights.

Since 1899 Japan has received a number of foreign applications for patent protection. (See Table 6.1.) The United States was particularly strong in electrical equipment and Germany was strong in weapon manufacturing and chemicals. British interests were diverse but particularly strong in shipbuilding.

TABLE 6.1

Foreign Holdings of Japanese Patents,
by Country, March 1910

Country	Number of Patents	Percent
United States	1,373	38.4
Britain	948	26.5
Germany	647	18.1
France	198	5.5
Italy	68	1.9
Sweden	56	1.6
Austria	47	1.3
Denmark	36	1.0
Hungary	36	1.0
Netherlands	33	0.9
Belgium	33	0.9
Switzerland	31	0.9
Russia	23	0.6
Norway	22	0.6
Other	29	0.8
Total	3,580	100.0

Source: Daigoji tokkyokyoku nenpo (Fifth annual
report of the Patent Office, Tokyo: Tokkyokyoku, 1911),
Table 12.

During the first seven years under the International Regulations,
Japan experienced a gradual invasion by foreign patent rights. In
1899, foreigners owned about 17 percent of the Japanese patents. In
1905, 28 percent of the patents granted were to foreigners.

The Patent Office had its own classification system, dividing
items into 136 kinds. Table 6.2 presents the major items and classi-
fies them into three categories, according to the degree of native
contribution.

In category A the native contribution is more than two-thirds.
The items are mostly small, useful gadgets for local use that were
developed in the practice of traditional craftsmanship, mostly in the
private sector, devised by those without college and science back-

TABLE 6.2

Degree of Native Japanese Patent Holdings, 1899–1905

Category	Number of Patents Native	Total	Percent Native
Category A			
Electric lamps	6	46	13.0
Electrochemical industry	5	34	14.7
Miscellaneous electrical applications	10	61	16.4
Steam motors	14	73	19.2
Metallurgy	6	31	19.4
Guns, bows	28	128	21.9
Telegraphy, telephony, electric signals	30	118	25.4
Regulating, distributing electricity	22	73	30.1
Category B			
Electric batteries	7	20	35.0
Steam generators	47	121	38.8
Chemicals and chemical products	36	86	41.9
Mechanisms and mill gearing	55	130	42.3
Mining machinery	14	33	42.4
Ships and boats	32	75	42.7
Signals and communicating apparatus	18	40	45.0
Furnaces and kilns	22	44	50.0
Rope tramways and the like	18	35	51.4
Carriages	73	138	52.9
Civil engineering	26	48	54.2
Printing machines and appliances	18	50	64.0
Miscellaneous chemical works	36	56	64.3
Spinning	27	42	64.3
Match manufacturing machines	24	37	64.9
Category C			
Miscellaneous manufacturing machines	199	295	67.5
Buildings and structures	34	49	69.4
Ceramic industry	10	14	71.4
Pumps, other means of raising liquids	48	67	71.6
Tools	71	94	75.5
Measuring, indicating, registering	84	111	75.7
Food, beverage, confectionary machines	48	63	76.2

(continued)

Table 6.2, continued

| | Number of Patents | | Percent |
Category	Native	Total	Native
Kitchen appliances	73	93	78.5
Matches	9	11	81.8
Medical instruments	34	38	89.5
Safes and locks	77	86	89.5
Illuminating apparatus	190	212	89.6
Dyestuffs	50	55	90.9
Paints, varnishes, lacquers	40	43	93.0
Toilet, hairdressing articles	56	60	93.3
Corn husking implements	58	62	93.5
Knitting and netting	89	95	93.7
Dyeing machines and appliances	60	64	93.8
Sanitary appliances	64	68	94.1
Furniture	99	105	94.3
Tobacco manufacture	39	41	95.1
Sugar manufacture	41	43	95.3
Looms and weaving	273	285	95.5
Tea processing machines	69	72	95.8
Winding and twisting machines	85	88	96.6
Traps	61	63	96.8
Umbrellas and walking sticks	34	35	97.1
Bedroom furniture	44	45	97.8
Stationery	158	161	98.1
Boots, shoes, wooden clogs	116	118	98.3
Agricultural appliances	238	242	98.3
Boxes, cases, trunks, bags	123	125	98.4
Grain polishing machines	69	71	98.6
Sericultural appliances	56	56	100.0
Weaving and knitting	59	59	100.0

Source: Daiichiji tokkyokyoku nenpo (First annual report of the Patent Office, Tokyo: Tokkyokyoku, 1907), Table 17.

grounds. Agricultural implements, such as a rice-polishing machine, belong to this category. Some technologies, such as that for making matches, had been completely transferred by the turn of the century.

In category B the native contribution is reckoned at between one-third and two-thirds. It includes Western technologies that are nearly assimilated but still being transferred, such as cotton-spinning machinery, shipbuilding technology, and printing machines. In these fields the native contribution is near the level of international tech-

In category C the native contribution is less than one-third. It includes new Western inventions, such as electrical equipment, and such large-scale technologies as metallurgy and weapon manufacturing. It is science-based and occurs mainly in the public sector, which has qualified scientific and technological personnel.

By selecting cases of agricultural technology, spinning and textile engineering, and military engineering and its associated mechanical engineering from categories A, B, and C, respectively, I shall try to show how the public/private and Western/traditional dichotomies were removed.

Agriculture

The most important traditional technology was, of course, agricultural technology. Most of the government revenue in the early Meiji period before industrialization came from a land tax on the agricultural sector. It was a prerequisite for modern economic growth that agricultural technology be improved and farm production increased, in order to use the agricultural surplus to defray the cost of modernization and capital investment for industrialization. Most agricultural technology was practiced in the private sector through the private initiative of experienced farmers, and the Meiji government tried to introduce its new agricultural experts trained in Western technology.

The new agriculturalists were educated at two agricultural colleges, one in Komaba (Tokyo) and the other in Sapporo (Hokkaido). Students were mostly sons of impoverished samurai who rushed to apply to the colleges, attracted by government scholarships that were better than in other schools. They had no previous background in agriculture and were trained by foreign teachers.

In the earliest period of the Komaba Agricultural College, which was founded in 1877, British and German agricultural scientists tried to introduce the European style of animal husbandry and to import Western plants, livestock, and farm implements. They had nothing to offer for the improvement of paddy-rice production, even though rice was the major crop raised in Japan. The Komaba College proved to be unsuccessful in the 1880s when farmers realized that Western techniques were difficult to adapt to traditional, small-scale, land-saving paddy-rice agriculture. They turned to the veteran farmers for instruction.

There was a struggle between college graduates and veteran farmers for leadership of the agrarian reform movements. [12] The policy of local and central governments to experiment with newly imported methods and techniques was for the most part abandoned in the 1880s. The application of the newly acquired technology of the college graduates now focused on paddy-rice production, but it was limited to such academic studies as chemical analysis and biological studies. [13] The initiative in technological improvement of production was seized by veteran farmers, who took advantage of some of the conclusions drawn by the new agricultural scientists and applied them on the basis of their own largely empirical knowledge, acquired through long experience. The new research findings were taken into consideration insofar as they supported the farmers' knowledge. An example is the chemical analysis of traditional fertilizer by O. Kellner, a German chemist hired by the Komaba Agricultural College. [14] Except for research on plant diseases, basic college agricultural research in the early Meiji period did not contribute to the reform of agricultural technology.

The exploitation of Hokkaido, on the other hand, led to successfull transfer of foreign methods. The government policy of establishing farms in the newly settled areas, designed to aid jobless samurai, encouraged the experimental introduction of farm appliances and new varieties, which were generally adopted by those unexperienced in conventional farming. [15]

In the early Meiji period rather superficial efforts were made to import the visible results of Western achievements, but without much of practical success; however, since the 1890s scientific research has made great contributions to Japanese agricultural production in two important fields: the promotion of soil fertilizer and the breeding of improved varieties. It was only in the twentieth century that academically originated research in soil chemistry and the application of empirical breeding knowledge and Mendelian genetics[16] yielded practical results. Japanese agricultural science and technology concentrated on the land-saving intensive growth of crops without radical land reform or a change in economic framework.

Indigenous Industry

In textile enterprises, college-trained engineers mainly served as official supervisors assessing the work of entrepreneurs. In 1903 a government official criticized their attitude in an address entitled "The Encouragement of Indigenous Industries":

> People and industrialists do not know yet how to utilize indigenous inventions. Most inventors are mechanics or industrialists, without adequate scientific background, or schoolteachers without technical experience. Those who had been trained in specialized disciplines are concerned only to import and imitate new industry. They have the mental habit of feeling ashamed to devote themselves to the reform of indigenous industry and stay out of it. [17]

He noted the manufacture of matches as a good example of an indigenous industry.

Though incomparably smaller than major industries like spinning, match manufacturing provides an exceptionally interesting case of technology transfer. It had no roots in traditional technology but was entirely imported from Europe in the early Meiji period. Since it did not require much capital stock, it achieved business success by acquiring technical know-how quickly, reaching export capability by the turn of the century. [18] As shown in Table 6.2, most of the patents on match manufacturing were native ones, and those foreigners who applied for Japanese patents were taking advantage of the Japanese business boom in the industry.

Textiles

The major industry of Meiji Japan was textiles (spinning, silk-reeling, and weaving), which, according to the statistics of 1900, employed 233,000 people and had machines of 39,000 horsepower—more than ten times the size of the government armament industry. [19] The government encouraged business by establishing pilot plants with imported spinning machines and then turning these into private businesses, which obtained government loans for further importation of machinery. At the turn of the century, while the quantity of imported machinery was rapidly increasing in government heavy industry, the importation of spinning machines decreased considerably as capital investment increased. Textile products were turned from imports to exports. [20] As indicated in Table 6.2, at this time native patents on spinning machines considerably exceeded foreign patents. This may be one indication of the "domestication" of spinning-machine technology, though wholly domestic production of modern spinning machines started only in the 1910s.

The textile industry is basically a profit-making activity of the private sector, and hence its technological advancement has been accomplished in that sphere. The technology was basically completed before modern engineering science was formulated in the middle of the nineteenth century. No course was given in the Imperial College of Engineering. [21] In the early Meiji period, spinning technology was looked down upon by college graduates, who considered it to be crafts-men's work and not the business of an intellectual class. [22]

The invention of the garaboki (a simple spinning machine) in 1876 was accomplished in the context of traditional home spinning, without any foreign influence. However low its manufacturing cost, its inability to produce quality-controlled fine yarn prevented its nationwide diffusion. In 1890 garaboki spinning reached its peak, 11.8 percent of total production. It declined soon after the importation of a ring spinning frame, which made possible both high speed and continuous production. [23]

In contrast with the spinning industry, which was still completely dependent on imported machinery, the hand loom was produced domestically and played a great part in promoting the textile industry. Compared with the spinning industry, a weaving business is too small to afford costly imported machinery. Also, it is more dependent on domestic technology to satisfy the local demand for varieties of fabric. Looms made mainly by local carpenters were basically of wood with metal fittings. Local ironsmiths and machine shops for farm implements could repair such simple machines. S. Toyota, a son of a carpenter, devised a power loom in 1890 that proved to be suitable for making cheap cloth of narrow breadth for export to Korea and Manchuria. In 1914 he again designed an automatic loom; but, mainly because of cheap and abundant labor available at the time, it was not widely employed until 1929, when a factory act to prohibit night labor was enforced. [24]

Military and Mechanical Engineering

It is a cliché among historians of Japanese technology to characterize prewar Japanese science and technology as "armament-centered and unbalanced." Yet we do not know what is balanced and what not. In the absence of an adequate international criterion, any account tends inevitably to be impressionistic. In addition, the Japanese military sector did not permit examination from outside. Hence, reliable data are still scarce.

Prevailing theory says that since private enterprises in the advanced capitalist countries are governed by the pursuit of profit, their interest in the armaments industry is indirect. In contrast with this, the Japanese industry was led by a government preoccupied with the goal of directly increasing the wealth and military strength of the nations through the efforts of government-owned industries. Japanese industry and technology were from the beginning strongly colored by the militaristic policy. For example, the exhibits at the National Industrial Promotion Fair held in 1877 were composed mainly of the best-quality facilities and machinery belonging to the army and navy arsenals, as well as of other government-operated plants. Exhibits from the privately owned plants were few in number and poor in quality.

Priority for military engineering was reflected in the departmental organization of the Imperial College of Engineering. The core of modern scientific technology was mechanical engineering, from which engineering science was formulated. Japan was not too backward to catch up with this rising science, which was systematized only in the nineteenth century. Mechanical engineering was also fundamental to naval architecture and military engineering. In the early Meiji period, however, the government shipbuilding industry dominated the job market for college graduates in mechanical engineering. It was wholly and directly connected with the armaments program of the government. Though it was not included in the original program of the college, drafted by Henry Dyer on the model of the Zurich Technische Hochschule, the new department of shipbuilding was added in 1882, at the request of the navy.

The Meiji government's interest in strengthening national power can be seen in the creation of special departments for the study of armaments at the university level, an act without precedent in advanced countries. In 1887, within the Imperial University, there were founded a department of arms technology, at the suggestion of the army, and a department of explosives, at the request of the navy. These were opened in the College of Engineering in order to secure engineering personnel for military arsenals. Since these departments did not involve existing university disciplines, no adequate teaching staffs were available at first. Only with the coming of the twentieth century did they have full-fledged professorships and regular students. In the meantime, the arms technology department developed a discipline of precision engineering, while the explosives department was led to chemical engineering. [25]

In 1887-96 many state-owned plants were transferred to private ownership, so that private capital increased steadily. However, there was no appreciable decrease in the total horsepower or the number of employees of governmental plants. [26]

On the other hand, the manufacture of machine tools, which formed the foundation of the machinery industry, was left in the hands of small and medium-size private enterprises, where no college graduate engineers were available until World War II. [27] This neglect revealed the basic weakness of Japanese technology during wartime, when high-quality machine tools ceased to be imported. Simple lathes and other standard machine tools had been homemade, but machine tools for automation and precision had been imported from Germany and the United States.

RISE OF TECHNOCRATS, WORLD WAR I AND AFTER

Start of Financing Scientific Research

Until the 1880s, the scientific institutions created by the government were mostly for geophysical survey work: the Navy Hydrographic Office (founded in 1871), the Tokyo Meteorological Observatory (1875), the Geological Survey (1882), the Army Ordnance Survey (1884), and the Tokyo Astronomical Observatory (1888). Their work was essential for a nonindustrial modern state. After the 1890s, however, many national research institutions were established for fostering industrial developments, such as the Electrical Laboratory (created in 1891), the Central Inspection Institution for Weights and Measures (1903), the Fermentation Laboratory (1903), and the Railway Research Institute (1907).

Before World War I, Japanese industrial laboratories existed only in the public sector, in order to give guidance in the technology needed by private enterprises. The mobilization of research in Europe and the United States during World War I provided great stimulus and opportunity for government and scientists in Japan to think about the financing of scientific research. The Japanese, who had stayed out of major World War I hostilities, took advantage of the wartime economic boom. Scientists changed from academic bureaucrats to technocratically oriented bureaucrats who advocated planning scientific research for national goals.[28] The latter were backed by the rising industrialist class. The creation of the Riken[29] (Institute for Physical and Chemical Research) in 1917 was a landmark in this change, since the major source of funds (85 percent) was the industrial sector rather than the government. During and after World War I private enterprises, notably in the chemical industry, established their own full-scale industrial laboratories. This signified the maturity of capital formation as well as the rise of interest in R&D in the private sector.

It was also during World War I that two other arrangements to promote scientific research were formed in the public sector: university-affiliated research institutes and governmental research funds.

Owing to increasing specialization and the advancement of research in science and technology, Humboldt's ideal of the unity of education and research at the university level was virtually destroyed in the early part of the twentieth century. The American solution of the problem was to create graduate schools, so that the unity was delayed from the undergraduate to the graduate level, where students could be educated by formal curriculum as well as by apprenticeship in research. The other solution was the German method of creating a governmental research institute, called Kaiser-Wilhelm-Gesellschaft, to liberate scientists from their educational burden. This model was followed by the Soviet Academy of Science. [30]

The Japanese tried to circumvent this difficulty by creating the Riken, modeled after the Kaiser-Wilhelm-Gesellschaft. Still more significant was the affiliation of research institutes to a government university, where professors were full-time researchers without formal obligations to teach while enjoying academic freedom and higher standing than those working in the government and industrial laboratories.

The Ministry of Agriculture and Commerce created an Invention Fund in 1917 to encourage inventive researchers and, later, an Industrial Research Fund. The Ministry of Education's Science Research Grant program was started in 1918. Other small philanthropic foundations were also created during the war to promote scientific research. This heyday of scientific research funding was rather short-lived; and when the economic recession started in 1920, the amount of government funds either remained the same or decreased.

The worldwide depression of 1929 had relatively little influence on Japanese business, however, because Japan's involvement in the Manchurian Incident of 1931 caused a munitions boom. A study of the industrial production indices of Japan and the United States shows that while the United States was unable to recover its 1920 level of prosperity until the start of World War II, Japan soon regained normality. Heavy industry was mainly responsible for the economic recovery. By about 1935, heavy and chemical industry production exceeded that of light industry.

With the aim of avoiding economic depression by means of scientific research and of promoting power of its industry to compete internationally, the Japan Foundation for the Promotion of Science was created in 1931. It arose from a proposal by scientists and started functioning in 1933 with government funds dramatically bigger (about ten times) than any ever contemplated before. Nationalization of science was a worldwide trend at this time, but Japanese science took particular advantage of this trend to reach international standards.

In spite of the heavy government intervention in scientific research, we should not overlook the trend in the machine industry, constant since the early Meiji period, toward moving industrial technology from the public sector to the private sector. As a 1931 survey indicates, nearly half of the researchers in engineering fields worked for private businesses.

Historians are concerned with finding an indicator of the military inclination of prewar Japanese scientific research; but in the absence of adequate data, we must be content with Hiroshige Tetu's rough estimate that in 1931 nearly half of the national research investment in engineering went to military research establishments. [31]

Thus, we may characterize the interwar period as one of involvement in the worldwide trend to nationalization and militarization of scientific research, superimposed on the background trend of the transfer of industry from the public to the private sector.

WORLD WAR II AND AFTER

After 1939 the wartime recognition of the need for science mobilization and armaments led to the expansion of education programs for top-level scientists and engineers (according to one estimate, three times the usual number of scientific manpower was produced). As the war intensified, students of the humanities were called to active military service, while students of science and technology were exempt.

Demands from the military and from business made engineering research particularly favored during the war. According to a questionnaire sent to engineering scientists after the war, their happiest time was that of the war mobilization, when they were given preferential treatment, in the form of abundant funds and materials. Furthermore, the complete isolation from Western scientific circles inevitably provided an opportunity for public recognition of those who had formerly been psychologically dependent on Western authority in their fields.

Immediately after the war, many Japanese spokesmen of science stated that the lack of recognition in prewar Japanese society of the importance of scientific research had contributed to the Japanese defeat. The experience of military supremacy, thought control, and the virtual failure of war mobilization must have given many scientists bitter memories and led to a negative assessment of the war effort. Wartime isolation especially resulted in a hiatus that presented the postwar generation with tremendous difficulties. Only recently have more critical attempts to reevaluate wartime scientific

FIGURE 6. 1

Japanese Research Expenditure, by Type of Institution, 1952–62

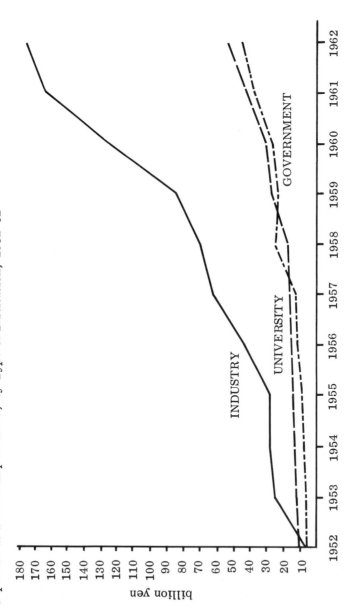

Source: Nihon no kagaku gi jutsu katsudo (Activities of Japanese science and technology) (Tokyo: Maruzen, 1965), p. 97.

225

TABLE 6.3

Economic and Functional Classification of
Japanese Public Expenditure, 1900 and 1960
(percent)

	1900	1960
Directly productive activities	20.78	3.63
Agriculture, fishery	1.32	2.24
Commerce, mining, manufacturing	2.54	0.80
Transport, communication	16.92	0.60
Education	2.30	11.33
Defense	45.29	8.50
Social welfare	1.93	21.82
Other (general and local administration)	29.70	54.72

Source: Choki keizai tokei, zaisei shishutsu (Long-term economic statistics, Public expenditure, Tokyo: Toyokeizai, 1966).

efforts as the source of the postwar boom in Japanese technological
advancement and economic growth appeared. [32]

During the occupation years some strategic research, such as
nuclear and airplane research, was forbidden by the occupation
forces; the two Japanese cyclotrons were thrown into the sea. Military research was almost totally absent. Demilitarization and the
demobilization cut most of the government support of science, which
may have been partially responsible for the postwar transfer of sci-
tific and technological leadership to the private sector.

After the war, governments in other advanced countries supplied
the largest share of funds for scientific research. In contrast, in
postwar Japan private enterprise, which formerly did not supply a
large proportion of research funds, underwrote a significant portion
of scientific research.

Japanese private enterprise has grown to the point where it
must, and can, sustain its own research. Against a universal trend
in the postwar period to move toward "government science" or
"nationally sponsored science," such as the major space and atomic
research programs, the center of gravity of Japanese science moved
from the public to the private sector, as is shown in Figure 6.1.

Table 6. 3 reveals that the prewar public spending for defense and industry has now moved to social welfare, development, and education. This trend is reflected in the structure of scientific and technological research in such a way that postwar Japn is said to have stayed out of big science in the public sector and to have concentrated on profit-making, economical development on the basis of imported know-how. However, Japan is now facing the problem of a laissez-faire science policy, which may be partly responsible for heavy pollution and other public nuisances. At the same time, when the source of technological borrowing is nearly exhausted, Japan may have to find new, costly ways of breaking through by itself. The Japanese government is beginning to take note of this reality, and the national budget for scientific research is being increased gradually and steadily.

CONCLUSION

Considering the past of Japan's science and technology, which began under government sponsorship in the early Meiji years, the direction of the scientific information flow has been always one-way, from abroad to the public sector, and in turn from public sector to private sector. The public sector has always turned Japanese science and technology in the directions that are its most serious concern: enriching and strengthening the country.

The course of this endeavor was first determined by the samurai intellectual class, which was seeking a new source of income after being deprived of family stipends. They took the initiative and were responsible for the systematic introduction of technology into public institutions. The process of modernization could, however, never be successfully completed unless the efforts in the public sector reached the private sector. Early enthusiasm for Western science led to the diffusion of the ideology of modernization to the social grass roots. At the same time this educational effort resulted in the liquidation of samurai/commoner class difference, to the extent that by the 1920s commoners had a higher representation among the students at the Imperial University.

As a device for explaining the transfer process, I introduce the following matrix:

	Western-origin	Tradition-bound
Public	I (military)	II (mining)
Private	III (matches/ textiles)	IV (agriculture/ brewing)

Technology of Western origin was usually practiced in the public sector, and tradition-bound technology in the private sector. In the process of the transfer from section I some adaptive technologies emerge in sections II and III. We have examined the success or failure of technology transfer from the imported section I to other sections.

In the early Meiji period, college graduates, mostly of samurai origin, were busy reading and translating Western scientific works, visiting and studying in the West, learning Western science, copying high technology, assisting Western advisers, teachers, and engineers in surveying and construction jobs, and planning for Japan to become an industrial nation, catching up with the most advanced high technology (section I). On the other hand, more practical inventions and improvements at the grass-roots level were made by nonsamurai, artisans, and experienced farmers often under the influence of the Westernization policy (section IV).

Government enterprises focused their efforts on areas where no existing system was available, such as railway construction and telecommunications (I to III) or where large-scale capital and mechanization were needed, such as government-owned mines (I to II). In such fields all essential aspects of enterprises were introduced in toto; both chief engineers and less skilled workers were recruited abroad. If production technology and the production system were radically different from the domestic ones, even where traditional domestic engineering and industry existed, as in shipbuilding (I to II) and the spinning industry (I to III), the mechanization system and the managerial form were imported as a single, inseparable unit.

In the Japanese experience, a special adaptive technology was selected in order to compete with more capital-intensive advanced countries by using Japan's cheap labor. This aspect was not consciously emphasized by the Meiji government, whose foreign advisers, with very few exceptions, lacked much imagination in adapting Western technology to local situations. Their samurai followers, without much experience in domestic technology, tried to copy and demonstrate the Western technology, often with disastrous results. A notable labor policy of the Meiji government was to provide jobless samurai with new means of livelihood, but in doing so the government did not seem to have conceived of labor-intensive intermediate technology. Rather, it directed them into more labor-saving new technologies, such as large-scale farming in the colonization of Hokkaido (I to III) and/or mechanized mining (I to II).

On the other hand, in purely domestic private-sector industries where no similar Western counterpart existed, such as paddy-rice agriculture, sake brewing, and lacquerware and ceramic production (IV), the innovation process was naturally very slow and gradual. Only the generation trained in science and technology under the modern

Western-style school curriculum began to utilize the knowledge acquired at school and to apply it, if practicable, in their domestic businesses.

In between these two extremes were many intermediate technologies. There were cases in which the production scale was too small for the enterprise to be in the public sector, yet local demands made the private sector quite receptive to the imported knowledge. Typical of this was the match-manufacturing industry, where at first cheap labor and match sticks were the only exploitable factors; later potassium chlorate and phosphorus became available domestically, thus making it possible to complete all processes in Japan. Silk reeling and textile manufacture were also such that imported machinery in government pilot plants served to demonstrate the way to technological progress, although for a considerable period production was carried out by means of homemade wooden machines and small-scale production systems. The local synthesis of the two extremes was often found in this area of the transfer section (I to III).

The most important technology—arms technology—should not be forgotten. It has always remained within the public sector (I). The major areas of mechanical engineering and naval architecture were monopolized by military arsenals and large government industries without much effect on the private sector. Machine-tool engineering, despite its basic character, was relatively neglected, and continued to be a small-scale private industry.

If we characterize 100 years of Japanese industrial development by one phrase, it was a constant process of transfer from the public to the private sector. This policy was publicly announced and was followed during the 1880s. It was basically, if not explicitly, continuous, although often interrupted by the effects of the government arms industry.

This effect totally disappeared after World War II. With economically motivated, private industry based on profit-making, a laissez-faire kind of science and technology became dominant in the budgeting structure of postwar Japanese R&D. This contrasted with other advanced nations, such as the United States and Britain, where public expenditure for science and technology became the major and predominant source of research budgets.

Postwar Japan also presents an example that conflicts with the prevailing belief that science and technology input contribute to economic growth in proportion to its size. Among advanced nations, Japan paid less for science and technology and gained more in economic growth. Unless we specify the quality of science and technology and the quality of economic growth, we may not be able to say anything definite; but the postwar private leadership in Japanese industry and technology may give a clue to this question.

On the level of catching up and profit-making, postwar Japan proved that borrowing is less expensive than spending on costly original research. In so doing, it needed more intermediate scientific manpower than highly talented Nobel laureates. This could be accomplished only through a social revolution in which vertical and horizontal mobilities should be accelerated for the recruitment of scientific and technological manpower.

Let me conclude with a metaphor. Science and technology have usually been received in the upper social strata in the developing countries, and then have penetrated the society as a whole. These two steps were unusually orderly in Japan. If a class system is inflexible, the new knowledge fails to reach the lower levels. If we could complete the above two steps at once, the growth of science and technology might be more efficient and far-reaching.

The introduction of science and technology into Japan has taken place with unprecedented speed, but our studies suggest that the transformation can take place even more rapidly and more thoroughly if the conditions are right.

NOTES

1. S. Nakayama, "Kokuei kagaku" (Nationalized science), in I. Sugimoto, ed. , Kagakushi (History of sciences) (Yokyo: Yamakawa, 1967), pp. 351 ff.

2. The original is reprinted in S. Nakayama et al. , eds. , Nihon kagaku gijutsushi taikei, kokusai (Source books of the history of science and technology in Japan: International relations) (Tokyo: Daiichi-hoki, 1968), pp. 35-36. Hereafter cited as NKGT.

3. NKGT: Kyoiku I (Education no. 1) (Tokyo: Daiichi-hoki, 1964), p. 353.

4. Ibid. , pp. 392-393.

5. Ibid. , p. 202.

6. NKGT: Tsushi I (Outline history, I, Tokyo: Daiichi hoki, 1964), pp. 180-86.

7. Tokyo Teikoku Daigaku gakujutsu taiken, Rigakubu (Survey of research activities in the Tokyo Imperial University, Faculty of Science, Tokyo: Tokyo Daigaku, 1942). In the chemistry department, the change from local science to internationally recognized subjects became noticeable in 1885.

8. T. Muramatsu, "Nihon no kogaku soseiki no jakkan no mondaiten" (Some problems of the early Japanese engineers), Kagakushi kenkyu (Journal of the history of science, no. 32, 1954): 8-14.

9. I. Amano, "Kindai nihon ni okeru koto kyoiku to shakai ido" (Higher education and social mobility in modern Japan"), Kyoiku shakaigaku kenkyu (Journal of educational sociology, no. 24 (1969): 84.

10. Takahashi Korekiyo jiden (Autobiography of Takahashi Korekiyo [founding president of the Japanese patent office]). The part relevant in this context is reprinted in NKGI: Tsushi I, pp. 339-40.

11. "Tokkyo dai-100 go no meisho oyobi shutsugannin ichiran" (Titles and applicants' names of patents no. 1 to no. 100), in ibid., pp. 334-36.

12. J. Iinuma, Nihon nogyo no saihakken (Rediscovery of Japanese agriculture) (Tokyo: NHY Books, 1975), ch. 3.

13. NKGT: Nogaku I (Agricultural sciences) (Tokyo: Daiichi hoki, 1967), p. 26.

14. Ibid., p. 14.

15. Ibid., pp. 122 ff.

16. NKGT: Seibutsu kagaku (Biological sciences) (Tokyo: Daiichi hoki, 1965), ch. 5, sec. 1.

17. NKGT: Tsushi II (Tokyo: Daiichi hoki, 1967), p. 240.

18. T. Furushima, Sangyoshi II (History of industry) (Tokyo: Yamakawa, 1966), pp. 474 ff.

19. T. Tsuchiya, Zoku nihon keizaishi gaiyo (Outline of economic history of Japan), II (Tokyo: Iwanami, 1939), pp. 180-81.

20. Ibid., pp. 178-79.

21. Tokyo Teikoku Daigaku gakujutsu taikan, Kogakubu (Survey of research activities in the Tokyo Imperial University, College of Engineering, Tokyo: Tokyo Daigaku, 1942), p. 91. The courses of instruction are given in English in The Engineer, December 3, 1897, p. 544.

22. NKGT: Tsushi I, p. 466.

23. NKGT: Kikai gijutsu (Mechanical engineering) (Tokyo: Daiichi hoki, 1966), pp. 119-20, 124-26.

24. Ibid., pp. 136-37.

25. Tokyo Teikoku Daigaku gakujutsu taikan, Kogakubu, chs. 6, 10; see also The Engineer, loc. cit.

26. Nihon keizai tokei soran (Survey of Japanese economic statistics) (Tokyo: Asahi Shinbun, 1930), p. 723.

27. Muramatsu, op. cit.

28. Hiroshige Tetu, Kagaku no shakaishi (Social history of science) (Tokyo: Chuokoron, 1973), ch. 3.

29. K. Itakura and E. Yagi, "The Japanese Research System and the Establishment of the Institute of Physical and Chemical Research," in S. Nakayama et al., eds., Science and Society in Modern Japan (Cambridge, Mass.: MIT Press; Tokyo: Tokyo University Press, 1974), pp. 158 ff.

30. Loren Graham, "The Formation of Soviet Research Institutes: A Combination of Revolutionary Innovation and International Borrowing," in XIVth International Congress of the History of Science, Proceedings, I (Tokyo: Science Council of Japan, 1974).

31. Hiroshige, op. cit. , p. 116.
32. Ibid. , pp. 216-20.

7

SCIENCE AND TECHNOLOGY IN BRAZILIAN DEVELOPMENT

José Pastore

Brazil is one of the largest and fastest-growing nations in the world. Almost as big as the 50 United States, it occupies fully half of South America. Its vastness and its climatic variations are especially noteworthy. The Brazilian Amazon region is considered one of the wettest, most impenetrable forests on earth. Less well known is the fact that if this region of Brazil were a separate country, it would be the ninth largest in the world. Western Brazil, mostly savanna, is also immense. Together these regions make up about half of the land surface of the country, yet they are almost uninhabited and are quite underdeveloped.

Practically all of the nation's 110 million people live within 1,000 kilometers of the Atlantic shore, most much closer than that. About a third of the population lives in the Northeast, a region of great poverty; the rest live in the Center-South. Here are concentrated most of the large urban areas—São Paulo, Rio de Janeiro, Belo Horizonte, Brasília, and Pôrto Alegre—as well as most of the nation's agricultural and industrial production. Indeed, with 20 percent of the people, the subtropical state of São Paulo alone currently produces about half of the nation's GNP. It is here, near the coast and close to the rich iron mines of Minas Gerais, that Latin America's most productive manufacturing center is located.

Brazil's economic development has been the most rapid in Latin America. The country has undergone a vast and far-reaching mechanization of manufacturing, transportation and communication. At present mechanization is beginning in agriculture. Science and technology also have progressed markedly in recent times.

In analyzing both Brazil's scientific and technological development and its economic and social development in historical terms, a hypothesis of one-way causation is untenable. In this study I hope to clarify to some extent the nature of the interaction between the two processes. My central hypothesis is that factors inhibiting or facilitating technological development are highly dependent on key economic forces and political interests.

More specifically, this implies that the development of Brazil's limited technological system was influenced by forces emanating from the production system, particularly by the shortage or availability of natural and human resources. This interaction occurred under changing conditions, during which the technological system responded in different ways and took different forms at different times. Most common has been the borrowing from abroad: internal technological innovation has been quite limited. The production sector has displayed some initiative in diffusing innovations among industries so as to make the necessary adaptations, and, to a much lesser extent in generating its own innovations and establishing research institutes.

Another central observation is that until the present, scientific and technological research undertakings have been small-scale enterprises in comparison with those of the wealthier nations. Underlying this is the hypothesis that the Brazilian colonial experience influenced the development of scientific and technological establishments.

It appears that Brazil has passed through two colonial stages and has now entered a third. [1] In all three stages the profit flow has been mostly from the colony to distant countries. In the first stage, which lasted more than three centuries, the Portuguese maintained both political and economic control, using Brazil as a profitable source of agricultural and mining products. This stage ended when the Portuguese royal family fled from Napoleon's armies and, in 1822, set up the Empire of Brazil. In the second stage, lasting until about 1930, Brazil maintained political independence but continued to be used by other countries as an agricultural and mining colony. The two phases were marked by an "extraction-export colonialism." Yet in the second stage Brazil was less dependent economically because it controlled its own huge coffee production. In the third stage, a kind of "industrial internal-market colonialism," Brazil has encouraged wealthier nations to set up manufacturing plants to produce goods for sale mostly in Brazil. This last stage is expected to generate trained personnel and training institutions, new manufacturing plants, viable organizational patterns, and other modernizing effects by providing equipment and skills that could ultimately lead to a more independent economy.

None of these variaties of colonialism, whatever its virtues, tends to encourage the development of an innovative technological research establishment. This is particularly true, in industry, for firms whose R&D centers are already available in other countries.

The main exception is agriculture, for two reasons. First, today's export business is mostly in the hands of Brazilians. Second, a great many of the plant varieties and plant diseases that flourish in Brazil are not those of the northern hemisphere. Thus the need for home-grown research establishments in agriculture is more obvious than in manufacturing. In this area science and technology have made some progress.

This paper presents a brief overview of the beginning of science and technology during the colonial period, then moves to a deeper analysis of the twentieth-century development, particularly since 1930. It then focuses on the basic mechanisms related to the develop-ment of agricultural and industrial technology and to the formation and the evaluation of the nation's modest scientific establishment. Finally, the present state of science policy in Brazil is analyzed.

THE FIRST SIGNS OF TECHNOLOGICAL DEVELOPMENT

During Brazil's first four centuries, progress in science and technology was modest at best. In this respect Brazil and the other Latin American countries were quite similar. The factors behind this stagnant situation may be partially located with the Iberian legacy.

The Portuguese performance in the sixteenth and seventeenth centuries showed many contrasts and contradictions. On one hand, it had a notable level of military and organizational efficiency, which had developed in its defeat of the Moors (thirteenth century) and a highly integrated sociopolitical system. On the other hand, during these centuries, it had no large and active class of entrepreneurs and innovators. Yet its great discoveries were certainly not random events, and indicate the existence of a vigorous social, economic, and political life, as well as a high inventive capacity in shipping and seamanship. On the whole, the Portuguese experience was one of fast, short-run economic progress paralleling long-run poverty.

Three factors can help explain the decline of Portugal. Eco-nomically, its affluence was based on an extreme form of mercantil-ism of gold and silver. Agriculture and manufacture were neglected, and the rapid increase in the prices of food and other essential com-modities made it highly dependent on foreign nations, particularly England and France. Socially, Portugal was rigidly organized: society was sharply divided into a small privileged, aristocratic elite and a mass of illiterate peasants with limited social and political roles. Culturally, society was dominated by and suffered from the rigidity of the Inquisition ideology, which suppressed all disagreement and free debate. [2] In short, Portugal's progress was limited by economic, social, and cultural rigidities.

This was obviously not an environment for intellectual development, but it was the only cultural dowry to be transplanted to Brazil. As a consequence, the new society was modeled on a predatory culture, with a heavy emphasis on mining and monoculture, and on a dualistic social structure that still pervades Brazil and most of Latin America. The technology required by the elite was extremely rudimentary, since land and minerals were abundant and Indian and black slave labor was cheap. For centuries, these conditions allowed agriculture and mining to be carried out with crude production methods.

In sharp contrast, England took the first decisive steps toward the Industrial Revolution, which was based on technological development and social change. Economically, the previous expansion of the agricultural sector of the British Empire (first half of the eighteenth century) provided a surplus of food, a capacity to support rapid urban population growth, and, consequently, an increasing domestic market. Later, the intense growth of export activities brought a fantastic market expansion. Optimistic business expectations, a decrease in interest rates, the rise in wage levels, and shortages of natural resources induced technological innovations (particularly in the cotton and iron industries) that, with increasing efficiency, generated economics of scale and flexibility in factor substitution. A sizable class of entrepreneurs was formed who quickly became an important element of political stabilization and played a decisive role, making economically rational decisions based on the free examination of conflicting alternatives and open criticism. [3]

The economic histories of England and Portugal highlight the difference between the two societies, one of which generated a sophisticated complex of technical and social innovations to underpin its modernizing historical role, and the other which made economic and political decisions leading to technological stagnation. In the first case, economic and political pressures tended to induce innovations and brought realistic hopes for the future; in the second, economic encouragement of productivity was practically nonexistent and political discontent was easily managed by means of rigid social controls. England's type of society strengthened its economic, technological, and political hegemony; but Portuguese society could hardly do other than to choose dependency. This dependence is manifested by socioeconomic stagnation; in others, it is marked by development based on heavy borrowing.

Brazil's progress has been conditioned by several forms of colonialism. The economic, social, and cultural background that inhibited intellectual development in Portugal was transplanted to Brazil. The rigidity of its social structure and the relative abundance of its natural resources gave its small elite all the conditions necessary to pursue the predatory style of mining and agriculture. As a conse-

quence, the country exhibited very little innovating vigor during most
of its colonial history.

The following two sections analyze the few technological innovations that occurred in agriculture and infrastructure during the nineteenth century and the beginning of the twentieth.

The Beginning of Agricultural Innovation

Along with mining, agriculture was the key economic activity
during the colonial period, and is still important today. For centuries,
economic cycles of the Brazilian economy have been based mostly on
sugarcane, mining, rubber, cotton, coffee, tobacco, and cocoa. Given
Brazil's 8.5 million square kilometers and slavery, land and labor
were abundant and cheap throughout the colonial period. There was no
pressure to develop or import biochemical innovations to use land
efficiently or mechanical technology to save labor. [4]

Even so, a few technological innovations appeared during the
colonial period. For example, Portugal demanded that animal manure
be used as fertilizer, as opposed to the customary "slash and burn"
agriculture. This was a political objective to reduce nomadism and to
establish population centers for defense purposes. In the South, economic objectives lay behind the introduction of a few European agricultural innovations at the end of the eighteenth century and the beginning of the nineteenth: the introduction of cotton seeds, the arrival
of a few agricultural technicians, and the publication of technical bro-
chures (manuais de lavoura) on cotton, tobacco and sugarcane cultivation techniques. During the first three decades of the nineteenth century, coffee farmers in São Paulo used fertilizers to revive "tired"
soils. In the middle of the century, the plow was introduced by American immigrants as a solution to labor shortages on the coffee plantations. [5]

The first attempts at practical agricultural training were made
at the beginning of the nineteenth century, although none of them lasted
long. Later, from 1859 to 1900, agricultural institutes were established in the states of Bahia, Pernambuco, Sergipe, and Rio Grande
do Sul. Most of these, too, were short-lived, due to a lack of steady
public support. [6] The first successful and lasting attempt was the
Agricultural Research Institute and Agriculture School of Bahia, created in Rio Grande do Sul in 1883. As in other areas, the first agricultural teaching units were devoted more to training than to research.
Most of the courses were based on imported knowledge of soils, plants,
and animals that were seldom adapted to the ecological peculiarities
of Brazil. A few scattered attempts at agricultural research were
limited to replication of well-known European experiments.

The first solid move toward the use of science and technology for agricultural development was the creation in 1887 of the agricultural Research Institute of Campinas, São Paulo, whose main purpose was the study of tropical crops. The Institute's first director, Franz W. Dafert, from Austria, immediately set up basic and applied research activities focusing on soils, cytology, genetics, plant pathology, and conservation techniques. Main emphasis was on coffee, cotton, sugarcane, corn, and citrus fruits. On the basis of the available research results, good soil, and management capability, São Paulo became the leading state in Brazil in terms of production of all these goods.

The success of the Agricultural Research Institute of Campinas encouraged the federal government to create research units in other Brazilian states, particularly Bahia, Minas Gerais, Rio de Janeiro, and Rio Grande do Sul. Equally important was the stimulus given to the foundation of many agricultural colleges at the beginning of the twentieth century: Piracicaba, São Paulo (1901), Lavras, Minas Gerais (1908), and Viçosa, Minas Gerais (1916).

By the turn of the century the country had undergone two major changes that played an extremely important role in the development of agricultural research: the shift of the economic center from the Northeast to the Center-South (Rio de Janeiro, Minas Gerais, and São Paulo); and the decline of traditional crops (sugarcane, cotton, and tobacco) and the rise of coffee. Coffee brought new strength to the Brazilian economy. As was mentioned, Brazil's economic situation oscillated frequently throughout the nineteenth century, especially in the first half: the country rarely had a favorable trade balance; manufacturing was practically nonexistent; and, to complicate the picture, European beet sugar was ruining Brazil's once-lucrative market for cane sugar. Also, cotton prices were falling fast. The whole economy was severely affected by the sugarcane and cotton problems. Coffee production emerged as a powerful alternative. Indeed, through the good price for coffee in the international markets, the nation managed to achieve a trade surplus that continued, with few interruptions, for more than 50 years.

Since coffee is a labor-intensive crop, the abolition of slavery helped to speed a vigorous migration from Europe to São Paulo. Immigrants, good soils, a fair climate, transportation facilities, and a large demand in the international market encouraged the expansion of Brazil's coffee production. By 1880, half of the total export revenue came from coffee. The coffee states had the most rapid rate of population growth (3. 2 percent per year), but even so they managed to keep income ahead of the population growth. Real income from exports grew at about 5 percent annually. A new class of entrepreneurs emerged, and after 1930 the profits from coffee financed a more consistent industrialization. [7]

The rise of coffee growing in the Center-South provided a substantial impulse to agricultural research. This development path illustrates the case in which the direction seems to have been from the economy to the research establishment. Coffee was a money-maker. It was important for the balance of payments and for the economy as a whole. As problems related to soil and labor emerged, Brazil responded by encouraging research. Later the agricultural research establishment expanded to other crops.

Yet noneconomic forces were also operating. It is important to understand them, in order to grasp the style of research that was initially implanted. The creation of agricultural research units in Brazil was greatly influenced by the prevailing European pattern. This led to a "diffuse model" in which each research unit tried to diversify its activities, focusing on many different products and attempting to generate a wide array of technologies. The proper choice of techniques and crops was left to the farmer.

This model may be efficient in an environment having adequate resources for agricultural research and a mass of well-educated and organized farmers, bringing their specific needs to researchers and administrators. Scientists' individual interests can be satisfied, given the wide range of choice in their respective areas of research. At the same time, the model guarantees that the desires of the majority of the farmers (particularly of those in a position to exert influence on the investigators) are satisfied too. As a consequence, a wide and diversified array of research results permits the development of a large number of specific production systems suited to the peculiar ecological and economic conditions of the farmers. [8]

The diffuse model, therefore, tends to meet the requirement of ecological and economic peculiarities in agricultural research that is extremely important, given the variety of soil, climate, and market conditions. But it is expensive. Moreover, the developing nations lack the two essential ingredients demanded by the diffuse model: abundant research resources and an educated mass of farmers. This double scarcity makes interaction difficult, especially in the absence of a well-organized agricultural extension service.

But even without these two crucial ingredients, Brazilian researchers followed the individualistic European tradition for many decades. As a result, the limited economic and human resources were more or less randomly spread among disparate activities, and contributed little to the existing production systems. Information tended to be dispersed, limited, and insufficient to be transformed into effective technological packages.

In the absence of a critical minimal number of active farmers, the model became institutionalized as an extremely individualistic activity. Research topics and methodology were viewed as the per-

sonal property of the investigators, even though research was entirely
publicly financed. Research priorities were viewed as sacred. Scarce
as they were, the resources were allocated to a wide variety of dis-
connected projects which, in turn, were carried out not by qualified
research teams but usually by an individual scientist working alone.
Little emphasis was placed on the training and preparation of new research generations. On the contrary, the few scientists trained abroad
tended to become isolated as soon as they returned.

This pattern existed throughout Brazil. However, the first exceptions arose in those states where land and labor were becoming
scarce and the farmers were better-educated and better-organized. [9]
A new system of forces emerged to forge more solidly the link between agricultural research and productivity. This was the pattern in
São Paulo and Rio Grande do Sul after 1930, with variations from
product to product.

Infrastructural Needs, Manufacturing, and the Role of Engineering

The concern with the colony's defense encouraged the development of military engineering in Brazil, especially in construction of
forts, canals, and bridges; production of weapons and munitions; and
cartography. The first weapons factory was built in 1762. To satisfy
these needs, training courses in "military sciences" were introduced.[10]
These courses slowly evolved into more formal training programs of
practical engineering (focused on artillery, mining, bridges, pontoons,
explosives, and forts) and, finally, into the establishment of military
engineering as a recognized profession.

This evolution seems to fit the classic pattern of development
of military engineering in many societies: from concrete needs the
country moved to practical training, and from practical training to a
formal theoretical approach. The establishment of military schools
in Brazil also performed several important broad functions in the
country's development process and led to the formation of engineering
as an autonomous discipline. Military engineers helped to speed the
transfer of technology from England, particularly railway, telegraph,
bridge, and canal construction,[11] and had a decisive impact on the
emergence of mathematical and physical sciences and civil engineering.

The most interesting influence in this respect was the role of
military engineering in helping to solve the prevailing mining problems
and in the creation of the first school of mines.[12] Until the first quarter of the eighteenth century, mining was relatively easy: the known
deposits of gold, iron, and diamonds were on the surface and readily
accessible. The extraction process was labor-intensive, and labor
was cheap. But this changed. Labor shortages and the depletion of the

superficial deposits induced mining companies to seek mechanical processes and to acquire a better knowledge of chemistry and mineralogy. For more than forty years the mining companies urged the government to create a mining school, which was finally established at Ouro Preto in 1876, under the leadership of a French mineralogist, Henri Gorciex, and Brazilian military engineers. Since then, a close relationship has been maintained between the military engineering schools and the School of Mines. Almost all the early graduates were absorbed by the military academics and by the government agencies in charge of road and bridge construction. Thus, the responsibility for infrastructure construction was shifted from the army and its military engineers to civil agencies and the new civil engineers. The School of Mines developed a diversified training program that prepared personnel for infrastructural development, and it initiated a long tradition of metallurgical engineering in Brazil.

A significant part of the history of Brazilian civil engineering related to the railway movement. The first railways entered Brazil in the middle of the nineteenth century, much earlier than the civil engineering schools;[13] and by 1889, Brazil had over 10,000 kilometers of railroads. The first railway, from Rio de Janeiro to Petrópolis, was built in 1854. The second, from São Paulo to Santos, was completed in 1867 and played a key role in bringing coffee from the hinterland to Santos for export. The building of the two railways, particularly the latter, was a challenging technical venture involving a sizable number of British and Brazilian technicians. To bring the railroads up from the coast, over the mountains, and onto the hilly interior plateau required considerable ingenuity. New bridging systems were created and new solutions for cable-tracking were found.

For the first time the nation used large quantities of iron. During the 30-year construction phase, many military engineers participated in the British-Brazilian teams. The expansion and maintenance phases brought about the establishment of several training programs and laboratories with the participation of the civil engineering schools. The School of Mines was the first Brazilian civil institution to become involved. After it came the São Paulo Engineering School, founded in 1894, with emphasis on road construction, mechanical engineering, resistance of materials, and electric power generation.[14]

From the beginning of the century there was a close relationship between the São Paulo Engineering School and the railway and electric sectors. It was not a coincidence that one of the school's first buildings was located adjacent to the São Paulo Railway's main station: the school's laboratories and railway shops were used interchangeably for material testing and training activities.[15] The same type of interface occurred with the São Paulo Tramways, Light and Power Company, responsible for electric generation and urban streetcars.[16] In particular, the school's Laboratory of Material Resistance played an important role in testing materials and equipment both for the railways and the power plants.

For many decades, even after the founding of the São Paulo Engineering School, the railways played an important role in the diffusion of techniques for foundry and lathe operations, steam-driven machinery, and train maintenance. São Paulo Railway maintenance shops did technical consulting for other industries (textile, steam sugar mills, foundries), resulting in some inventions. There were many cases in the history of Brazilian manufacturing where the entrepreneur-inventor took his idea to the railways shops for testing and improvement. One of the most remarkable examples was the baron of Mauá, who was responsible for the completion of the first two railways and had a great deal to do with the introduction of new technology for steamships, shipyards, pipe laying and manufacturing, iron processing, cranes, gas street lamps, the telegraph, and electric-power generation. [17]

In contrast with infrastructure, the relationship between engineering and industrial development was weak until 1930. During the nineteenth century, attempts at industrialization in Brazil were shaken by numerous oscillations in the country's economic policies. In addition, the technological advances of European industry and the small size of the domestic market severely limited Brazilian industrial development. [18] The economy was dependent largely on export of primary products (particularly coffee, rubber, and cocoa). But the abolition of slavery at the end of the century brought new difficulties to the prevailing traditional agriculture and, as a consequence, to the public budgets. The government's expenses were higher than its tax revenue. Coffee was in serious oversupply. The economy was extremely unstable, and in 1929 it virtually collapsed. This fact is usually taken as the key catalyst that moved the nation to a new import-substitution phase (cement, paper, metal, textiles, and chemicals) and a steadier pace of industrialization. [19]

During the early years of industrialization, when plants were few and the economy was plagued by long and frequent periods of instability, industry did not demand large numbers of engineers. Little research capability was required. From 1930 on, however, the import-substitution strategy raised the demand for professionals and technicians (but not for researchers). The engineering schools, particularly in São Paulo, Rio de Janeiro, and Minas Gerais, managed to maintain a reasonably high standard of professional training. The first clear signs of research activity devoted to industrial development came after 1934, with the transformation of the Laboratory of Material Resistance of the São Paulo Engineering School into the Technological Research Institute. Its main aim was to perform experiments and to provide technical assistance to the nascent construction and mechanical industries.

The first two Brazilian specialists, Paulo Souza and Victor Freire, immediately realized that, given the expanding role of the laboratory, there was need for additional manpower. As a consequence, the Engineering School imported Wilhelm Fischer from the University of Vienna (1903), who came to play an important role in the training of Brazilian technicians. From its beginning the laboratory had a practical orientation, focusing upon the testing of imported and native materials and providing information to the production sector. The railways and the electric-power companies depended heavily on this activity; in fact, they gave the laboratory the responsibility for testing most of the key materials, including cement, tracks, pipes for electric plants, and railroad ties. The institute quickly became well-known and began enlarging its staff, training its major research personnel abroad and strengthening its relation with the industry sector through contracted research. [20]

As a general appraisal, it can be said that despite the existence of several training institutions and agriculture engineering, the scientific and technological achievements of Brazil were rather meager until 1930. Schools at the university level (there were no universities) were few and were directed to professionalization rather than to research. The scientific establishment was practically nonexistent; and originality in science was still a result of individual effort, European training, and (often) private wealth. [21] The few technological innovations in agriculture, infrastructure, and manufacturing were accomplished quite independently of a scientific or technological establishment. Borrowing was the rule; industry-to-industry diffusion and training opportunities abroad were the key to improvement of the agricultural and engineering sectors.

THE IMPACT OF AGRICULTURAL RESEARCH ON AGRICULTURAL PRODUCTIVITY

Thus far I have employed the induced-innovation hypothesis as a key to explaining forward and backward movements of the technological and development processes. Applied to agricultural research, this hypothesis argues that the crucial elements in determining the application of research resources are the scarcity or availability of the key production factors (land and labor) and the interplay between farmers and researchers: the stronger the interaction, the more dynamic the innovation process. [22]

The central idea of this section is to analyze the importance of land and labor prices and the role of the interaction component in

Brazilian agricultural research after 1930. To what extent was this stimulus-response mechanism responsible for the research results? Was the sequence only from market to research, or were there other patterns? How did the dialectic process vary from commodity to commodity, from institution to institution, and from period to period?

This paper hypothesizes that the importance of the stimulus-response mechanism and the sequential direction in agricultural research in Brazil have been highly dependent on the nature of the commodities market. To test this proposition, six important agricultural commodities are analyzed: coffee, sugarcane, and cotton, which are export products; and rice, beans, and corn, which are domestic products. It is hypothesized that the research environment related to export products in Brazil was systematically more favorable than that of the domestic commodities. It is expected that, as a consequence, interaction between farmers and researchers was relatively intense and that the research establishment produced substantial results. Conversely, unfavorable market conditions for the domestic products would inhibit innovation and diffusion.

The basic strategy is to relate research emphasis on each of the six products to its productivity pattern. Next, the social matrix in which advancements were made is analyzed in terms of the market conditions, with focus on the interaction between research and the market, on the question of continuity and discontinuity of the research work, and on the quality of the research personnel. In this way it is hoped to progress beyond the usual cost/benefit analyses and to gain a broader view of the relation between agricultural research and agricultural productivity in Brazil.

Export Products

Coffee, sugarcane, and cotton have long played extremely important roles in the Brazilian economy; until very recently, when soybeans became a major export commodity, these three accounted for about two-thirds of all exports. For centuries, Brazil's balance of payments was greatly dependent on their performance. As a consequence, reasonable attention was always placed on the production and trade of these commodities, and a series of protective policies was developed to stimulate production and to support prices.

Coffee

Coffee was introduced into northern Brazil in 1727. By the end of the eighteenth century, it had been shifted to better soils in the South. [23] This shift permitted the country to increase its production and export capacities. In the first half of the nineteenth century, Brazil produced about 40 percent of the world supply, and coffee was responsible for 64 percent of Brazilian exports. [24] However, production expansion depended primarily upon land expansion until almost the middle of the twentieth century, when a combination of worn-out soils and severe foreign competition called further attention to the need for coffee research. Research on coffee had, in fact, been underway since the founding of the Agricultural Research Institute of Campinas in São Paulo. [25] During its first 10-15 years, under the influence of Franz Dafert, the institute dedicated a series of studies to the chemical constitution of coffee trees, comparisons of productivity levels between the traditional and recently imported varieties, organic and chemical fertilizer experiments, techniques for transplanting, soil studies, and many other investigations. Support from the government and farmers was quickly obtained, and the prestige of the coffee research teams was readily established within the institute. A school of outstanding investigators was brought together and an extensive research program on varieties, taxonomy, genetics, soils, and fertilization was consolidated between 1930 and 1940. By 1935 demonstration farms in São Paulo had increased their productivity 40 percent by using chemical fertilizers developed by the institute.

In addition to the institute's work, the state government in 1927 created another important research unit devoted primarily to plant and animal pathology, the Biological Institute of São Paulo, which attracted able young researchers in plant pathology, physiology, and other areas. Also, extension and sanitation programs were created and operated jointly by the Biological Institute and the Agricultural Research Institute.

The 1930-40 period brought a clear awareness that the coffee industry could no longer rely on land expansion alone. During the first three decades of the century, there was a dramatic decrease in coffee productivity (see Figure 7.1). By 1930 the crucial and most substantial research effort had been initiated. Consolidation of coffee research teams, stimulation of research in plant pathology, and establishment of extension and sanitation programs can be taken as clear indications that by midcentury Brazil had become concerned with the future of the crop and with the economy as a whole. The most impressive results came by the end of the decade, and the first signs of productivity gains were seen in 1943-47. These gains are also associated with the exploitation of new soils in Paraná and the use of chemical pesticides.

At the beginning of the 1940s, the Campinas Institute's efforts were extended to include soil erosion. Two important lines of research were inaugurated on soil conservation and plant varieties. An intensive research program was dedicated to the development and adaptation of a new variety, Mundo Novo, which had been created empirically by

FIGURE 7.1

Coffee Productivity

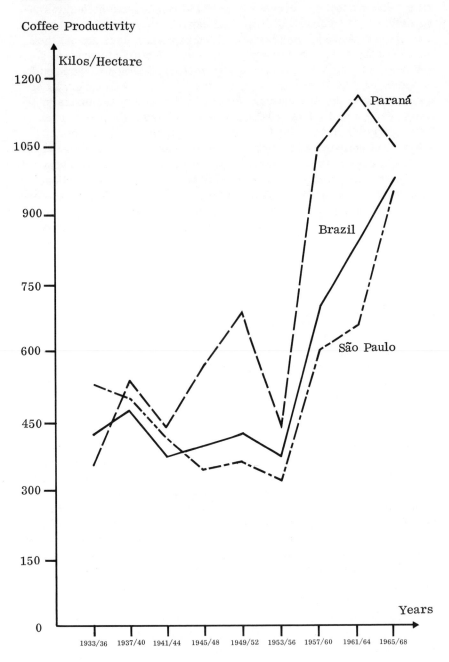

Source: Instituto de Economia Agricola.

private farmers. Efforts to improve and to make regional adaptations of the new variety marked the institute's activity during the 1940s. Improved seeds were first distributed on an experimental basis in 1948, with extremely good results: Mundo Novo produced 240 percent more than the initial typical variety or Arabica coffee. By 1952 more than 50 percent of the coffee plants (more than 400 million) were of the Mundo Novo variety.[26] From that year on, productivity climbed from about 350 kilos/hectare to 490 in 1958, 590 in 1963, and 680 in 1968. This dramatic increase in the aggregate productivity level for the country as a whole was largely due to the success of the research effort. Continuity in the coffee experiments, top-quality personnel, and steady government support were the prime reasons for numerous positive results in terms of varieties, fertilization, and soil techniques. Good management on the part of most coffee producers also helped, but research results were a necessary condition.

Sugarcane

Sugarcane has been one of the most important products in Brazil since 1534, when it was introduced in São Vicente (São Paulo) and rapidly spread to Rio de Janeiro and the Northeast. The crop grew well on large properties, was cultivated by slave labor, and required only redimentary technology.[27] This pattern prevailed for centuries, particularly in the Northeast, where land and labor were quite abundant. In other areas, such as Bahia, Rio de Janeiro, and São Paulo, attempts were made to increase land and labor productivity as early as the beginning of the nineteenth century.[28]

Productivity gains in sugarcane in Brazil are closely related to research on plant improvement, which began with importation of a better variety (Caiana) from French Guiana in 1810. Initially this variety had four important advantages over the prevailing one (Crioua): shorter growth and maturation period; much higher resistance to climatic variations; higher juice and sugar yield; ease of processing. It was, however, severely affected by the "Gomose" disease (Xanthomonas vascularum), which dramatically reduced the country's production in the middle of the nineteenth century. The response to this problem was to import other varieties, mainly from Java, and, for the first time, to use a variety native to Brazil, Cristalina. Attention to new varieties was accompanied by attention to new practices, including the introduction of mechanization of a greater and more even spacing between plants and the use of green fertilizer made from sugarcane bagasse and other plants.

Research on sugarcane was intensified in the twentieth century, particularly after 1930, when a series of international price crises began pressuring producers to push for better quality and better productivity. Research activities initiated at Campos (Rio de Janeiro) focused upon improving the Javanese varieties (POJ) and creating a new national variety (CB). These varieties solved one of the key problems by presenting a higher resistance to prevailing diseases, especially the "mosaico" (Marmor sacchari holmes). Also, the CB variety yielded a much higher level of sucrose. In 1935 a sugarcane department was created at the Agricultural Research Institute of Campinas; together with the Campos station, it provided a substantial capability for research on new varieties, spacing and density, cultivation practices, and fertilization, as well as construction and maintenance of modern sugar mills. The research teams were small but of top quality. Their efforts were greatly appreciated by farmers and industrialists; consequently they were able to conduct their experiments on the farms, thus quickly obtaining feedback. Interaction and research continuity were assured. [29]

Figure 7.2 shows the trend of sugarcane productivity in Brazil, the Northeast, and São Paulo. Before the research on varieties, spacing, cultivation practices, and fertilization, São Paulo produced less than 15,000 kilos/hectare. After five years of research, land productivity practically doubled (almost 30,000 kilos/hectare); and by 1943-47 the state was producing 43,000 kilos/hectare. Constant gains continued until the beginning of the 1960s. Recently, the more modern establishments have been producing 100,000 kilos/hectare. Today the existing research apparatus and the available good soils are being used almost to full capacity, and the savanna soils are beginning to be cultivated. To meet the requirements of those poorer soils, there is a new line of research emphasizing fertilization, irrigation, climatology, sucrose level, entomology, and plant pathology, with special attention to two diseases, "broca" (Diatrea sachoralis) and "raquitismo" (virus disease).

It is important to reemphasize that because of sugarcane's agroindustrial character, farmers and industrialists were always involved in its research. New varieties were tested directly by the private sector, which provided useful feedback to the research stations. In fact, farmers and those in the sugar industry tended to put pressure on the stations for better and more useful research results. Today, the more modern farmers and sugar mills are organized into a strong cooperative (COPERSUCAR) that not only provides information to the research agencies, but also has started its own research, with encouraging results. In addition, a special program to improve sugarcane (PLANALSUCAR) was created in 1971, with a strong and decentralized research effort.

In short, for more than 50 years sugarcane, one of the most important export products of Brazil, has received special research attention from the public and private sectors. Research teams have

FIGURE 7. 2

Sugarcane Productivity

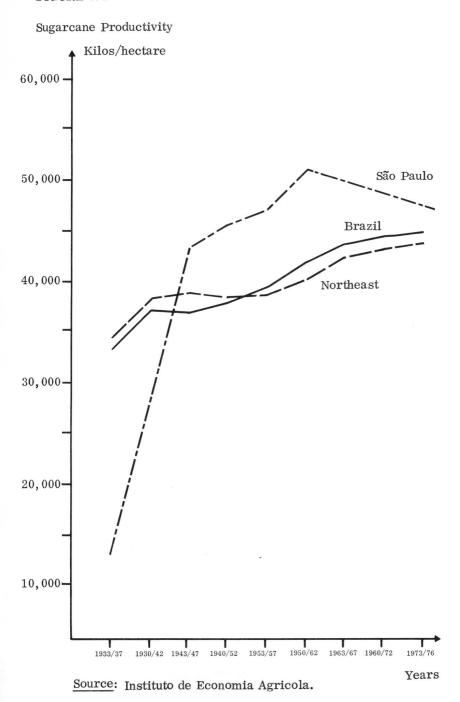

enjoyed reasonably stable support, interaction with farmers and industry has been considerable, and the impact on productivity, particularly in the South, has been notable.

Cotton

Cotton, another important Brazilian export crop, has also received considerable research attention. The history of cotton can be divided into two distinct periods. In the first, from the mid-eighteenth century to 1930, the textile mills of the industrial nations provided a voracious international market for Brazil's cotton. During this time the crop expanded rapidly in the Northeast, where rudimentary techniques and considerable land and labor were used. The end of the American Civil War (1865) gave an extra stimulus to the Brazilian cotton crop. Cotton spread to the South, particularly São Paulo, where land was not being fully used because of the spread of coffee diseases. American immigrants, well-to-do refugees from the American South who were experienced cotton planters, introduced new seeds, plows, and Whitney-type cotton gins.

Cotton research did not begin until the early twentieth century, when an integrated system of research, extension, and subsidies was created. The foreign literature was surveyed. New research was conducted on plant varieties, fertilizing, plowing, and other aspects of cotton production. Again, the Agricultural Research Institute of Campinas contributed a great deal to knowledge regarding the crop.

The second period started in 1930, when São Paulo became the main producer of cotton and Brazil began to participate even more actively in the international market. Additional stimuli to this shift were the 1939 economic and coffee crises and the reactivation of the textile industry in São Paulo. Scientific knowledge accumulated in São Paulo provided an additional basis for development. Both the Agricultural Research Institute of Campinas and the Biological Institute gained extra support to strengthen their research activities on cotton plant improvement and plant pathology. Their main lines of research were establishing collections of good varieties, based on imported species; checking selection techniques ("roughing") and large multiplication fields; selecting demonstration farms for variety testing and substitution; creating new varieties more adapted to Brazilian conditions; making studies of hybrids to improve yields and increase the length of fibers; controlling of pests and diseases, especially "broca" (Entinobothus brasiliensis), "pulgão" (Aphis gossypii), "percevejo" (Horcias nobilellus), "lagarta" (Platyedra gossypiella), and "murcha" (Fusarium oxysporum). [30]

The enormous impact of research on São Paulo's cotton productivity has been documented by Harry Ayer and G. E. Schuh, who

showed that it yielded one of the highest rates of return in the world. [31]
Figure 7. 3 shows that before the concentrated research effort, São
Paulo had a very low land productivity for cotton, less than 400 kilos/
hectare in 1943. This level has practically doubled every ten years:
800 kilos/hectare in the 1950s and 1,200 in the 1960s. Recently, São
Paulo has averaged more than 1,500 kilos/hectare and has made a
substantial improvement in fiber quality. [32] Cotton research there has
been conducted with stable support. Research teams have recruited
well-prepared personnel. Also, with the creation of the integrated
extension-research system, interaction between cotton researchers
and cotton growers has become the rule. Seed production has been
under a state monopoly strictly controlled by the research units.

Research in the Northeast, on the other hand, has been marked
by timidity and discontinuity. Emphasis, such as it is, has been placed
almost exclusively on the fiber length, whereas São Paulo focused on
the length-productivity binomial. The physiology of the Northeast cotton plant is poorly understood, as is its most common pathology. Since
the 1950s the region has lost more than 10 percent in yield, which was
already low.

Domestic Products

Rice and beans are the basic items in the Brazilian diet. Corn,
the most widely cultivated commodity, is used to feeds humans and
animals. Together these three products form the basis of the domestic food supply. [33] Yet, despite their importance, rice, beans,
and corn are among the most poorly understood crops in Brazil.
Neither science nor technology has done much to increase their production and quality.

Rice

Rice was introduced into Brazil as an export crop in the eighteenth century, yet it was never significant. In the mid-nineteenth century it came to be part of the daily diet, and Brazil began to import
large quantities from India. Production expanded considerably in the
twentieth century in the South-Center, particularly São Paulo (Vale
do Paraíba) and Rio Grande do Sul (Pelotas), where it was based on
small plots, natural irrigation, and rudimentary technology. [34] Rice
imports declined from 70 percent of consumption in 1910 to 10 percent in 1916, and to practically none today.

Brazil's rice yield per hectare, however, is still one of the
lowest in the world. Productivity levels have been quite unstable,
extremely dependent on climatic variations, and subject to insect

FIGURE 7.3

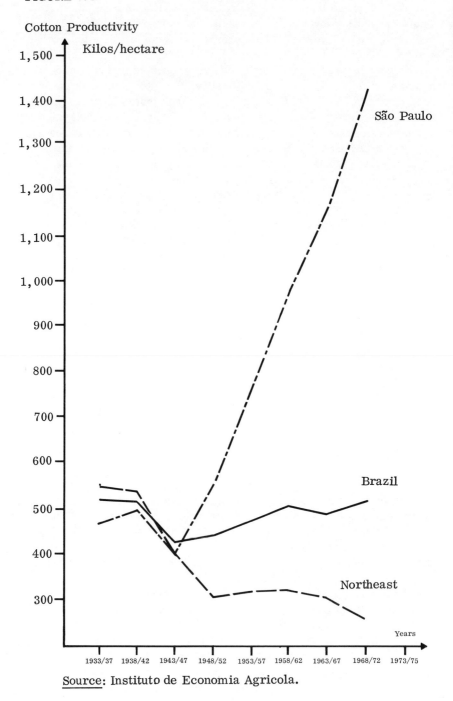

Cotton Productivity

Source: Instituto de Economia Agricola.

252

infestations and diseases. In 1938-42, as shown in Figure 7.4, the average productivity was 1,600 kilos/hectare. Rice was grown on the land with the best soils and good water conditions. With the exhaustion of the good soils and water facilities, productivity declined in almost every region: in 1968-70 the average yield was 1,400 kilos/hectare.

Until very recently, rice research was conducted only at the Agricultural Research Institute of Campinas and the Southern Rice Institute. Main research areas have been development of local varieties from imported rice; sprout-multiplication techniques; dry rice hybrids; resistance to diseases and insects; and production of basic seeds. [35] Funds for rice research have been unstable in comparison with such "noble" products as coffee, sugarcane, and cotton. Also, interaction between researchers and farmers has been much more limited than in the previous cases, due to the funding instability and to the extreme dispersion of rice producers. The lack of an organized market, domestic or international, means that there is little encouragement for research; as a result, rice research is especially vuluerable to the budget crises that sweep the agricultural research stations from time to time. Researchers have never been forced by the market or the government to concentrate their efforts on rice. The first direct attempt to stabilize and to concentrate talents on rice research was made by the federal government through the Brazilian Agricultural Research Center in Goiás, which brought in highly qualified personnel and established close ties to the International Rice Research Institute and other recognized research institutions.

Beans

Brazil inherited beans from the aborigines. From earliest times, beans have been cultivated as a subsistence crop, using the poorest natural and human resources. Productivity levels in Brazil are extremely low compared with the United States, Japan, or Mexico. Moreover, Figure 7.5 shows a dramatic drop in Brazilian productivity: from 869 kilos/hectare in 1933 to 631 in 1973. In the Northeast this decline was still more severe: from 897 kilos/hectare in 1933 to 493 in 1970.

This productivity loss can be attributed to two basic factors. First, the market for beans has been quite small and confined to limited areas of the country. They are considered a secondary product, usually intercropped with corn, and other products under primitive techniques. Second, the instability of research teams has deterred the development of seeds with high yield potential and high resistance to insect infestation and diseases.

Research on beans has never received high priority in Brazil. The two scattered studies from the 1930s and the 1950s were limited

FIGURE 7.4

Rice Productivity
Kilos/hectare

Source: Instituto de Economia Agricola.

Years

254

FIGURE 7. 5

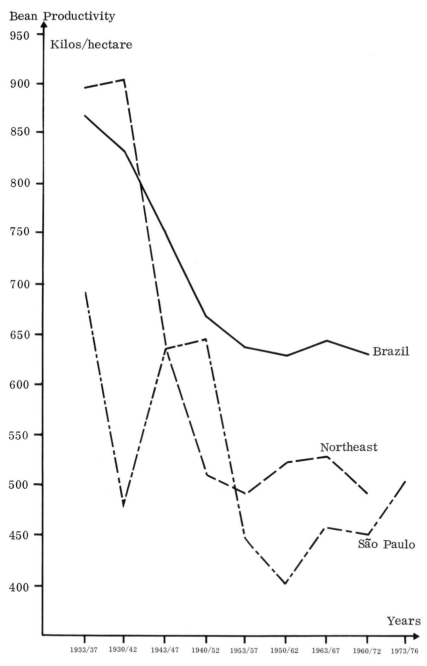

Source: Instituto de Economia Agricola.

to diagnosing the factors to which the plant is sensitive. [36] However, little is known about ecological and climatic adaptability or ways to control pests and diseases.

These needs were identified long ago, but research institutions have always encountered extreme difficulties in attracting the financial resources and professional talent. Bean research teams have been the first to be dissolved during crises. Beans are a case in which a subsistence crop is paralleled by subsistence research. As in the case of rice, only recently (in 1975) did the federal government begin, through the National Center for Rice and Beans in Goiás, to investigate the genetics, disease resistance, and economic profitability of the plant.

Corn

Corn, the most widely planted crop in Brazil, is a staple food for both animals and humans. As in the case of beans, Brazil inherited corn from the aborigines. Today more land is devoted to corn than to other crops, and it is the second most important commodity in terms of production value. [37] Yet corn productivity is extremely low, about one-fourth of the average in the United States or France. Figure 7. 6 shows that some productivity gains have occurred in São Paulo since 1953, reaching about 2,000 kilos/hectare in 1970. Conversely, in the Northeast production has been declining and is at an extremely low level in absolute terms (about 700 kilos/hectare). However, even in the most developed state (São Paulo), corn productivity is about one-third that of advanced countries. [38]

The nation has given reasonable support to corn research. Experiments on fertilization, soil adaptability, plant improvement, disease control, hybrid seeds, and other problems have been conducted continuously since 1934 by capable personnel. In this respect, corn research is quite similar to that of the export products. In addition, private firms became active in seed production after 1950. Under experimental conditions, in both public research institutes and private firms, corn productivity is as high as in the most developed countries. Improved varieties, adequate fertilization and soil preparation schemes, and appropriate cultivation techniques have raised corn productivity under experimental conditions to 6,000 kilos/hectare. The country's average productivity, however, remains insensitive to these research advances.

Corn seems to be a deviant case in which the production sector has not responded to research productivity gains. This has posed a complex question to research policy agencies: how to evaluate a line of research that unquestionably represents a positive contribution to physical productivity, but has practically no impact on the country's average performance. Perhaps the research gains are not transferable

FIGURE 7. 6

Corn Productivity

Kilos/hectare

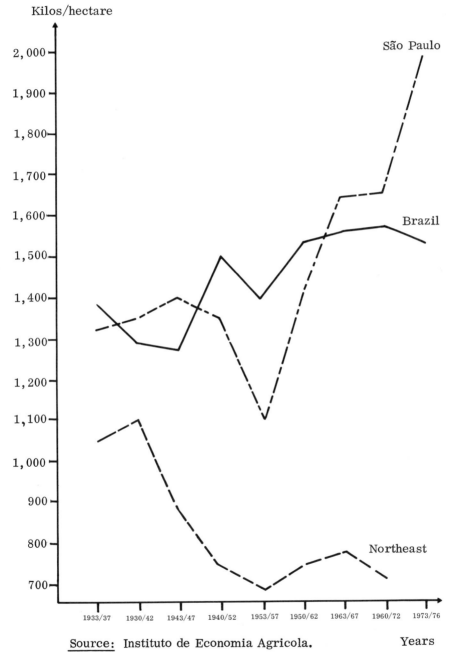

257

to concrete situations because unstable prices do not permit the adoption of the techniques used by researchers. It is intriguing, however, that some simple and inexpensive changes in spacing would substantially increase corn productivity, even under the most favorable ecological conditions.

Yet, there is another possibility. Since corn is a secondary product for most farmers because of unstable market conditions, the research effort may not have been sufficiently imaginative. Even more modern farmers, who are in a more favorable market position, do not show any substantial improvement in corn productivity. This brings into question both the idea that good research results are blocked by bad market conditions and the actual value of those results. This type of concern is behind the wide-ranging policies of the newly created National Corn Research Center (in Minas Gerais), which emphasizes simple technologies that seem to be more appropriate for the corn producers.

These six case studies illustrate that the performance of a particular research area in agriculture is highly dependent on the nature of the market and the efficacy of supporting policies for these commodities. Positive impact is closely related to research continuity, interaction between researchers and farmers, and the quality of research personnel. Yet the Brazilian situation shows that research continuity and the quality of the personnel are more important than large amounts of funds. Sugarcane, an export crop, is a case in which little money and a small but highly qualified research group provided one of the most substantial impacts on Brazilian productivity. Corn illustrates a situation where massive efforts produced practically no impact in terms of the country's productivity. These results suggest that a methodological strategy aimed at affecting quality of the research teams, the nature of personnel recruitment and promotion, and the nature of the market seems to be the answer for providing additional insights to the current cost/benefit analysis.

Internationally, Brazil has been in a good position to grow and sell coffee, cotton, and sugarcane. From these come the country's foreign earnings. [39] Behind recent gains in the productivity of these crops lies a stable, well-funded, and innovative research organization that was initiated because Brazilian growers and distributors realized they needed greater knowledge of agricultural technologies appropriate to the nation's special conditions, and that much of this information could not be imported from abroad. Despite its practical effectiveness, this establishment has not yet contributed much to basic scientific theory. Perhaps the needs of the technological researchers will encourage investment in basic research, which in turn may lead to new contributions to basic theory.

Research on domestic crops that yield no foreign earnings has either languished (rice and beans) or been ineffective in practice (corn). Because of the success of research on export crops, the government is finally attempting to improve domestic crops. Other requirements necessary are dependable markets and distribution systems, and considerable interaction between these and the researchers. If available, these staple foods should find a rich domestic market. It seems reasonable to expect that new research efforts can bring domestic food benefits corresponding to those of the export crops. If this happens, the food industry will no doubt continue to encourage research on these and other staples; and eventually these efforts might contribute to or stimulate the growth of basic science.

TECHNOLOGY IN BRAZILIAN INDUSTRIAL DEVELOPMENT

Modern Brazil has one of the largest and most diversified industrial sectors in Latin America. The present share of industry in the total output is more than 40 percent, which means an average rate of growth of about 8 percent per year. The industrial sector employs more than 30 percent of the Brazilian labor force.

The beginnings in industrialization in Brazil came during the Empire, which lasted until 1889. The earliest examples were based on the country's low production costs and comparative advantages in food products, construction materials, iron and iron products, and textiles. However, the sector did not develop rapidly during the nineteenth century. Continuing political control by landowners, the small domestic market, competition from European industry, and Brazil's strong comparative advantage in primary products caused the slow and oscillating development of industry.

Import Substitution Policies

The rate of industrialization in Brazil during the twentieth century was accelerated by three marked waves of import-substitution policies. The first was connected with World War I, when the interruption of supplies from overseas eliminated foreign competition, and many new industries were created to fill the gap and even to supply overseas markets. [40] Industrial production grew at over 25 percent per year between 1914 and 1919. [41] The second wave came after the 1929 collapse: because of the severe drop in coffee export earnings, rigid restrictions were placed on imports and again domestic industries were stimulated. This second import-substitution phase was supported by President Getúlio Vargas, who had no ties with the backward landowners and was sympathetic to industrial development. The third import-substitution stimulus occurred with World War II, when import difficulties again encouraged the installation and development of the steel industry, and factories for machinery, electrical equipment, and similar goods. By the time the war ended, about 20 percent of the GNP was produced by industry. [42] At that time Brazil had large foreign-exchange reserves; but by the end of 1946, after one year of unrestricted imports, these were wiped out. Import controls and internal inducements to industrialization were chosen as the key mechanisms by which to equilibrate the balance of payments. However, more substantial industrialization did not take place until Vargas returned to office (1951-54). The enthusiastic Juscelino Kubitschek (who was president from 1956 to 1960) provided the actual support for industrial development. [43]

The restriction on imports induced foreign firms to establish branch factories in Brazil, so as not to lose the sizable Brazilian market. In addition, the government provided subsidies and special treatment to encourage foreign investment: for example, special treatment for imports of capital goods and the unrestricted repatriation of earnings and capital. Under these conditions, foreign corporations set up branch plants in Brazil during the 1950s and 1960s. They provided dynamic and capital-intensive factors in areas such as automobile production, automotive equipment, chemicals, and pharmaceuticals. They also invested over $100 million per year, or roughly 10 percent of the total investment in manufacturing. As a result, many complementary industries grew up, forming a constellation of foreign assemblers and Brazilian suppliers that acted as a self-reinforcing process based on a growing demand for both intermediate and capital goods. Import-substitution industrialization has spurred production enough so that Brazil is beginning to export manufactured goods, especially automobiles.

Import Substitution and Technological Inducements

Brazilian industrial technology has developed almost exclusively by borrowing. In the first and second stages of import substitution, technology came embodied in the capital goods. In other words, the country imported machinery in order to substitute for finished consumer products that would otherwise have been imported. In addition, the first two phases of Brazilian industrialization were oriented to the demands of wealthier groups that were used to sophisticated imported goods. A marked similarity between the domestic and the external know-how was thus established. The national technological and scientific establishments were not thought to be prepared for this style of industrialization and, as a result, were not asked to participate intensively. But machinery needs attention. Slowly, Brazil's industrial technicians and entrepreneurs developed an intimate knowledge of the imported machines. Modifications were sometimes introduced to allow for differences in local raw materials. Technical literature and contracts with immigrant engineers and technicians were other sources of know-how for local personnel. In fact, during the first half of the twentieth century, Brazil accepted more than 2 million immigrants from Europe, Japan, and Mediterranean countries, most of them with more education and industrial experience than the average Brazilian worker.

Entrepreneurship came from immigrants, especially São Paulo's Italians and native-born Brazilians, descendents of the coffee elites who had gone to the engineering schools in the 1920s and 1930s.[44] The engineering schools established at the end of the nineteenth century were adequately prepared to supply several industrial sectors with needed engineers. At the beginning they worked with immigrants. When the more ambitious projects were established, such as capital goods, electrical equipment, and (later) steel, Brazilian engineers were involved in their installation, maintenance, and (to some extent) modification and adaptation. The Institute of Technological Research of São Paulo responded to the new needs by developing domestic capabilities in chemistry, metallurgy, civil construction, and geology.

Consequently, the growth of industry brought a new emphasis on pragmatic and technical education, in contrast with the former literary and humanistic tradition. Thus educational values shifted, particularly in São Paulo, the main industrial center. Also, the early stages of industrial development generated a new "industrial mentality" described as

> . . . an ideology which emphasized a willingness to engage in the day-to-day details of technical work in order to participate in the world of modern technology and to help establish it in Brazil. . . . Strongly nationalistic, this ideology also stressed the importance of import substitution as a means toward national economic and political progress.[45]

To sum up, the first two stages of the import-substitution strategy (1914-19, 1930-39) were marked by generous protective mechanisms, heavy borrowing of "embodied technology," and the generation of modes of thought appropriate to industry. The domestic technological and scientific establishments were not encouraged to innovate. Rather, their contributions were limited to supplying professionals and to modest joint ventures regarding the adaptation and modification of imported equipment. If, on the one hand, protective polities provided seeds for Brazilian industrialization, on the other, the heavy emphasis on borrowing encouraged continuation of long-run technological inefficiency. One can argue with Joel Bergsman that

> Brazil would have been better off if infrastructure, including education, market organization, (and research capacity), had been better built and better operated; and if in addition direct subsidies to wage and other costs had been employed—as much as possible to pay for training, apprenticeship programmes, re-equipment, and other real-cost-reducing action. However, to recommend such a programme as a complete substitute for protection would be nonsense. The principal reason that such a programme would be economically justified in Brazil is the very reason why it would not be done: Brazil has an infant economy; underdeveloped; poor in institutions and human resources. [46]

During the third wave of import substitution (1947–61) the industrial sector grew 262 percent, while agriculture grew 87 percent. The need for industrial development began to go beyond the simple importing of capital goods. As a consequence, the country began to establish a broader industrial complex. A series of more complicated manufactures entered the country, including automotive goods, chemicals, capital goods, and pharmaceuticals, all requiring a high degree of technological skill. Local personnel were not prepared to supply this, although the university system and a few technological institutes were growing. Emphasis continued to be on professionalization rather than on R&D. Together then, technological requirements and government incentives allowed foreign companies to establish supremacy. By 1960, more than 50 percent of the capital-goods companies were local subsidiaries of overseas corporations; chemicals (except petrochemicals) reached 70 percent; pharmaceuticals were approaching 90 percent. The automobile sector was just beginning, with all the factories run by foreign companies.

Technological Transfer and Domestic Capabilities

The demand for more complex technology was largely met by plant-to-plant transfer, technical assistance contracts, licensing and patenting, and foreign engineering consulting. Foreign companies merely used the designs and the know-how from their home plants. Brazilian-owned firms were not very different. They tended to enter into technical assistance and licensing contracts. Domestic innovations and intramural R&D could not keep up with the nation's and government's insistence on fast industrial development, which was needed to cope with the rapid population growth. The domestic technological and scientific establishment was not prepared for such a fast rate of industrialization. [47]

Reasons for this relatively modest degree of domestic research activity are usually thought to be connected with the opportunity costs of generating local technology in contrast with the less complicated, cheaper, and quicker technological transfer. Yet perhaps more important than opportunity costs were managerial passivity and the country's permissiveness with respect to technological transfer. This point will be briefly expanded below.

In a country as large and diverse as Brazil, complete knowledge of the local environment and of factor prices is impossible for foreign companies. In addition, since costs of generating new information are high, managers tend to exploit what they already know and to gather more information grudgingly. They use the methods familiar to them before their arrival in Brazil. Their R&D is done in the home country, not Brazil.

Reinforcing this laissez-faire attitude, Brazil has until recently been permissive about technological transfer. Its rapid economic growth created such a fantastic seller's market that price competition was no longer a limiting factor. As a result, erroneous technological decisions did not often result in substantial losses or even profit decreases. Special market conditions and generous government incentives permitted inefficiencies in the choice of technology, and generated a kind of passivity in searching for local know-how. [48]

What is the pattern of this technological dependence in recent years? A pioneering empirical analysis of the origin and development of industrial plants in Brazil shows the following: (1) During their installation, 62 percent of the firms acquired their "know-how" from abroad. (2) This dependence is a little more pronounced for firms established after 1965 than for those installed somewhat earlier (1930-45). (3) Research activity is limited to making minor adaptations of foreign technology to fit a few peculiarities of local inputs. (4) About half the Brazilian firms maintain permanent contracts with foreign organizations for use of external know-how, technical assistance, and licensing. In the case of foreign-owned plants, most of these contracts are with their main offices abroad. (5) Only 10 percent of the Brazilian firms, predominantly large national plants, show some signs of innovating activities. [49]

During the three import-substitution phases, industrialization was quite dependent on foreign consulting firms. National consulting firms in engineering and project design were always scarce in Brazil. The first large industrial plants (steel, automobiles, electricity, and chemicals) were completely designed by foreign consulting firms. The next phase, operation and maintenance, brought foreign subsidiaries to Brazil. Today more than 70 percent of engineering activities in Brazil are performed by foreign firms or subsidiaries, which currently concentrate on large projects (including government contracts). Among the reasons given by the local contractors for selecting foreign firms are limited local capability, lower risk with foreign engineering, better information capability, and ease in obtaining funds from international agencies. [50]

Foreign technological participation may imply a considerable political cost, but that multinational corporations have played an important role in hastening the industrialization process, in permitting some competition in the international markets, in elevating the technical capacity of small and medium-size Brazilian suppliers, in broadening the demand for highly educated technicians, and—as a consequence—in stimulating university and middle-level technical training.

More recent industrial development (1960-75) has been marked, in general, by a large expansion and diversification. The disturbingly high inflation of the late 1950s and the political instabilities of the early 1960s have been considered prime causes of the decline of industrialization. Brazil had an average rate of industrial growth of 10 percent in 1950-61; but after 1962 the country experienced declining rates, reaching 5 percent in 1965. The rapid pace of industrial development was regained (10 percent average) from 1968 until the oil crisis of 1974. Highest rates were in capital goods, automobiles, durable consumer goods, and chemicals. The infrastructure also grew quickly, especially highways, electric power, ports, communications, and education.

The Roles of Foreign Companies and the State

Industrial growth in the 1960-75 period was marked by increase in participation of foreign manufacturing firms and of the state. Today these sources jointly have the major share of participation in most of the dynamic industrial sectors. Domestic firms today maintain preeminence in the traditional areas of wood products, furniture, beverages, printing, shoes, and textiles.

An intensification of technological transfer has been noted not only among the foreign firms but also among the state-owned firms; both are interested in fast growth at low costs. Although expenditure on technological transfer has increased, Brazil seems to be spending modestly (0.3 percent of GNP) in comparison with other countries that have gone through a rapid industrialization and have relied on foreign technology, such as Japan, France, and Mexico. Overall, current estimates indicate that an increasing proportion of GNP is being spent on technology transfer, possibly reaching 0.8 percent by 1980. [51] Can some of this dependence be offset by Brazilian research institutes?

An attempt to reverse the present dependence on imported know-how will take time and will require a combination of three interrelated mechanisms: new mechanisms by which to select the technologies to be imported; an increase of imports in selected areas for the next decade or two; a substantial strengthening of the scientific and technological infrastructure of the country during the next 10 or 20 years. This will require an elaborate and sophisticated policy of technology importation and innovation. The major stance at present, which is wholly permissive in some areas and exactly the reverse in others, simply is not viable. In some places there is already a serious concern for balancing domestic capabilities and foreign know-how. Where Brazil has confronted the problem of factor scarcity with flexible decision-making mechanisms, participation of domestic know-how is higher than where administrative rigidity has been the rule. Specifically, this paper hypothesizes that when administrative flexibility has been dominant, expansion and innovation (at least by borrowing) have occurred; conversely, when rigidity has prevailed, the result has been stagnation. Moreover, it appears that when flexibility prevailed in a sector with an abundance of production factors, expansion involved considerable transfer of technology with few domestic innovations; however, when production factors were scarce, administrative flexibility induced joint innovations between foreign and Brazilian technicians. The steel, cement, and oil industries illustrate these hypotheses.

Steel

Steel production was negligible at the end of the nineteenth century and the first quarter of the twentieth century. Bridges, railway tracks, building materials, and machinery were mostly imported. A key problem was the scarcity of good metallurgical coke. In the early 1930s a Brazilian-Belgian company in Minas Gerais (Companhia Belgo-Mineira) began to use charcoal to produce steel and introduced several changes in technique, such as the improvement of the furnace productivity through the use of synthesis, decreased energy consumption, and improvements in the furnaces. These advances permitted the multiplication of small private plants but did not solve the critical steel shortage in the country. [52]

The recognition of the importance of a large and strong national steel industry in Brazil was the key issue of numerous political battles during most of the 1930s and 1940s. The first large steel plant (Companhia Siderurgica Nacional, Volta Redonda, Rio de Janeiro) was constructed in 1942 and went into operation in 1947. Today Brazil's production practically meets the domestic demand for nonflat products and special steels, although the country continues to import flat-steel products. [53] The enormous expansion of the automobile industry in the 1960s and the rapid economic growth of the country in the 1970s have generated a dramatic increase in the demand for steel, now at 18 percent per year. In the 1960s the country produced about 2 million tons; in the late 1970s total production is about 12 million tons. New plants are being built to raise this to about 18 million in 1980 and 33 million in 1985.

This experience illustrates the successful and rapid deployment of a relatively advanced technology. The installation of a steel complex in Brazil faced a scarcity of capital for heavy investment as well as the shortage of coke. In this case government policies allowed the private sector to make its own decisions and, as a result, the sector expanded. State intervention and public monopoly were considered as an alternative, but the final solution provided a flexible joint-venture scheme, with large investments of public capital and use of foreign technology. At first technology was imported; later it was adapted to Brazilian conditions, mostly with the cooperation of American and Japanese technicians. But the direct participation of foreign operating personnel did not last very long. Brazilian engineers and metallurgists quickly learned the operating and maintenance techniques, due in part to the long training tradition in mining, foundry, and steel processing that had been under way in Brazil since the end of the nineteenth century. [54] Today there are five major steel plants, practically all with their own research centers staffed by highly qualified personnel. They have made improvements regarding raw materials and blast-furnace operations, and have brought in new ideas from abroad. Foreign participation is still predominant in the design and installation of new mills. In this case, therefore, factor scarcity and administrative flexibility combined to yield a dynamic industry that has stimulated its own domestic research programs.

Cement

The combination of the rapid expansion of population, the weakness of the steel sector for many decades, and the relative abundance of limestone near the main urban centers led to rapid development of the cement industry in Brazil. The country now has one of the most advanced cement engineering sectors (buildings, bridges, roads, and other construction) in Latin America.

Since its beginning the cement industry has had no interference from the government. All the plants were private, and were given considerable autonomy to import technology and manpower. The initial efforts in the 1930s were based entirely on imported equipment and European technicians. During the 1940s the cement sector expanded greatly, and by the mid-1950s and 1960s the country became self-sufficient. It was able to build the new capital city of Brasîlia, add 5,000 kilometers of paved roads per year, and provide the basis for rapid urban development.

This expansion was due largely to the addition of new plants, with practically no new technological development. The abundance of limestone and its proximity to the main urban centers did not encourage innovation. The sector imported its entire "technological package" from Denmark in the 1940s and has not changed it very much, although in recent years prestressed concrete has been introduced in building and bridge construction. The cement industry is still importing basic equipment, although, because of the petroleum crisis, oil processes are being replaced by dry techniques.

Thus, with an abundance of raw material and with administrative flexibility, the sector imported a complete operating package that is still in use. The demand has been met with a low level of technological innovation; very little research has been done on cement in Brazil. Yet the country is self-sufficient, and there are no economic stresses in sight that might lead to technical innovation.

Oil

The oil sector is a different story. Economic tensions are already present. The country is importing more than two-thirds of its domestic fuel needs. Brazil's walls are losing productivity at a rate of about 3 percent a year, while the demand is increasing at 5 percent and prices are rising dramatically.

In the mid-1950s rigid legislation was passed establishing a government monopoly in prospecting, research, refining, and transportation of petroleum and its derivatives. Government investments have been massive in this sector; including in research. Today the Brazilian Petroleum Company (PETROBRAS) has the largest budget in the country. For more than ten years it has supported an intramural research center that now has a $10 million annual budget and more than 500 technicians. Yet not much oil is being produced. The nation is doing poorly in terms of oil prospecting, and the performance of the research center falls far short of expectations in spite of its financial resources. [55] Administrative rigidity seems a weak strategy to deal with technological problems under severe factor scarcity. Economic imbalances generated political tensions that finally induced the country to relax the monopolistic privileges of PETROBRAS in 1975 and to open prospecting to private companies. In this sense the oil industry is the exact opposite of the steel industry, where resource scarcity was met with administrative flexibility that permitted joint ventures to acquire and develop the needed technology. With oil, non-economic factors, such as national security, generated administrative rigidity that severely hampered the free search for technological solutions.

In most cases the participation of domestic agencies in the industrialization process was limited, as was pointed out, to operating and maintenance activities (except in steel, in which a little R&D was conducted by Brazilian technicians). Yet during these years of experience with industrialization, quite a few research institutions were created. Today, Brazil has about 50 institutions that devote part of their time to industrial technological studies. About 60 percent are under the direct administration of the federal government and 20 percent come under state government control. They employ more than 6,000 people with university and other technical backgrounds. Half of these research units are located in universities. Despite this relatively large number of industry-oriented institutions, their interface with industry is limited. About 20 percent of them have no contact at all with industry. Sixty percent limit their diffusion activity to publications. Only 10 percent have a deliberate policy of establishing closer relations with industry. [56] The most active institution is the Technological Research Institute of São Paulo. Its initial responsibilities, at the end of the nineteenth century, were mainly to provide technical assistance to the railway and electric-power companies. In the 1930s it responded to new needs by setting up other research programs and helping to create domestic technical centers for work in metallurgy, wood and paper, ceramics, chemical engineering, and many other fields. [57] The Institute was transformed into a public corporation in 1976 so as to become more flexible. It now has 450 researchers and a 12-million-dollar annual budget.

DEVELOPMENT OF THE SCIENTIFIC ESTABLISHMENT

Science in Brazil has developed only recently. Throughout the colonial period, the development of a solid scientific establishment was blocked by the economic, social, and cultural backwardness stemming from the Portuguese legacy. Portugal's own scientific establishment was weak and severely isolated from the mainstream of European science for several centuries. Even the diffusion of elementary education in Brazil was impeded by Portugal during the colonial period: in the early twentieth century more than 80 percent of the Brazilian population was illiterate. In addition, science was utterly unimportant to the economic and religious elites of colonial Brazil.

Occasionally during the seventeenth to nineteenth centuries, attempts to conduct scientific investigations in Brazil were made by foreign naturalists interested in gathering data on the fauna, flora, and natives, and on raw materials of possible commercial value. These "scientific explorations" were quite independent of Brazilian institutions. [58] Their works were published abroad and placed in foreign libraries. Consequently, a knowledge of several aspects of Brazilian society could be obtained only through foreign literature.

The move of the Portuguese royal family to Brazil gave a bit of encouragement to the development of scientific training. Two main concerns were the shortage of medical doctors and the lack of engineers for the army. These fields remained weak for 50 years, and their benefits were restricted to a small minority of the population. Serious epidemics at the end of the nineteenth century stimulated the medical schools to expand beyond professional training into research on tropical diseases. Engineering became more research-oriented with the creation of the Minas Gerais School of Mines and the São Paulo School of Engineering at that time.

Brazilian society continued to be predominantly agricultural, hierarchical, and patriarchical. Secondary education was a privilege enjoyed by the few, and the educational emphasis continued to be literary rather than scientific. Little value was placed on the pursuit of science for its own sake. [59] University training, limited to a very small minority, was obtained in Europe.

At the end of the nineteenth century, those who came to Brazil from Europe after being trained in the natural sciences began to urge the nation to set up its own scientific centers. The movement found support within the medical and engineering schools. Government or private support, however, was limited. Brazilian society recognized a special role neither for scientists nor for the institutions they would require.

The first moves to establish scientific research in Brazil came with the beginning of the Federal Republic (1889) and the first timid steps toward industrialization at the turn of the century. The national demands for agriculture and industrial production, as well as transportation and communications, generated a need for more educated workers. Changes in the secondary level of the educational system began to take place, especially in São Paulo. This was influenced by American urban values, which, contrary to the French tradition, stressed experimentation and practical work. * Also important was

---

*The leaders of the secondary-education reform—Caetano de Campos, Marcia Browne, Horacio Lane, and Anisio Teixeira, strongly influenced by John Dewey—had studied in the United States (1895-1915).

the role of the São Paulo Secondary Teaching School, where, in 1933, the first research-oriented institution for higher education was born: the College of Philosophy, Sciences, and Arts. Educational movements from 1890 to 1920 were marked by an urban and positivist orientation that clearly departed from the prevailing literary and legalistic emphasis that predominated in law schools.

The positivist orientation was associated with an increasing rural-urban movement. Rich farmers and new industrialists established their homes in the urban centers. From 1922 to 1930, the country witnessed the emergence of a small middle class. By 1930 the public came to support the expansion of both educational opportunities and scientific research. The initial legal basis for the organization of the Brazilian university system was approved in 1931. Brazil's first university, the University of São Paulo, was created in 1934 from the already existing College of Philosophy, Sciences, and Arts. The natural scientists' earlier attempts to institutionalize scientific research were successful. Perhaps even more important, the new university quickly attracted a number of brilliant European scientists who saw in it an opportunity to develop a vigorous center for research in the natural and social sciences. Interestingly, although American ideas lay behind the earliest phases in the emergence of the university centers (based on a secondary education system), the first Brazilian university was wholly European in tradition. [60] France, Germany, Austria, and Italy provided the initial talents for both the University of São Paulo (1935-45) and the other Brazilian universities that came into being shortly afterward.

In the universities, scientific units emerged and developed within a liberal, almost laissez-faire, environment in which the state took on faith the ability of European scientists to define and develop appropriate research problems. In other words, the state has generally supported a liberal and individualistic style of scientific work.[61] This is sometimes viewed as evidence of the state's indifference, rather than of its support. Though they were urged to concentrate on national needs, they had complete academic freedom to design research and training programs during those early years.

Under these conditions, the earlier generations of university scientists were mainly concerned with the general theoretical goals of their fields rather than with the application of scientific knowledge to production technologies. This raises the question of how to appraise the role of science in a developing society. To what extent should the scientific efforts be directed to the goals of the economic and social development of the country? The answer to this complex question involves yet another: Can science really be effective in a developing country?

Attempts to answer this question in the more advanced societies have shown that the impact of science on technology and economic development is recent and limited. Technological progress in industry and agriculture seems to have relied mostly on the accumulation of practical improvements, carried out by workers and farmers. Only recently has modern technology begun to rely on science; but even this dependence should not be exaggerated. [62] "The statement that rich countries have more science because they can afford more science may be as true as the statement that the rich countries are rich because of their sciences."[63]

Yet even if it is true that a satisfactory answer to the question requires a causal theory (which is not available in the social sciences), there is still enough known to permit a bit of preliminary thinking. First of all, scientific research as practiced in the universities does have an important role in training new generations. This function may be more crucial in developing societies than in advanced ones, given the precarious state of knowledge in the former. Second, a strong scientific establishment is necessary in order to understand transnational knowledge, both in science and in technology. The better the national science establishment, the faster a nation can put available knowledge to work. Finally, the exercise of free scientific thinking may be a good way to develop logical and critical perspectives on man and society. Thus, even admitting the relative inefficacy of science in generating technology and development, the scientific establishment may still be beneficial in a developing society.

The performance of the Brazilian scientific disciplines during their short existence has varied from field to field. The following paragraphs present a brief analysis of the contribution of the more established basic sciences.

Biological Sciences

Biological sciences developed in Brazil in close connection with the medical and agricultural sciences. Much research activity has been done within technological institutes devoted to health and agricultural problems. Although the field had its start with foreign help, several purely Brazilian efforts achieved key breakthroughs, especially in biochemistry. [64] Oswaldo Cruz did vital work on the control of several epidemic diseases; Carlos Chagas discovered Tripanosoma crucis, the vector of Chagas' disease (1909); and José Baeta Vianna (in Minas Gerais) contributed to the biochemical studies of goiter. There also were pioneering research efforts on snake venom by João Batista de Lacerda (1881), K. Slotta and Courant (1939), C. Klobustsky and A. Konig (1939), Otto G. Bier (1947), and Henriques-

Henriques (1958), and on the biochemistry of yellow fever,* beriberi, and Hansen's disease (G. G. Villela, 1928).

Contributions of biochemical research to crucial national problems is clear not only in medicine but also in agriculture (improvement of coffee, cocoa, soybeans, and tea) and in industry (food technology, vitamins, vaccines, enzymes, and other products).

Biochemistry courses are taught in medical, agricultural, and arts and sciences programs. In addition, the discipline has matured enough for the awarding of 40 master's degrees and 15 doctorates annually; most of the recipients are hired by the higher education institutions. Biochemistry today constitutes a small but relatively active scientific community in Brazil and generates a volume of research that has gained international recognition. [65]

Genetics also has developed rapidly, especially as applied to the agricultural problems of plant improvement, cytogenetics, genetics of microorganisms, quantitative methods, and animal improvement. † Like biochemistry, it has been oriented to Brazilian problems and has provided many useful results for corn, vegetables, and improved seeds in general.

The discipline has attained some autonomy. Introduced in the College of Philosophy, Sciences, and Arts by a French geneticist, André Dreyfus, in 1934, it was further strengthened by Theodore Dobzansky, from Columbia University, beginning in 1943. It is perhaps best known today for its contributions on chromosome physiology and genetic counseling. Today geneticists teach and conduct research in more than 20 university units in Brazil. Graduate programs in genetics now produce 60 new geneticists each year, half with master's degrees and half with doctorates. [66]

---

*The Yellow Fever Institute was founded in 1900. During the previous decade about 21,000 people died from the disease. This figure dropped to about 2,600 during the next decade, largely as a consequence of the institute's work.

†Genetics was first taught at the College of Agriculture of São Paulo in 1918 by A. Teixeira Mendes and Octavio Dominques, and later by the German geneticist Friedrich G. Brieger. It was introduced in the Agriculture Research Institute of Campinas by Carlos A. Krug, who completed his graduate work at Cornell University in 1932. Foreign professors played a decisive role in the formation of an influential group of Brazilian geneticists: Clodovaldo Pavan, Eduardo Brito da Cunha, Edmundo Magalhaes, Oswaldo Froto Pessoa, and many other professors who organized research and training programs throughout the country.

In the biological sciences it is important to stress the role of three applied-research institutes: the Biological Institute, the Butantan Institute and the Adolfo Lutz Institute. The Biological Institute was founded in 1927 as a response to a disastrous disease of the coffee tree (Hypothenemus lampei) that destroyed one-fifth of the state's billion coffee trees. Its first director, Arthur Neiva, had studied under Oswaldo Cruz. Under his direction the institute assumed several responsibilities for basic and applied research on plant protection, vaccine production, entomology, plant pathology, and bacteriology. [67]

In spite of its applied interests, the institute rapidly became an important center for basic research. In 1934, with the foundation of the University of São Paulo, the institute served as a training and research unit. Its most innovative contributions were on pests and diseases of coffee, citrus fruits, sugarcane, and bananas, and on animal diseases, particularly brucellosis, cattle tuberculosis, aphtous fever, hydrophobia, and poultry pathologies.

The Butantan Institute was founded at the end of the nineteenth century to produce vaccines against bubonic plague. [68] Its first director, Vital Brasil, was a physician. He devoted his entire life to the study and production of snakebite serum. The institute is now widely known for its contributions on bubonic plague, smallpox, tetanus, typhoid fever, diphtheria, tuberculosis, and snake venom. In the latter area Butantan has identified more than 70 new snake species from which vaccines have been produced for practically all of Latin America. Equally innovative was its research on the organic chemistry of different types of animal poisons, pharmacology, endocrinology, immunology, and the bacteriology of snakes and spiders.

In the area of public health, the São Paulo Bacteriology Institute and Food Chemistry Institutes—founded in 1892 under the indirect influence of Louis Pasteur and in 1940 merged into the Adolfo Lutz Institute—played an important role in the control of many epidemic diseases. [69] Due largely to its close relations with research centers in France and in the United States, the institute has made a number of discoveries concerning typhoid fever, smallpox, cholera, meningitis, encephalitis, and hepatitis. Today it works closely with the Emilio Ribas Epidemic Hospital and continues to focus on tropical diseases. It also provides training for students in colleges of medicine and public health schools.

To sum up, the emergence and development of biological sciences in Brazil involved a high degree of interaction with the production sectors and with social welfare. Biochemistry, genetics, and applied biology have produced only two generations of Brazilian scientists, but they are effective research workers who are quite responsive to the needs of the nation. These sciences improved the quality of life by helping the country cope with serious epidemic diseases and other public health problems.

Chemistry and Physics

Chemistry and physics have also attained a reasonable effectiveness both in training and in research, although their main emphasis has been on activities directed by foreign scientists at the College of Philosophy, Sciences, and Arts of São Paulo: in chemistry, Heinrich Rheinbolt and Heinrich Hauptman; in physics, Gleb Wataghin, Guido Beck, David Bohm, and Hans Stammreich.

In chemistry, the first generations of Brazilian chemists—Ernesto Giesbrect, Madalaine Perrier, Simao Mathias, Paschoal Senise, Blanka Wladislaw, Giuseppe Cilento, Geraldo Vicentini, Walter Borzani, Persio de Souza Santos, Giovanni Brunello—have managed to organize one of the strongest teaching and research programs in the country. Chemistry has an active graduate training program that has been graduating 50 new doctoral students annually and twice that many at the master's level. Although the discipline is mostly academic, various programs have started new lines of research on problems of production including petrochemistry, water and food chemistry, electrochemistry, Raman sprectroscopy, and monazitic sand studies. [70] The discipline is taught and researched in several university schools, but the most influential units remain in São Paulo and Rio de Janeiro.

Physics, like chemistry, made its first appearance as an applied science in the schools of engineering. [71] During the 1930s and 1940s foreign professors at the College of Philosophy, Sciences, and Arts and the São Paulo School of Engineering were important in this field. This pattern continued during World War II, especially with personnel from the United States; the emphasis was on experimental physics. This period was marked by a series of research advances, some of which were quite fundamental, such as the discovery and artificial production of the pi-meson by Cesar Lattes and his associates. After the war, theoretical physics was again emphasized and foreign physicists were involved: Marcelo Damy de Souza Santos (experimental physics), Mario Schenberg (theoretical physics), Cesar Lattes (nuclear physics), José Goldemberg (nuclear physics), José Leite Lopes (theoretical physics), Abrabão de Moraes (mathematical physics), Oscar Sala (van der Graaf accelerator). The close connection with the international community was further stressed with the organization of the Brazilian Center for Physical Research in 1949. With the financial support of the National Research Council, the center enjoyed the cooperation of many foreign physicists, including Richard P. Feynman, C. N. Yang, and J. Robert Oppenheimer, who came to Brazil, usually for short periods to participate in training programs. During 1950-70 the emphasis on experimental physics was reestablished, especially in nuclear physics, with the installation of several nuclear accelerators that permitted original advances in electromagnetic radiation research. In the 1970s the State University of Campinas was founded. Today it has the country's largest staff devoted to high-energy studies, solar energy, and theoretical physics.

Physics has a more modest graduate training program than does chemistry: it awards 20 doctorates and 40 master's degrees per year, mainly in São Paulo and Rio de Janeiro. As in chemistry, the importance of physics goes beyond the campus. The needs of industry are creating a rising demand for engineers, which in turn places a pressure on physicists not only for training activities but also to maintain complex equipment, to set quality-control standards, and to advise manufacturers of new instruments including, transistors and computers. By current estimates, the demand for well-trained physicists will be greater than the supply until the 1980s. [72]

To sum up, chemistry and physics have attained a reasonable degree of scientific maturity. They have sometimes been criticized as being too theoretical and devoted to problems that are of more interest to academic communities than to national needs. The heavy participation of foreign scientists is thought to be the prime cause of this situation. Yet the fact is that the fields are young in Brazil; visiting foreign scientists have helped to save many steps by training the first Brazilian cohorts. The country is still forming its critical mass who will, it is hoped, be able to contribute to basic knowledge and to the technological needs of production.

Other Natural Sciences

In spite of Brazil's vastness and its urgent need for energy and raw materials, the sciences related to the study and exploration of natural resources are much less developed than the basic sciences described above. In the area of mineral resources the country had an early start with the emphasis on iron and gold in the eighteenth century, but so far it has accomplished little in terms of increasing its capability to map and survey its geological resources. The geological peculiarities of Brazilian mineral deposits (zinc, manganese, uranium, thorium, and iron) limit the usefulness of imported know-how and demand a great deal of locally oriented research. [73]

Equally important is the exploitation of water resources for agricultural and industrial purposes. Brazil has a large coast and great rivers that provide a variety of opportunities for new material and animal resources. In the area of sea and river resources, however, Brazil is just beginning its activities. Oceanography is a new science in the country and much knowledge is still required, for instance, on fishing, including the identification of new species, and investigation of deep sea fishing and freezing and storage. Biological productivity of the sea and rivers is an area still to be explored.

Regarding energy resources, Brazil is also paying a price for its small investment in science. The rapid economic development of the country is creating an increasing demand for all kinds of energy: hydroelectric power, oil, natural gas, coal, and wood, as well as nuclear and solar energy. The country's dependence on imported energy is increasing dramatically. A recent study shows that Brazil had increased its total energy imports from 17 percent in 1960 to 27 percent in 1970, with estimates of more than 50 percent by 1985. [74] Consequently, domestic production of commercial energy will not be sufficient to meet the expanding demand in the near future.

A reasonable effort has been made in the area of hydroelectric power: Brazil had a fairly large program in hydraulic engineering at the São Paulo Engineering School and has managed to adapt and develop new engineering techniques in building large electric power plants. However, in petroleum, natural gas, and coal the country has accomplished little in spite of long-term work and heavy research support directly given to the Brazilian Petroleum Public Company. For both these conventional and for nonconventional (nuclear) energy sources, Brazil is still dependent on foreign know-how. The Brazilian-German agreement on nuclear energy anticipates the building of eight large nuclear plants (10 million kilowatts each) and the establishment of the technological infrastructure for further plants. Programs of research and education of technical personnel are being developed by the nuclear research institutes of São Paulo, Rio de Janeiro, and Minas Gerais; and the State University of Campinas is concentrating efforts on solar and helium energy. [75]

Social Sciences

The study of human behavior has also been severely affected by intermittant support and lack of systematic empirical research. However, there are two fields in which the emphasis on applied research has encouraged the development of social technology. Economics and sociology, although modestly developed, have played important roles in shaping many of the country's policies since the 1950s.

Modern economics began in the 1950s with a research effort initiated at the University of São Paulo and the Getúlio Vargas Foundation in Rio de Janeiro. In the mid-1960s schools of economics in other states invested a great deal in graduate training, at home and abroad, thereby providing a new generation of economists who have been quite influential in formulating and administering the economic policies of the country. Involvement with the policy-making process has yielded a great deal of applied and some basic research.

Sociology was brought to Brazil primarily from Europe. At first it was the province of individual scholars such as Oliveira Viana and Gilberto Freyre, both of whom played pioneering roles early in this century.[76] However, as an institutionally organized science, sociology gained importance only later at the University of São Paulo and the School of Sociology and Political Science. Again foreign professors were crucial for the establishment of the discipline, emphasizing studies of racial prejudice, class struggle, sociology of education, and rural sociology.* Rural sociology is becoming an important specialization in the country. Sociological knowledge is systematically applied to introduce a series of changes in order to increase the efficiency of agricultural extension systems, agricultural research organizations, colonization, and irrigation programs. In this sense, as with economics, sociology has been used in Brazil as a social technology, although a series of political discontinuities interrupted this emerging pattern in the mid-1960s.

To sum up, science is new in Brazil, having started after 1930. The diversification of the production system, the beginning of industrialization, the urban concentration, and the relative opening of the Brazilian social structure were the basic ingredients in the emergence of university education in the country.

Until 1930 the great bulk of university education was limited to the law and literature; professional training was emphasized and practically no attention was given to research. After 1930 the first initiatives for science and "uncommitted research" helped to form a new intellectual class.

The opening of the first universities after 1930 was initiated by the state, but within a fairly liberal atmosphere in which the organization of science was entrusted to scientists themselves. The period 1934-45 marked a phase in which the nation made decisive investments in higher education, including importation of foreign personnel, establishment of a few active research units, and the development of new courses on basic sciences. The next 30 years were dedicated to the consolidation of the few more advanced universities and to a rapid expansion of the higher-education system. This expansion—more than 900 percent increase in enrollment—has been largely to meet the intense social pressure for higher education in the country. On the whole, graduate training and research have been less developed, and

---

*Roger Bastide, Claude Lévi-Strauss, Paul Arbousse-Bastide, Donald Pierson, Horace Davis, and T. Lynn Smith provided the intellectual support for influential Brazilian scholars, especially Fernando de Azevedo and Florestan Fernandes.

quality has not accompanied the booming expansion of the university system. Yet certain well-consolidated research groups have already begun to make basic contributions to scientific knowledge.

CONCLUSION

Brazil has shown the fastest rate of growth in Latin America in recent decades. Particularly since the end of World War II the country has maintained an average rate of growth of 7 percent annually, a relatively high figure by international standards. Industry has grown at a 9 percent rate, while agriculture has stayed below 5 percent. The per capita benefits of this growth have been limited by the nation's rapid rate of population growth. Even so, it has grown from $300 to $800 since World War II. [77]

Social indicators also display reasonable levels of improvement, although the rates of social development are far more modest than economic growth. Health, education, and employment have gradually progressed since the war. General mortality declined from 20 to 9 per 1,000. Infant mortality was reduced from 202 to 85 per 1,000 live births, and life expectancy climbed from 43 to 61 years. [78] The percentage of children 7-14 years old in school went from 35 percent to nearly 85 percent; opportunities for secondary education rose by 300 per cent during the period; and university education expanded by more than 1,000 percent. [79] Employment opportunities were generated to the extent that more than 95 percent of the population of working age has a job, although underemployment has remained high. More than two-thirds of the jobs are in industry and services. [80] The distribution of personal income is a different story. Rapidly increasing numbers of jobs for a small proportion of highly educated workers is a contributing, though possibly a transitional, cause of the increasing trend of income concentration. [81]

There are estimates that the investment in primary and other levels of education has been responsible for more than 30 percent of the growth accomplished. [82] This means that about one-third of the annual rate of growth is due to the nation's maintaining and improving the quality of manpower. In addition to that factor, the population increased during the period, contributing another 17 percent to its economic growth. Therefore, manpower factors together explained 47 percent of recent growth. Investment in physical capital contributed 32 percent. The residual, including measurement errors and technological progress, had an impact of 21 percent. [83]

This aggregative approach to analyzing the net effect of different factors on economic growth provides a reasonable overall view of the process, but gives a poor picture of the qualitative impacts of science and technology in the development of the production sector, in particular, and in the socioeconomic development process, in general. This paper has examined, in a more specific manner, the role of technological development in Brazilian agriculture and industry, as well as the expansion and contribution of science to the progress of technical knowledge and the general welfare.

As has been shown, Brazilian science and technology illustrate a typical case of colonial development, marked by historical dependence in terms of technology. The recent acceleration in the growth of science has so far been unable to provide the advances needed for production technology. Technology for industrial development, and to some extent agriculture, was largely imported. Industrial technology has been mainly imported in ready-made form, although some of it was modified to meet the country's needs. For export products, agricultural innovations have been demanded, and on the whole provided, but not in the domestic sector. Basic science developed more autonomously and is more directly linked with the transnational scientific community.

The intense borrowing and the laissez-faire style of scientific development have restricted the development of large national "critical masses" for science and technology. Yet technology transfer has contributed greatly to the improvement of Brazil's industry and agriculture, and indirectly to the establishment of new schools and research centers.

The development of the national scientific and technological capability has also been inhibited by a weak exchange network of personnel, ideas, and resources among the governmental, educational, and production sectors. [84] The technological infrastructure has rarely been strong enough to function productively and to establish close institutional links with the production sectors and the political and administrative structures. The scientific infrastructure, on the other hand, has generally concentrated its attention on training university students; consequently, few have been engaged in research on problems relevant to the country's development. [85]

In the government, especially in civil service, the low salaries have not been attractive to outstanding scientists. On the production side, given the country's historical permissiveness, multinational corporations dominate the market, particularly in industry. They usually prefer to carry out their R&D activities abroad.

Altogether, these factors make for very fragile links between local personnel, ideas, and resources; as a consequence, they tend to restrict the development of national capabilities. [86] If one believes, with Graham Jones, that "a satisfactory strategy for technical progress must involve an optimum combination of assimilating scientific and technological advances from other countries, with an indigenous organization for research, development, innovation, and diffusion,"[87] Brazil seems to have succeeded in performing the former role but has a long way to go in terms of generating its domestic capability.

However, Brazil's modest domestic research establishment has made a few contributions, some of which have been highlighted here. They include the development of biochemical innovations related to agricultural export commodities, and innumerable adaptations in the industrial sectors, especially in steel, construction, and transportation. Biochemistry and applied biology have also made important contributions to the control of critical epidemic diseases.

One must recognize, however, that the adverse socioeconomic and political conditions prevailing in Brazil for centuries cannot be fully overcome in a few decades. Scientific and technological establishments are nascent and still small, given the size and heterogeneity of the country. Investments in R&D are still limited, and probably could not be much greater, given the relative scarcity of university-educated personnel. Recent expansion of the university system has not been followed by substantial improvements in graduate training and sound research. [88] This means that for two or three decades the country will be concentrating on the raw material for more effective R&D activities: the preparation of top-quality personnel at the undergraduate and graduate levels.

Explicit support for scientific and technological development in Brazil is also a recent phenomenon. Since the 1950s the country has had six development plans, but most of them were quite timid in formulating national policies for science and technology. The first specific reference to the issue came with the Strategic Development Program for 1968-70, which devoted two chapters to science and technology and stressed that the general import-substitution strategy should be followed by a technology-substitution process. In this regard the Basic Plan for Scientific and Technological Research was formulated, and a series of centralizing mechanisms was created to command the allocation of the federal funds. However, the high speed of the country's development did not permit waiting for the slow action of a heavy and centralized bureaucratic apparatus. Very few of the plans became concrete accomplishments.

Science and technology also received specific attention in the next two national development plans. For 1972-74 period the First Basic Plan for Scientific and Technological Development was drawn up with a heavy emphasis on such sophisticated goals as nuclear energy, space research, and technology-intensive industries, as well as on infrastructure technology and agricultural research. The plan provided a reasonable budget that, however, was never spent for lack of organized research institutions and an excess of governmental centralism. For 1975-79 the country is pursuing its Second Basic Plan, approved by the federal government early in 1976. This plan provides a much more explicit policy for technological transfer and for the creation of national capabilities, especially in agriculture and natural resources; it also emphasizes environmental research, energy, new technologies, and human resources. The National Research Council was made the coordinator of the entire scientific and technological system. Unquestionably, the Second Basic Plan is a long list of relevant prospects for scientific development.

If one makes a comparative analysis, there is no doubt that Brazil has improved substantially in terms of plan-making exercises. Financial resources, however, are still modest: $3 billion for 1975-77. Also, reality is teaching Brazilians that there is not yet a "critical mass" available to absorb even those limited resources. The country lacks researchers, both in quality and in quantity. The plan recognizes this fact, and has allocated 26 percent of its funds to the development of human resources. This means that before 2000 the nation's ambitious plans may become realities.

Brazil has established a variety of financial mechanisms such as the National Fund for Scientific and Technological Development, many research financial agencies such as the Financing Agency for Project Development and Fellowship and Grant Agency, and several state research foundations. Still in short supply are capable personnel, flexible research institutions, and efficient mechanisms to articulate government efforts in educational agencies and the production sector.

The most ambitious current attempt to coordinate research and economic development is sponsored by the São Paulo State Council for Science and Technology, in which three mechanisms facilitate interaction among key components of technological development. First, research institutions are becoming more flexible and have been encouraged to operate under research contracts in the industrial and agricultural sectors. [89] Second, a special financial mechanism supports these sectors so that they may contract research with local agencies at subsidized rates. Finally, an external technical assistance, supported by a U. S. Agency for International Development loan, is reinforcing the local research agencies in terms of additional qualified personnel, transfer of new technologies, and support of joint-venture research projects. This model has already proved to be sufficiently flexible and decentralized to bring together the consumers and the producers interested in technical knowledge.

Such social innovations seem to be more urgent than the simple increase in available funds. The organization of an articulated system or subsystem for science and technology in Brazil is much more complex than the solution of a mere technical problem. The operation of a subsystem always involves questions related to the organizational capacity of the society, the changing of many economic and political factors, and the transformation of other subsystems such as education, communications, and some of the social and cultural traditions. Historically, these changes have occurred spontaneously in many societies; however, this does not mean that they cannot be deliberately induced.

The inducement of desired changes requires an optimum combination of a few centralized activities with myriad decentralized mechanisms through which consumers and producers of technical and scientific knowledge can exchange their needs. The flexibility of the subsystem is the key guarantor of successful development. For example, there are areas in which additional research would be more useful than strengthening the supporting service activities or raising the general level of productive competence.[90] Under other conditions, concentration on key research topics is crucial to take advantage of the country's resources. Yet, in other situations, transfer of ready-made knowledge is the best choice for the nation. For this reason a technical-scientific subsystem must be sufficiently flexible to be able to assess needs and obtain feedback from all involved actors: farmers, industrialists, consumers, researchers, research organizations, financial agencies, universities, and the government. Developing nations tend to require more intervention than the developed ones in order to harmonize this complex orchestra; however, they must avoid moving from a chaotic laissez-faire state to a stagnant centralized establishment.

Future possibilities for an efficient technological and scientific subsystem in Brazil depend on the nation's ability and willingness to provide a rapidly decentralized and financially secure system for encouraging the application of considerable amounts of trained talent to problems of the economy, including those dictated by the domestic market, and to basic research problems dictated by theoretical questions.

NOTES

1. George Basalia, "The Spread of Western Science," Science 156 (1967): 611-22.

2. Helio Jaguaribe, "Ciencia e technologia no quadro socio-político da América Latina" (Rio de Janeiro: Conjunto Universitario Candido Mendes, 1974). (Mimeographed.)

3. P. Deane, The Industrial Revolution in England: 1700-1914 (London: Collins Clear-Type Press, 1969).

4. Caio Prado, Jr., Historia económica do Brasil (São Paulo: Editora Brasiliense, 1967).

5. M. Luiza Marcilio, "Crescimento demográfico e evolução agraria paulista: 1700-1836" (São Paulo: Tesa de Livre Docencia, 1974). (Mimeographed.)

6. José Pinto Lima, Técnicos para o desenvolvimento da agricultura (Rio de Janeiro: ABCAR, 1961).

7. Roberto C. Simonsen, Aspectos da historia econômica do café (São Paulo: Companhia Editora Nacional, 1940).

8. José Pastore and Eliseu Alves, "Uma nova abordagem para a pesquisa agricola no Brasil" São Paulo: FIPE/USP, 1975). (Mimeographed.)

9. Instituto Agronómico de Campinas, "Informaçoes sobre o instituto agronómico, 1887-1972" (Campinas: IAC, 1975). (Mimeographed.)

10. Augusto Fragoso, "Os engenheiros militares no Brasil," Revista militar brasileira 85, no. 3 (June-September): 7-34.

11. João Baptista Magalhaes, A evolução militar no Brasil (Rio de Janeiro: Biblioteca do Exército Editora, 1958).

12. For a detailed historical analysis of the school of mines, see A escola de minas: 1876-1966 (Ouro Preto: Escola Federal de Minas de Ouro Preto, 1966).

13. Odilon Nogueira de Matos, Café e ferrovias (São Paulo: Editora Alfa Omega, 1974).

14. Escola Politécnica, Anuario (São Paulo: Universidade de São Paulo, 1946).

15. Alexandre d'Alessandro, "A escola politécnica de São Paulo: A Historia de sua historia," Revista dos tribunais 3 (1943).

16. João Luis Meiller and Francisco I. A. Silva, Meio século de tecnologia (1899-1949) (São Paulo: Instituto de Pesquisas Tecnológcas, 1949).

17. Irineu Evangelista Souza, Autobiografia (Rio de Janeiro: Edicoes de Ouro, 1964).

18. Nicia Vilela Luz, A Lutz Pela industrialização no Brasil (São Paulo: Editora Alfa Omega, 1975).

19. Anibal Villela and Wilson Suzigan, Política do governo e crescimento da economia brasileira: 1889-1945 (Rio de Janeiro: Instituto de Pesquisas Econômicas e Sociais, 1975).

20. Instituto de Pesquisas Econômicas e Sociais, Pesquisa Tecnológica no Brasil: Analise de cinco institutos oficais (Rio de Janeiro: IPES, 1971).

21. Nancy Stepan, "Emerging Role of the Scientific Establishment in Brazil" (New Haven: Yale University, 1973). (Mimeographed.)

22. Yujiro Hayami and Vernon Ruttan, Agricultural Development: An International Perspective (Baltimore: Johns Hopkins Press, 1971).

23. Simonsen, op. cit.

24. Affonso E. Taunay, Historia do café no Brasil (Rio de Janeiro: Departamento Nacional do Café, 1939).

25. Alcides Carvalho, "Café" (Campinas: Instituto Agronómico de Campinas, 1974). (Mimeographed.)

26. Ibid.

27. Patricia D. Ferreira, Lucila R. Brioschi, and Maria Luiza P. S. Mendes, "Desempenho da pesquisa agrícola no Brasil—produtos de exportação vs. produtos domésticos" (São Paulo: FIPE/USP, 1976). (Mimeographed.)

28. Alice P. Canabrava, "A grande propriedade rural," in Sergio B. Holanda, ed. , Historia de civilização brasileira, II (São Paulo: Difusão Europeia do Livro, 1960): 423-78.

29. Oriovaldo Queda, "A intervenção do estado no industria acucareira" (Piracicaba: ESALQ, 1972). (Mimeographed.)

30. Instituto Brasileiro de Potassa, Cultura e adubação do algodoeiro (São Paulo: IBP, 1965).

31. Harry W. Ayer and G. Edward Schuh, "Social Rates of Return and Other Aspects of Agriculture Research: The Case of Cotton Research in São Paulo, Brazil," American Journal of Agricultural Economics 54 (1972): 557-69.

32. Ibid.

33. SUPLAN, Subsecretaria de Planejamento e Orcamento Milho, Arroz e Feijão, serie subsidios para o II PND (Brasília: Ministerio da Agricultura, 1974).

34. Antonio Bayma, Arroz (Rio de Janeiro: Serviço de Informação Agrícolas, 1961).

35. Instituto Agronómico de Campinas, Melhoramento do arroz no Instituto Agronómico (Campinas: IAC, 1972).

36. Universidade Federal de Vicosa, Anais do I simposio brasileiro do feijão (Vicosa: Imprensa Universitaria, 1972).

37. Armando Conagin and B. B. Junqueira, "O milho no Brasil," in Cultura e adubação de milho (São Paulo: Instituto Brasileiro de Potassa, 1966).

38. SUPLAN, Produção e abastecimento/Perspectivas e proposiçoes, 1975/76, milho (Brasília: Ministerio da Agricultura, 1975); and Subsidios aos planos anuais de produção e abastecimento—milho (Brasília: Ministerio da Agricultura, 1975).

39. SUPLAN, Produção e abastecimento/Perspectivas e proposiçoes, 1975/76, arroz (Brasília: Ministerio da Agricultura, 1975).

40. Werner Baer, Industrialization and Economic Development in Brazil (New York: Irwin, 1965).

41. Alfredo Ellis, Jr. , A evolução da economia paulista e suas causas (São Paulo: Companhia Editora Nacional, 1937).

42. Joel Bergsman, Brazil: Industrialization and Trade Policies (London: Oxford University Press, 1970).

43. Ibid.

44. Nathaniel H. Leff, The Brazilian Capital Goods Industry: 1929-1964 (Cambridge, Mass.: Harvard University Press, 1968).

45. Ibid., p. 17.

46. Bergsman, op. cit., p. 180.

47. Francisco A. Biato, "Política tecnologica e política industrial," Revista de administracão de empresas 14 (1974): 61-66.

48. Samuel A. Morley and Gordon W. Smith, "Managerial Discretion and the Choice of Technology by Multinational Firms in Brazil" (Houston: Rice University, Program of Development Studies, 1974). (Mimeographed.)

49. Francisco A. Biato, Eduardo A. Guimaraes, and Maria Helena P. Figueiredo, Potencial de pesquisa tecnólogica no Brasil (Rio de Janeiro: Instituto de Pesquisas Económicas e Sociais, 1971).

50. Sergio Francisco Alves and Ecila M. Ford, "O comportamento tecnológico das empresas estatais" (Rio de Janeiro: FINEP, 1975).

51. Nuno Fidelino de Figueiredo, A transferencia de tecnologia no desenvolvimento industrial do Brasil (Rio de Janeiro: Instituto de Pesquisas Económicas e Sociais, 1972).

52. Bruno Leuschner, Transferencia de tecnologia na industria siderurgica (São Paulo: IPE, 1971).

53. Instituto Brasileiro de Siderurgia, "Subsidios para a definição da política siderurgica brasileina" (Rio de Janeiro: IBS, 1975). (Mimeographed.)

54. Werner Baer, The Development of the Brazilian Steel Industry (Nashville, Tenn.: Vanderbilt University Press, 1969).

55. Financiadora de Estudos e Projectos, Pesquisa tecnológica em empresas estatais (Rio de Janeiro: FINEP, 1975).

56. Biato et al., op. cit.

57. Meiller and Silva, op. cit.

58. Rodolfo Garcia, "Historia das exploraçoes cientfficas," in Dicionário histórico geográfico e etnográphico do Brasil (Rio de Janeiro: Imprensa Nacional, 1922).

59. Stepan, op. cit.

60. Fernando de Azevedo, ed., As ciencias no Brasil (São Paulo: Editora Melhoramentos, 1955).

61. Vanya M. Sant'Anna, "Ciencia e sociedade no Brasil" (São Paulo: Universidade de São Paulo, tese de Mestrado, 1974). (Mimeographed.)

62. Nathan Rosenberg, "The Role of Science and Technology in the National Development of the United States," ch. 4 of this book.

63. Richard Nelson, "Less Developed Countries—Technology Transfer and Adaptation: The Role of the Indigenous Science Community," Economic Development and Cultural Change 23 (1974): 61-78.

64. Leal Prado, "A evolução da bioquímica no Brasil," O Estado de São Paulo, Suplemento do Centenario no. 15, April 12, 1975.

65. Ibid.

66. Warwick Kerr, "Historia da genética no Brasil," O Estado de São Paulo, Suplemento do Centenario no. 45, November 8, 1975.

67. José Reis, "Rocha lima, a obra e o homen," O Estado de São Paulo, Suplemento do Centenario no. 27, July 5, 1975.

68. Flavio da Fonseca, "Instituto Butanta: Orígem, desenvolvimento e contribuição," O Estado de São Paulo, Suplemento do Centenario no. 29, July 19, 1975.

69. Instituto Adolpho Lutz, Revista Commentario do Instituto Adolpho Lutz 1 (1941): 1-40.

70. Simão Mathias, "Cem anos de química no Brasil," O Estado de São Paulo, Suplemento do Centenario no. 6, February 8, 1975.

71. José Goldemberg, "Cem anos de física no Brasil," O Estado de São Paulo, Suplemento do Centenario no. 10, March 8, 1975.

72. Ibid.

73. DNPM, Anuario mineral brasileiro: Departamento Nacional da Produção Mineral (Brasília: Ministerio das Minas e Energia, 1974).

74. Elias Paladino, "Uma política de energia para o Brasil" (São Paulo: Universidade de São Paulo, 1976). (Mimeographed.)

75. Sociedade Brasileira de Pesqisas Físicas, "Recomendações para o ensino da física no Brasil" (Rio de Janeiro: SBPF, 1975). (Mimeographed.)

76. Florestan Fernandes, A etnologia e a sociologia no Brasil (São Paulo: Editora Anhembi, 1958).

77. Carlos Geraldo Langoni, A economia da transformação (Rio de Janeiro: Livraria José Olympio, 1975).

78. João Yunes and Vera S. C. Rochenzel, "Evolução do mortalidade geral, infantil e proporcional no Brasil," Revista brasileira de saude pública, special issue (1974); Instituto de Pesquisas Econômicas e Sociais, "A situação da saude no Brasil" (Brasília: IPES, 1975). (Mimeographed.)

79. CNRH, "Diagnóstico social do Brasil" (Brasília: Instituto de Pesquisas Econômicas e Sociais, 1975). (Mimeographed.)

80. Conselho de Desenvolvimento Social, "Brasil: Indicadores sociais: 1960-1970 (Brasilia: CDS, 1975). (Mimeographed.)

81. Carlos Geraldo Langoni, Distribuição de renda e desenvolvimento no Brasil (Rio de Janeiro: Editora Expressão e Cultura, 1973).

82. Carlos Geraldo Langoni, As causas do crescimento econômico no Brasil (Rio de Janeiro: APEC Editora, 1973).

83. Langoni, A economia de transformação.

84. Diana Crane, "An Interorganizational Approach to the Development of Indigenous Technological Capabilities" (Paris: OECD, 1974). (Mimeographed.)

85. James P. Blackledge, "The University as an Adaptor of Technology in a Developing Country," unpublished paper (Denver: Denver Research Institute, 1972).

86. Crane, op. cit.

87. Graham Jones, The Role of Science and Technology in Developing Countries (London: Oxford University Press, 1971).

88. José Pastore, O ensino superior em São Paulo (São Paulo: Companhia Editora Nacional, 1972).

89. José Pastore, "Alguns principios para a modernização tecnológica no Brasil," Revista de administração de empresas 14 (1974): 67-71.

90. Jones, op. cit.

8

THE ROLE OF SCIENCE AND TECHNOLOGY IN THE ECONOMIC DEVELOPMENT OF GHANA

Edward S. Ayensu

A modest scientific and technological development holds the key and the most promise for the future of Ghana. My evaluation is based on the conviction that Ghana needs to engage in judicious economic planning by exploiting the best of its human and natural resources. In addition, the country should make intelligent use of its limited financial resources. While no one can deny the advantages of obtaining external help from Ghana's friends, the burden of fulfilling the country's economic and social destiny can be sustained only by Ghanaians. But above all, the country must have political direction to enable the government and people to focus on the kind of economic development that will be in the national interest.

The person who has had the most profound influence on the overall development of Ghana was the late President Kwame Nkrumah. While one may criticize him and his government for many errors in economic judgment and political activities, it is only fitting that his era should form the reference point in any discussion of the development of Ghana. It was certainly Nkrumah's foresight and leadership ability that aided Ghana in breaking away from colonial rule. It was he who foresaw the significance of science and technology in the economic development of Ghana. This essay, therefore, will examine Ghana in the period before independence was attained, the Nkrumah era, and the post-Nkrumah era. * The short- and long-term roles that

---

*In order that the Ghana situation may be clearly understood, socio-political background is provided throughout this paper to explain the overall general pattern of the country's economic development.

science and technology have played in the economic and social development of Ghana will be analyzed for each period.

HISTORICAL PERSPECTIVE

Before the Europeans arrived in the Gold Coast (Ghana), a good measure of scientific and technological skills was in evidence in the daily activities of the people. Archaeological evidence indicates that the natural resources of the Gold Coast were in constant use. As in most West African countries, stone tools manufactured by the earliest hunters and farmers have been discovered. Metalwork of all kinds has been found, especially in gold, with which Ghana is well endowed. Unfortunately, because a written history is lacking, the full extent of the scientific and technological practices of the prehistoric people cannot be known. The Gold Coast's scientific and technological history has been recorded only since the beginning of European contact some 500 years ago.

It was not until 1901, when the country was designated a colony, that the British government decided that its economic and social development was to be handled by the colonial government in the Gold Coast. The first major preoccupation of the Gold Coast government was to find funds to expand and improve the existing infrastructure of roads, railways and communications. At first the government thought that funds could be obtained from the gold-mining industry, since the early outlook of this industry seemed prosperous. More than 20 companies operated in the administrative divisions of the colony and Asante. Unfortunately, the large financial returns expected from the gold-mining companies did not materialize and the government turned its attention to agriculture, the next obvious money-making activity to finance the development of the Gold Coast.

As a first step, eight botanical and agricultural stations were established by the colonial government to engage in research on cocoa, rubber, cotton, coconuts, and other cash crops. The major botanical garden, established at Aburi in 1890, was already in use as an agricultural experiment station. The government tried to encourage foreign companies, such as the British Cotton Association and F & A Swanzy, Ltd., to establish plantations of various cash crops; but, because of the failure of an earlier cotton plantation scheme organized by the government, the companies were not enthusiastic about funding plantation projects.

The point to be noted is that it was left to the ordinary Gold Coast farmer to develop agriculture in the country. The farmers took up the challenge and started to cultivate cotton, coffee, cola, rubber, and oil palms. It was soon realized that these commodities did not yield enough funds, and the farmers then turned their efforts to the cultivation of cocoa. The export of the first 80 pounds of cocoa to England in 1891 brought the government a total of £4. As is evident even today, this crop's impact on the overall development of the country has been striking. The impact of the cocoa industry on economic development had started when the Basel missionaries, who were based in Akropong, began experimenting with cocoa seedlings and distributing them to their stations in Aburi, Krobo Odumase, and Mampong. Their operation was short-lived, however, because of the Asante invasion of the Lower Volta states between 1869 and 1871. The lasting interest in cocoa cultivation was engendered by the introduction in 1878 of cocoa pods from Fernando Po to the Gold Coast by Tetteh Quashie. Quashie succeeded in growing the crop, and sold his harvest to farmers in Akwapim to start their own plantations.

Further impetus to the cocoa industry was given by Sir William B. Griffith, governor of the Gold Coast from 1880 to 1896. He imported cocoa pods from São Tomé in 1886 and established the Aburi nursery, which supplied cocoa seedlings to all farmers interested in its cultivation. Sir William also encouraged the publication of scientific information on the botany and cultivation of cocoa for the guidance of the farmers who wanted to take the industry seriously. By 1907 the cocoa industry was booming and, as noted by F. Agbodeka:

> The amount of energy and enterprise displayed in this
> work all over the country was amazing; and all eyewit-
> nesses including the Curator of Botanical Gardens and
> officials from D. Cs. [district commissioners] right up
> to the Governor himself could not but admire the Gold
> Coast farmers who, single-handed, brought about this
> fantastic agricultural revolution. [1]

The kinds of technology imported into the Gold Coast during the colonial era were principally those designed for extractive activities: simple farm implements such as hoes, cutlasses, and the odd tractor; fishing nets, fishing hooks, lines, and sinkers; some gold-mining equipment; and a few bulldozers and timber trucks. These basic types of equipment were used in conjunction with the large available body of unskilled labor and the relatively small amount of skilled labor produced by the colonial education system.

The scientific and technological contributions made by the colonial government were directed toward the increased production of raw materials for export; there was almost no development of an integrated economic strategy focused on improving the lot of the country. The total contribution of science and technology in the economic growth of the Gold Coast was very modest indeed. One gets the impression from the overall performance of the country that the colonial government was interested in contributing the minimal amount of science and technology for maximal financial gains. We should not, however, ignore the fact that the establishment of botanical gardens and specialized nurseries with research facilities was an important long-range contribution to the economy of the country, particularly with respect to the selected export crops.

The colonial government was not particularly interested in promoting industrialization in the Gold Coast. If this had been the intention, one would have expected it to develop and diversify the agricultural program of the country, so that the economy would not be dependent on one or two crops. Furthermore, it seems that for Britain to obtain maximum benefit from all its colonies, it was much more economical to encourage each colony to produce the commodities that had the greatest promise, rather than to encourage diversification. This attitude is still held by many underdeveloped and developed countries.

One major difference between the colonial period and the present is that during the colonial days the world prices of commodities were essentially controlled and influenced in a way that brought about gradual but steady economic growth for the colonies, since it was in the interest of the colonial government to maintain a reasonable state of economic health. For example, since foreign trade figures began to be recorded in 1856, Ghana always maintained a healthy balance of payments by importing less and exporting more. In 1856 Ghana imported goods amounting to £105,634. In the same year it exported goods totaling £180,359. In 1870 it imported £253,398 worth of goods and exported goods amounting to £510,015. Similar relationships obtained until after independence, when imports began to outstrip exports and the Ghanaian economy began to be dependent upon the outside world. For example, in 1961 exports of goods and nonfactor services made up almost 29 percent, while the value of imports was equivalent to 34 percent, of the gross domestic product (GDP).

Another major concern of the colonial government was to train people to become civil servants who could help it administer the country. The person most instrumental in the advancement of good education was Sir Gordon Guggisberg, governor of the Gold Coast from 1919 to 1927. In 1925 he presented his educational plan to the Gold Coast Legislative Assembly. Among his major aims were thoroughness in primary education, the acquisition of the highest-quality teaching staff, equal educational opportunities for boys and girls, and the desirability of a coeducational structure. While English had to be taught in all schools, he also emphasized the need for a solid base of the various vernaculars spoken in the country. Guggisberg also requested that a sufficient staff of African school inspectors be trained and maintained to oversee the primary and secondary schools. He further suggested that trade schools be provided to offer a technical and literary education that would fit young men to become skilled craftsmen and useful citizens. While most of these ideas were implemented, it was clear that science education (mainly nature study and hygiene) was principally limited to subjects of theoretical interest.

It should be borne in mind that as far back as 1841 the British educational system that had taken root in West Africa stressed the basic skills in reading and writing as well as the elements of arithmetic, the English language, and geography. However, the main emphasis of the school curriculum was on religious instruction. Of the 42 schools established in the West African colonies, 28 were missionary schools and 14 were local government establishments. The educational program of the period was summarized by Eric Ashby as follows:

> But the main emphasis was on religious instruction to the exclusion, the inspector reported, of every other species of learning. There had been attempts to vary the pattern on the part of individual governors and missionaries. A Mr. Weeks of the CMS [Church Missionary Society] had contrived to combine a system of industrial training with the current type of scriptural education. Boys at his school learned trades and were taught to cultivate cotton and coffee; girls were instructed in housecraft and needlework. H. D. Campbell had admired the system and established a series of government schools on similar lines during his governorship in the thirties. But these experiments had largely disappeared with their authors. [2]

Governor Guggisberg's educational policies continued to develop, however, after his term ended in 1927. The major setback that education encountered was in the 1930s, when the economic depression drastically curtailed the financial resources needed to further education. The value of cocoa, which had developed rapidly as the major foreign-exchange earner for the Gold Coast, decreased from £50 per ton in 1929 to £20 per ton in 1930. Nevertheless, interest in education had already been generated to the extent that in the face of these financial stringencies, an engineering school was opened at Achimota College in 1931. It was at this school that the first 17 Gold Coast engineers were trained.

With the economic slump, technical education in the country suffered greatly. Although the University College of the Gold Coast (now the University of Ghana) was established in 1948, its academic activities tended to be more theoretical than technical. No substantial improvement in technical education was made until 1951, when the

Kumasi College of Technology (now the University of Science and Technology) was established. The year 1951 is significant in the history of education in the Gold Coast. It was the year in which Prime Minister Kwame Nkrumah's Convention People's Party came into power and established the Accelerated Development Plan for Education.

During the colonial era the economy of the Gold Coast was in the hands of foreign companies. After it became obvious that the local farmers were doing well for themselves, the colonial government, in collaboration with the foreign companies, planned, for all intents and purposes, to obtain complete control over the agricultural industries, particularly cocoa. The government announced its intention to introduce legislation requesting the improvement of the quality of cocoa, and to provide inspectors to ensure adherence to "proper" farming practices. It also announced plans to establish fixed produce markets and to license produce brokers. The Gold Coast farmers quickly realized that the government, strongly influenced by the alliance of foreign companies, was trying to squeeze them out of the cash-crop market by lowering the price of cocoa. Furthermore, the foreign companies, wishing to take over the cocoa industry, asked the government to stop its service of shipping cured cocoa to England for the farmers. The foreign companies (Messrs. F & A Swanzy, Ltd., Miller Brothers & Company) and the African Association went so far as to form a cartel in order to buy cocoa and other agricultural produce at fixed prices, and thus eliminate any competition.

After a long struggle the farmers were able to convince the governor of the Gold Coast, John P. Rodger, not to force them to submit to the demands of the foreign companies. The farmers then solved their transportation problems by constructing bush paths along which they could roll casks of cocoa from the interior to such major buying centers as Akuse and Accra.

The grip of foreign companies on the Gold Coast economy during the colonial period was also manifested by the practice of importing manufactured goods, instead of encouraging the development of local industries and locally manufactured consumer goods.

THE NKRUMAH ERA

After Ghana attained its independence in 1957, there was an expansion of the public and private sectors of the economy. As in most newly developing countries, the Ghanaian economy required large capital investments, particularly in infrastructural programs, as the major activity during the first phase of progress toward industrialization. Although the government was relatively well-off financially, it soon became apparent that the development and expansion of the economy would still have to depend principally on foreign companies. Some of the companies decided to divert their investments to particular kinds of industrial development while continuing to sell large quantities of foreign-made consumer goods in the Ghanaian markets. Some of the investments were made as joint efforts between the Ghana government and foreign governments. Investments also came from multinational corporations and directly from local private business concerns in Ghana.

The Volta River Project

By far the most important large-scale industrial venture jointly undertaken by the Ghana government and a foreign corporation is the Volta River Project, undertaken with the Kaiser Aluminum Company. The tenth anniversary of the completion of the Akosombo Dam on the Volta River, the most impressive physical asset Kwame Nkrumah established for Ghana, was celebrated in February 1976. This project, which was negatively and pejoratively described as "prestigious" by many short-sighted economists and political opponents of Nkrumah, has proved to be Ghana's salvation in the energy field. At present 90 percent of all electricity generated in Ghana comes from the installation at Akosombo. The powerhouse produces 912 megawatts. Seventy-five percent of all the electricity generated at Akosombo is now consumed by the aluminum smelter at Tema. Most of the rest is used by the mining industries, the Electricity Corporation of Ghana, and by distribution centers in the towns and villages. In 1972 the Volta River Authority also began to supply electric power to the republics of Togo and Benin (Dahomey).

The establishment of the Volta River Project was not without some serious difficulties, both technical and financial. The idea of establishing an aluminum industry using the Volta River as the hydroelectric-power base dates back to 1925. In 1945 a private company, West African Aluminum, Ltd., was established to investigate the economic possibilities. In 1949 the Gold Coast government appointed William Halcrow and Partners to examine the economic potential of the Volta River. This was followed by a series of negotiations between the Gold Coast and the British governments. The original arrangement was that the capital cost of the project, estimated at between £110 and £144 million, would be shared between the Gold Coast and British governments and the aluminum companies. By 1956 the original estimates had increased to such an extent that the British government and the aluminum companies began to lose interest in the project. Soon after, West African Aluminum Ltd. announced that it could not finance its part of the project. Nkrumah approached the U.S. government for possible financial assistance. He followed this with a personal visit to Washington and contacted the Kaiser Corporation, which ultimately spearheaded a consortium to bring the project to fruition. Although some may argue that the Ghana government did not do well with the Kaiser agreement because of its weak bargaining position, the hard reality is that without the Volta River Project, Ghana would be in an even more difficult financial position. What the project set out to do has, by and large, been achieved.

From the very beginning, Nkrumah realized that the power released from the Akosombo Dam would not be sufficient if full-scale industrialization took place in Ghana. Therefore, from the outset, preliminary studies were made concerning the future generation of hydroelectricity at Kpong and at Bui. The Volta River Authority is currently completing arrangements to start the Kpong hydropower development, which, it is estimated, will add some 160 megawatts to Ghana's hydroelectricity production.

Since the development of the Akosombo Dam, most of the electric power has been used for low-cost conversion of imported alumina into aluminum. Unfortunately the vast bauxite deposits available in Kibi and at Nyinahin in Ashanti are not being used by the Volta Aluminum Company. The reason given is that since the production of alumina from bauxite is mostly tied to specific uses, any attempt at overproduction would be financially uneconomical. The Volta River Project's major concern, however, is to produce electric power. Were it not for this project, Ghana would have been totally crippled during the recent world oil crisis.

The construction of the rock-filled Akosombo Dam has added a new dimension to the ecology of Ghana, Volta Lake. The dam rises 440 feet from the bedrock of the river, with a crest length of 2,200 feet. The reservoir behind the dam is 250 miles long and has a surface area of 3,275 square miles, making it the fourth largest man-made lake in the world. During planning of the Akosombo Dam construction, serious consideration was given to the effect that the creation of Volta Lake would have on water transportation, agriculture, and fisheries development, as well as to the implications of the displacement and subsequent resettlement of more than 80,000 people in new communities and new agricultural vocations.

The Ghana Academy of Sciences

During the Nkrumah era the Ghana government realized that the future development of the country depended principally on the application of science and technology. It observed that financial investment in the promotion of science and technology had aided the developed sectors of the economy of some developing countries to obtain substantial economic returns from those investments. In his enthusiasm to ensure that Ghana developed quickly, Nkrumah decided that the best way to mobilize the scientific manpower in Ghana was to put the scientists in close proximity to each other. This was reflected in a speech he delivered on November 25, 1964, at the laying of the cornerstone of Ghana's atomic reactor. It was at this ceremony that he made public the idea of establishing a "science city":

> The Science City will accommodate a number of special research institutes and will be a centre where the Academy [the Ghana Academy of Sciences] will undertake pilot industries based on its discoveries, that when the Academy recommends the setting up of any full scale industry, it will be in a position not only to give expert advice on the type of industrial plants to be established, but to make the necessary economic appraisal of the proposed industry. [3]

There is no question that Nkrumah knew very well what direction Ghana should take to develop a well-founded science and technology infrastructure. The Ghana Academy of Sciences was then a corporate body responsible for the organization and coordination of all scientific research in all branches of knowledge in Ghana.

The government showed a keen interest in the affairs of the academy, to the point that some critics felt that the institution was becoming too politically orientated. This observation was not entirely wrong, for many facets of the Ghana Academy of Sciences were fashioned after the Soviet Academy of Sciences. But the point to be remembered concerning the Nkrumah era is that it was during this period that the country, for the first time, took bold steps to infuse science education into all levels of the school system. On science education Nkrumah stated:

> There are not nearly enough Ghanaian scientists and Ghanaian technicians of all kinds for the work we have on hand. Even the finest laboratory, the best equipment or indeed the best reactor, will not produce scientific work of their own accord. Only men and women can do that and only after long and highly specialized training in scientific techniques. Hence we need to press on with the greatest urgency the scientific and technological training of young Ghanaians. Every boy or girl who shows talent in this direction must be encouraged and helped, because such talent is especially precious to us and we must foster and guard it. Our Universities, the various Institutes of the Ghana Academy

of Sciences, the University College of Science Education
in Cape Coast [University of Cape Coast], our newly established Medical School and the proposed University College of Agriculture will help in providing the training. [4]

It was during Nkrumah's presidency that the most remarkable
acceleration in the training of scientists and technologists occurred.
The efforts he made to support science education have so far not been
equaled. Science education was part of his Accelerated Development
Plan for Education, which resulted in an unprecedented increase in
the number of primary and secondary schools. From the time Nkrumah
took over the government of Ghana in 1957, until 1962, enrollment in
public schools increased some 120 percent. During that period he also
brought about the establishment of 2,500 primary schools and 374 middle schools.

Before Ghana became independent, the only introduction to science education in elementary school was a course in nature study and
hygiene. As part of a general feeling that progress in all aspects of
Ghana's development must be based on science and technology, the
Ministry of Education designed a method of briefing elementary school
teachers on how science courses should be introduced. In addition to
the development of new science curricula, the government established
the National Science Museum to provide both permanent and short-
term exhibitions aimed at arousing public interest in the sciences.

State Enterprises

The Nkrumah government further realized that Ghana could not
achieve economic progress without industrialization, and therefore
established a number of state enterprises covering a wide range of
industries. It was aware that the foundations of industry were in technology, and therefore encouraged the incorporation of science and
technology into the design of industrial products. One such industry
that I examined in some detail as it relates to science and technology
is the Steel Works Division at Tema. Soon after independence the government felt that a modest steel industry was a necessary component
of the overall development plan. Geological mapping surveys and
sampling, as well as ore estimates, had been prepared in the late
1920s and again in 1952-57, in keeping with the scientific interests of
the colonial era, which were essentially to make scientific surveys to
facilitate more effective exploitation.

Further studies were undertaken in 1962-65 by the government.
These studies showed that there are three main iron ore deposits of
commercial importance to the country. Although Ghana has no coal

from which coke can be obtained for iron reduction, it was felt that an integrated program involving the use of charcoal and the abundant hydroelectric power supply from the Volta River Dam, could realistically make the iron and steel industry an important component of the economy. The government's aim was to produce iron rods and steel bars for the building industries, flat steel and drawn wire to feed the nail factories, agricultural implements, and spare parts and components such as car mufflers to help the transportation industries. This would eliminate the necessity of importing such essential items from abroad.

The country's wide range of mineral resources was also considered in the establishment of subsidiary industries. For example, the government was aware of the several limestone deposits that could be used in the production of cement products and of lime, which is used in the metallurgical and chemical industries and in agriculture.

National Development Plans

During the Nkrumah era there was widespread realization in the government that the natural resources of the country had to be exploited to bring about industrialization. But doing this efficiently depended upon the establishment of an industrial policy. During 1951-59 the government made the first major attempt to formulate its plans. The First Five-Year Development Plan (1951-57) was drawn up at a time when the Gold Coast was engaged in internal self-government. It was initiated to provide a basic infrastructure to support the exploitation of natural resources and the establishment of agricultural activities.

When the country became independent in March 1957, Prime Minister Nkrumah declared that the first two years (1957-59) would be designated as a period of consolidation. This decision was taken to ensure that the framework necessary for economic development would be established.

The Second Five-Year Development Plan (1959-63) was given the sole objective of improving upon the First Five-Year Development Plan and the period of consolidation. During this plan period, provision was made for a definite program of industrialization and a certain degree of expansion in the agricultural sector. The major objective of the Second Five-Year Development Plan was embodied in a famous speech that Prime Minister Kwame Nkrumah delivered to the Ghana Parliament on March 4, 1959. In his speech he pointed out that although it was important that the country obtain political independence, such should not be an end in itself.

Having achieved political independence our next objective
is to consolidate that independence and lay the economic
foundation to sustain our national independence. . . . It
is the aim of my Government to create the means for the
good life, and create a society in which everybody in Ghana
can enjoy the fruits of his labour and raise the economic
and social standards of the people. This aim cannot be
attained all at once but it can be realized through a series
of five-year plans like the one we are about to embark upon. . . . The Ghana Government has also signed an agreement with the Government of the United States of America
guaranteeing the rights of American investors in Ghana.
It is willing to sign similar agreements with any other
country. [5]

To this end the Nkrumah government designed three major programs. The first involved industrial activities that were wholly private. The second was based on the concept of mixed enterprise, in
which the government owned part of the business. The third concerned
wholly governmental enterprises. Realizing the importance of the private sector of the economy, Nkrumah pointed out that his government
had informed the Industrial Development Corporation that the number
of enterprises owned completely by the government must be kept at a
minimum. Unfortunately this concept did not materialize, because the
more the government talked about it, the larger the governmental
enterprise became. There was no doubt that the government was on a
socialist path allowing little room for the free-enterprise system in
major industrial concerns.

The most comprehensive plan that Ghana put into operation was
the Seven-Year Development Plan (1963-70), which ended abruptly with
the 1966 change of government. It has been pointed out by various analysts and critics that this plan was by far the most thorough and con-
ceptually the most ideal plan drawn up by any African country, before
or after independence. Certain key elements in it were partly motivated by factors in the African political scene and partly by the socialist desires of the government. In the foreword to the Seven-Year Development Plan, President Nkrumah noted that the plan

. . . is designed to provide the basis not only of our national progress and prosperity, but also of our ability to
contribute to the advancement of the African Continent.
. . . Apart from the welfare of the individual Ghanaian,
we intend to use the resources which our participation
in the productive economy yields to the state to promote
the economic independence of Ghana and the unity of Africa. [6]

Inherent in the plan was the growing realization that some firm action was needed to encourage scientific research and technological advancement as the only means to raise the standard of living of Ghana and the rest of Africa. Another feature of the plan was the government's desire to transform Ghana from a low-energy economy into what Nkrumah described as a "strong, industrialized, socialist economy and society."

The Seven-Year Development Plan did not, however, work out as its architects had intended. First, it was assumed that the Ghanaian public was ready to live the kind of life characteristic of a socialist state. Furthermore, no one actually knew whether the individual Ghanaian was ready to place the eternal well-being of the country before his private interests and concerns. Second, the plan assumed that credit from within the country and from abroad would be forthcoming to help move an economy by means of a plan that had a price tag of no less than £1,016 million. In addition, it was assumed that the growth rate of the country would be approximately 5.5 percent annually. The plan partially succeeded because of the enterprise of Nkrumah, but unfortunately the kind of leadership that was needed at every level of the economy was lacking. Hence it was impossible to implement the elaborate yet comprehensive plan in a coordinated fashion. By and large, the Seven-Year Development Plan resulted chiefly in the establishment of many agroindustrial activities.

One major drawback that affected the total success of the Seven-Year Development Plan was that although the objectives of the government were clearly conceived, they were not realistically related to the availability of financial resources. For example, it was assumed that the export of cocoa could bring in enough capital to finance major portions of the plan. The steady fall of the world price of cocoa undermined the integrity of the plan as a whole.

The pattern of dependence on one crop had been recognized as dangerous even during the colonial era. In 1919 the governor of the Gold Coast, Sir Gordon Guggisberg, asked the Legislative Council: "We have all our eggs in one basket. The cocoa baskets are full—what about the other baskets if anything goes wrong with the cocoa crop or the cocoa market?"

There is no doubt that the Nkrumah government had many good ideas but its performance reflected the basic weakness in management that seems to be characteristic of many less developed and developing countries. For example, the idea of establishing state farms to feed the people and the government's determination to see agroindustries established were sound, but the actual implementation of these ideas and development plans left much to be desired.

The State Farms Corporation, established in 1962 by the Nkrumah government, consisted initially of 42 farms. By 1966 the number had increased to 105. The corporation employed some 22,000 permanent workers on 59,580 acres of fields. Of the total, 21,500 acres were allocated to food production (corn, rice, cassava, groundnuts), and 480 acres were devoted to vegetables (peppers, tomatoes, and onions.) Crops such as oil palm, rubber, bananas, coconuts, cola, and citrus fruits were established on 34,490 acres of land. About 810 acres were devoted to plantain and pineapple cultivation. Fiber crops such as cotton and urena lobata covered some 1,400 acres. About 800 acres were allocated to tobacco. In addition to the above crops, the corporation also undertook the rearing of 260,000 chickens, 2,300 cattle, 1,300 sheep, and 2,700 pigs.

The poor performance of agricultural output during this period was not the result of poor land use by the farmers (although this might have been true in some cases) or poor availability of seeds (this might also have been true in some cases); it was due almost entirely to poor administration and serious weaknesses in the organizational structure of the Ministry of Agriculture. The state farms were infested with corrupt personnel who managed to line their pockets with the proceeds of the harvests. In addition, there was an unholy alliance between the Nkrumah political functionaries and the United Ghana Farmers Council in a joint attempt to undermine the financial position of the state farms and a number of private farms. The entire agricultural sector of the economy was unduly politicized during the Nkrumah era. Many farmers were unhappy because the financial proceeds from the farms (which formed a substantial portion of the GNP) were appropriated by the government, the bulk of the funds being used to improve the cities, with very little returning to the rural areas for development. Similar observations can be made of other subsidiary activities of the government.

THE POST-NKRUMAH ERA

The Seven-Year Development Plan did not run its full course. The change in government in 1966 terminated the plan, which had been scheduled to end in 1970. From the time of the 1966 coup to 1967, the National Liberation Council sought to reassess all the projects and enterprises instituted by the Nkrumah government. On July 1, 1968, it issued a decree (no. 207) that established the Ghana Industrial Holding Corporation to carry on the industrial and manufacturing activities of the government. One of the major reasons for this decree was the fact that many of the enterprises established by the Nkrumah government were very poorly run, although the ideas behind their establishment were well intentioned. The net accumulated losses of the 20 divisions of the enterprises amounted to some 15 million cedis (in 1976, 1 cedi

= U. S. $0. 8667). To improve the economic performance of the corporation, an attempt was made to recruit the best managerial and technical manpower to run the viable divisions. The Sheet Metal Division was sold to a Ghanaian company. The Cocoa Products Division was transferred to the Cocoa Marketing Board, and the Sugar Products Division was reconstituted into the Sugar Estates Limited. Most of the changes were made to improve the management of the organizations.

Perhaps the major reason for the failure of these enterprises during the Nkrumah era was that soon after independence, when the government decided to industrialize, little thought was given to sustained supplies of both local and imported raw materials to feed the industries. The Nkrumah government (and subsequent governments) gave special priority to the state-owned industries but paid little attention to the flow of raw materials. Hence, as soon as the industries began production, it became evident that the government had to import, for example, large quantities of raw cotton for the mills to supply the booming textile industries. Similarly, imports of tallow and palm oil were needed for the soap factories. For Ghana to be self-sufficient in its supply of sacks for the cocoa industry, the government quickly established a sack factory without advance assurance of a local supply of raw jute. The two sugar factories were in a similar predicament. The local supply of sugarcane was so inadequate that the modern sugar factories were idle for three months a year.

It should be emphasized that the Nkrumah government was not unaware of the need for a local supply of raw materials. On the contrary, a systematic effort was made to establish state farms to produce such agroindustrial crops as palm oil, tobacco, cotton, rubber, sugarcane, groundnuts, sisal, and citrus fruits. The failure of these farms to meet their commitments was due mainly to poor management and, to a certain extent, lack of capital.

The scientific and technological input into the state farms was considerable. The government acquired large parcels of land, and imported many tractors and other farm equipment from both Western and Eastern European countries. Because the concept of technological transfer was not clearly perceived, the wrong kinds of tilling equipment were imported from some Western countries. The Eastern European countries, on the other hand, supplied the wrong types of tractors. Furthermore, the farm equipment was often left in the hands of political functionaries who indulged in corrupt practices. Most of the tractors from East European countries presented a special problem because no firm arrangements were made to obtain a regular supply of spare parts and replacement parts. As a result, only a few of the tractors were still in use by the end of the 1960s. The activities of many of the "modern" state farms therefore ended in total disaster.

It fell upon the small private farms using the "village" type of technology to supply most of the raw materials needed by the agroindustries. The small-scale, individual farms now constitute the most important agricultural enterprise in the country and account for over 80 percent of the total output of agricultural produce.

Import Substitution

During the colonial era the concept of import substitution was not given much consideration because the economic system operating at the time did not call for the kinds of industrialization that encouraged local industrial complexes, let alone the development of indigenous technologies that are central to the production of local food items. Virtually all the food products—canned meats, fruits, and vegetables—and numerous manufactured goods were imported while the country's efforts were directed to the production of cash crops for export.

After Ghana attained its independence, the government decided to encourage the public and private sectors to diversify their industrial activities by concentrating on the production of manufactured goods to replace the large imports of food and clothing. The major problem was to obtain the kind of financing that would make possible the importation of industrial equipment needed by local factories. A number of the state enterprises were established by the Nkrumah government under the policy of import substitution. The Ghana Industrial Holding Corporation, for example, was established to consolidate and make more efficient the manufacture and distribution of locally produced goods.

Yet over the years, and certainly in the Nkrumah and post-Nkrumah years, successive Ghanaian governments have misunderstood the role that import substitution can play in a growing economy. It has always seemed logical that for a country like Ghana to achieve economic independence, it is necessary that it rely principally on its own agricultural output to satisfy its basic food and industrial needs, and that it should rely on imports only when such items would help to improve local production capabilities. For some reason Ghana has seen fit to rely on importing manufactured goods, as well as unprocessed foods, as a normal component of economic activity. Ironically, a number of the raw materials needed to manufacture many food items are readily available in the country or could be used as substitutes for the imported manufactured items.

Another area that will lead to the rapid development of Ghana is the promotion of domestic manufacture of consumer goods. The textile industry is an example. The increased demand for material caused the textile industry to develop at a rapid rate. As a result Ghana has virtually eliminated the import of cloth. The balance between mass production and consumption in the textile industry suggests a certain amount of technological maturity.

It is disturbing to note, however, that in the year after the 1966 coup, the new government imported raw and processed agricultural products to the value of 50 million cedis, nearly one-third of the total value of Ghana's agricultural export at the time.

A major problem encountered during the initial production of manufactured goods was the adverse attitude of the Ghanaian populace toward the finished products. The reaction of many consumers has been one of strong bias in favor of imported goods. It is only in recent years, and certainly since the government policy of self-reliance and Operation Feed Yourself, that the snobbish attitude toward locally produced foods has somewhat diminished.

Operation Feed Yourself

Since the overthrow of the Nkrumah government in 1966, two successive governments have attempted to increase activity in the agricultural sector. It was, however, quickly realized that little could be done without a massive investment of capital and proper management. Soon after the current government took office in 1972, the National Redemption Council (NRC) launched a self-reliance program named Operation Feed Yourself. The aim of the government was to attain self-sufficiency by 1974 in the production of such basic food items as cassava, corn, rice, plantains, yams, groundnuts, millet, and vegetables.

Much progress was made in the production of cassava, maize, and, to a certain extent, rice. Disappointing results, however, were realized for plantains, yams, and millet. Although the government of Kofi Busia tried to promote agriculture by increasing the acreage of farm land, (see Table 8.1), it was unable to assign the high priority that agriculture deserved in Ghana. It was left to the present government to assign the highest priority to the agrarian sector of the economy. The NRC's 1972 budget statement indicated that the government was willing to increase production of such staple foods as corn, rice, yams, and other crops. In an attempt to solve the difficulties of the high cost of foreign food items, the NRC began to subsidize the prices. The government soon realized subsidies were expensive and were not the real answer to Ghana's economic problem. The NRC also revived factory projects in the agricultural and industrial sectors that had been initiated by the Nkrumah government but abandoned by subsequent ones. In the 1973-74 budget the NRC allocated nearly 15 million cedis for agricultural development, an increase of 6 million cedis over the

TABLE 8.1

Area and Production of Major Staples, Ghana, 1968-72
(area in 1,000 acres; production in 1,000 long tons)

		1968	1969	1970	1971	1972
Corn	Area	671	680	989	851	960
	Production	296	303	435	378	396
Sorghum	Area	373	370	387	404	425
	Production	82	81	85	89	93
Millet	Area	346	432	458	468	483
	Production	72	86	92	94	97
Rice	Area	113	120	128	136	138
	Production	64	60	64	68	69
Groundnuts	Area	150	183	181	200	220
	Production	37	37	36	49	54
Cassava	Area	425	450	476	524	602
	Production	1,423	1,485	1,618	1,782	2,047
Cocoyam	Area	313	320	352	387	426
	Production	961	983	1,082	1,189	1,309
Yam	Area	293	290	319	325	331
	Production	1,330	1,305	1,456	1,463	1,490
Plantain	Area	326	350	400	463	508
	Production	755	805	920	1,065	1,168

Notes: Area is the estimated number of acres fully under any particular crop. This is calculated as the sum of the acres fully under, for example, corn and a proportion of the acres under corn and other crops mixed. Production is calculated as estimated area in acres multiplied by estimated yield per acre.

Data prior to 1970 are not strictly comparable with figures for later years.

Source: Ghana Ministry of Agriculture.

previous year's budget. Substantial portions of the funds were allocated to the livestock, irrigation, and mechanization sector of the economy, as well as to improving supplies of grains and fertilizer. The NRC also realized the importance of storage facilities. The government arranged for the construction of more than 100 grain silos in 1972, to cut down the enormous losses from spoilage that had plagued the grain harvests.

The NRC continues to stress and to encourage agricultural production. Although much progress has been made since the introduction of Operation Feed Yourself, much more must be done to meet all the local and export demands of the country. The Ministry of Agriculture should undergo further drastic reorganization before the long-term ideas of the government can be brought to fruition. Another drawback in the ministry's activities has been the periodic change of top administrators. In order for the long- and short-range goals of the agricultural programs to be reached, it is important that an essentially stable corps of top-class administrators be associated with the ministry. Since the NRC government was established in 1972, the Ministry of Agriculture has passed through the hands of four commissioners. While some of the changes in commissioners have been unavoidable, it would seem that the government could make it possible for the appointment of someone over a sustained period in this important ministry. It is essential to note that this lack of sustained leadership in the Ministry of Agriculture is not new. From the time that Ghana gained its independence in 1957 until the 1966 change of government, the ministry passed through the hands of seven ministers of agriculture.

Impact of Technology on Sector Development

Cocoa

Cocoa has been Ghana's economic mainstay since 1910, when it became the world's leading cocoa-producing country. Over many years the cocoa industry has earned the largest amounts of foreign exchange for Ghana's economic development, contributing 62 percent of the foreign-exchange earnings and 33 percent of the local revenue in 1975. Current figures show that about 2.5 million Ghanaians are involved in cocoa production.

During the 1950s there was an increase in cocoa production partly because of the increase in new plantings and partly because of the use of insecticides. However it became quite apparent in the 1960s that continued dependence on cocoa would prove unhealthy to Ghana's economic growth because of the persistently poor performance of the industry, especially when a three-year moving-average production fell from a peak of 472,000 tons in 1963/64 to a low of 376,000 tons in 1969/70.

Apart from weather conditions that lower the general production of cocoa from time to time, the major consequence of the low prices of the early and middle 1960s was a reduction in new plantings and subsequent increase in the average age of existing cocoa trees.

The cocoa industry has suffered over the years because of two major diseases: swollen shoot, caused by a virus transmitted by the mealy bug, and capsid damage, which results from a fungus infection that develops after capsid flies have pierced the cocoa pods. The major crop failures of the 1960s were mostly attributed to these two diseases. However, the symptons of swollen shoot were first noticed in Ghana in the early 1920s. The disease was recognized in 1936 and the virus origin of it was pinpointed in 1940. During the colonial days the British established the West African Cocoa Research Institute at Tafo, Ghana, to deal with all aspects of cocoa biology. Later, when the other English-speaking West African countries attained their independence, the Ghana Cocoa Research Institute was established under the Ghana Academy of Sciences and the Council for Scientific and Industrial Research. In the early 1970s the institute was placed under the Ghana Cocoa Marketing Board, which is now under the Ministry of Cocoa Affairs. The Cocoa Research Institute has been conducting breeding programs using disease-resistant and high-yield varieties. There are about 700 nurseries in Ghana that supply seedlings for replanting. During 1972/73 nearly 3 million seedlings of the high-yield varieties were planted for farmers, free of charge. To control the capsids, farmers have been encouraged to obtain spraying machines and insecticides. But the most effective method for suppressing the disease has been the cutting of infected trees. The cutting program, begun in 1946, has been described as the longest, the most complicated, and the most expensive effort ever undertaken to eradicate a plant virus disease. Over 150 million diseased trees have been destroyed in Ghana. This figure does not include the many millions of trees that have died naturally. Although some progress has been made in the suppression of the disease, a major research effort is needed to understand the complex behavior of the vectors.

There is no doubt that a serious scientific and technological effort has to be made by the Ghana government and the cocoa industry in order to have any success in eradicating this disease. Unfortunately the manpower necessary for field and laboratory work is quite inadequate at the moment. For example, the number of cocoa entomologists in the country is so small that one wonders if the scientific community is effectively communicating its needs and obligations to the government and to the cocoa industry as a whole. One would assume that since the cocoa industry is the main earner of funds for the country, the government would not hesitate to invest in sound research programs to improve cocoa production by eradicating the disease and breeding high-yield varieties.

Forestry

After cocoa, Ghana's most important export commodity is timber. During the 1975 financial year the logs exported and processed locally amounted to 1.5240 million cubic meters. This figure is reported to be the lowest production in the period 1971-75. In 1973 the output was 2.074 million cubic meters, which was an increase of 197,000 cubic meters over the 1972 output. The unprecedented world demand for timber in 1973 and the rising prices increased Ghana's earnings from timber from 36.8 million cedis in 1971 to 143.2 million cedis in 1973 (see Table 8.2).

Of the nearly 200 exploitable species of timber trees available in the country, only about 30 have been harvested commercially. [7] The species that have featured prominently in the export market include wawa, utile, sapele, mahogany, edinam, makore, amire, kokrodua, and oprono. It is interesting to note that these nine species accounted for 94 percent of the total log export of the country. Italy was the leading buyer during the 1974-75 financial year (21.8 percent). Other buyers included West Germany (15.5 percent), the Soviet Union (13 percent), the Netherlands (9.8 percent), France (2.4 percent), the People's Republic of China (2.2 percent), Egypt (1.8 percent), and Japan (1.7 percent).

The timber industry has a long history of association with science and technology. In recent years several well-equipped plywood and veneer industries have been established in the country in addition to the many sawmills. The technological innovations in the timber industry since Ghana achieved its independence have improved production immensely.

Two of the most efficient pieces of equipment used in the timber industry are the Caterpillar 988 Pay Loader and the Timber Jack 550. The Pay Loader is a rugged machine that has a lifting capacity of 20 tons in seven to ten minutes, and loads an average of about 500 tons a day. This machine does little damage to the trunks when loading and is economical to operate. During a good season it can load 10,000 tons of logs a month, an economically efficient loading performance in comparison to former techniques. The Timber Jack 550 is most useful in extracting logs from the bush and taking them to the loading stations. It is one-third cheaper to operate than the ordinary Caterpillar, and its maintenance cost is 50 percent lower. It can haul 12 to 15 tons of logs from the bush per trip, and its extraction performance ranges from 2,000 to 2,500 tons per month.

TABLE 8.2

Output and Export of Logs, Timber, Veneer, and Plywood, Ghana, 1968-72

	1968	1969	1970	1971	1972
	Value (million cedis)				
Log exports	16.26	24.12	19.88	20.54	42.29
Sawn timber	17.83	21.80	22.71	20.85	32.80
Exports	12.30	14.96	17.09	12.22	21.17
Local sales	4.39	4.33	4.50	6.27	12.86
Own use*	1.12	1.10	1.20	1.24	3.34
Change in stocks	-0.31	1.07	-0.30	0.88	-4.80
Pitsawing	0.33	0.34	0.22	0.24	0.23
Veneer	0.10	0.18	0.08	0.23	0.29
Exports	0.02	0.15	0.08	0.24	0.21
Local sales	0.01	0.01	—	—	—
Own use*	0.07	—	—	—	—
Change in stocks	—	0.12	—	-0.01	0.08
Plywood	3.98	4.14	5.20	7.08	8.73
Exports	2.41	2.66	3.04	3.82	4.50
Local sales	1.56	1.41	1.99	3.27	3.40
Own use*	0.07	0.01	0.01	0.03	0.03
Change in stocks	-0.06	0.06	0.16	-0.04	0.80
Total value of output	38.17	50.34	47.87	48.70	84.11
Total exports	30.99	41.89	40.09	36.82	68.17
Total local sales	5.96	5.75	6.49	9.54	16.26
Total own use*	1.26	1.11	1.21	1.27	3.37
Total change in stocks	-0.37	1.25	-0.14	0.83	-3.92
Total pitsawing	0.33	0.34	0.22	0.24	0.23
	Volume (million cubic feet)				
Logs	49.05	56.92	55.24	57.40	66.33
Exports	20.10	24.60	21.22	24.95	33.60
Sawmill intake	25.65	29.10	30.94	29.03	28.14
Veneer/plywood intake	2.80	2.72	2.68	3.02	4.29
Pitsawing	0.50	0.50	0.40	0.40	0.30
Sawn timber	11.85	12.90	12.68	12.19	12.28
Exports	7.60	7.73	8.50	6.56	8.84
Local sales	3.38	3.29	3.32	3.93	4.02
Own use*	0.86	0.82	0.89	0.87	1.67
Change in stocks	-0.24	0.81	-0.23	0.63	-2.40
Pitsawing	0.25	0.25	0.20	0.20	0.15
Veneer	0.02	0.02	0.02	0.03	0.03
Exports	0.01	0.02	0.02	0.04	0.02
Local sales	—	—	—	—	—
Own use*	0.01	—	—	—	—
Change in stocks	—	—	—	-0.01	0.01
Plywood	0.96	0.89	1.11	1.38	1.65
Exports	0.73	0.67	0.77	0.91	0.89
Local sales	0.22	0.20	0.28	0.47	0.48
Own use*	0.02	—	—	0.01	0.01
Change in stocks	-0.02	0.02	0.05	-0.01	0.27

*Consumption by mills and mines.

Sources: Ministry of Lands and Mineral Resources, Forestry Department, Economic Survey, 1971; information provided by the Central Bureau of Statistics.

The future growth of the industry will depend, however, on several other factors. It has been shown that the timber from the reserved areas (which constitute 20 percent of the total forest area in Ghana) can be exploited by investing in sound forest management and reforestation programs. Most forest technologists agree that the establishment of a 25-30-year felling cycle will stimulate growth. Since few companies in Ghana have processing facilities other than sawmills, there is a tendency on the part of many timber companies (particularly the small logging companies) to leave unexportable logs of poor quality to rot on the forest floor. With the increasing desire by the government to encourage the local processing of logs, it would seem appropriate for the government to enforce the regulation of issuing operating permits to only those companies that can process all logs felled into veneers, plywood, and other wood products.

A tremendous amount of wood is left to waste in the concession areas after a few choice logs are extracted from the forest. It would seem logical to turn this substantial amount of unused wood into charcoal. At present Ghana produces 250,000 tons of charcoal per year from more than 2.5 million tons of firewood, using very primitive methods of operation. Naturally a much more efficient conversion method, using modern science and technology, is needed to reduce the available firewood to charcoal and useful by-products. In April 1975, the Forestry Department of Ghana, working with the United Nations Industrial Development Organization and the Food and Agricultural Organization, initiated a charcoal production and utilization project. This project is aimed at converting firewood into charcoal to meet domestic needs and also at developing more profitable industrial uses, such as the production of cement, calcium, and carbide and the processing of iron ores and ilmenite. It is also planned that the large-scale destructive-distillation installations will make possible the recovery of wood tar and pyroligneous acid, which can economically be put to good use. The wood tar can be processed into products to replace presently imported products, such as creosote, for many uses, including wood impregnation. The pyroligneous acid can be further fractionated into valuable chemicals, such as acetic acid and methanol.

The modern technological procedures for producing charcoal will be more economical than the primitive method of carbonization, which yields only 10 percent or less charcoal from a given volume of firewood. Production in portable kilns, on the other hand, has a conversion rate of 25 percent in addition to the recovery of various chemicals.

The long-range question that should be posed in the light of the rapid expansion of the timber industry is how long the timber resources will last. As intimated earlier, most of the extractable timber has been obtained from the nonreserve areas of the forest. Because of the lack of reforestation programs in these areas, major portions of the cleared lands have been converted into farming areas. Furthermore, the relatively extensive construction of rough roads in the concession areas makes the closed forest zones accessible to human habitation and use. Such areas will therefore be lost to the timber industry within the next 15-20 years. This practice seems to indicate that the timber industry will have to rely on the forest reserve areas to sustain the industry. The future of the timber industry will therefore depend on the strict management of the forest reserve areas of the country.

Fisheries

Before independence the fishing industry was considered marginal at best, because the fishing boats had fishing rights in a limited area. Furthermore, the boats were small and did not have the capacity to store fish for long periods. The fishing industry began to enjoy a new boom when the first motorized vessels were introduced in the 1950s. By the early 1960s it exhibited considerable potential in meeting Ghana's needs. From 1957 to 1967 fish production from marine and inland sources quadrupled to 110,000 metric tons with a value of more than $17 million. Most of the increase in catch was generated by the use of motorized boats.

This technological activity in the fishing industry has proved to be of great importance in increasing Ghana's ambition of reaching self-sufficiency in this sector of the economy. Unfortunately, several factors have affected the continued growth of the industry. In 1972 the marine fishing capabilities were seriously impaired by managerial problems. First, the government decided to promote Ghanaian ownership of the fishing vessels, which automatically resulted in the restriction of contract awards to foreign concerns. In 1967, for example, the Department of Fisheries records indicate that about 26.7 percent of all marine fish landings were made by individual Ghanaians using modern motorized boats and canoes fitted with outboard motors. The private Ghanaian fishing companies landed about 27.6 percent of the total catch. The State Fishing Corporation, using only motorized vessels, accounted for 14.8 percent. The foreign companies under contract, including groups from Japan, the Soviet Union, France, and Poland, contributed 16 percent of the total. In addition, 14.9 percent of the fish were caught by foreign companies that were not under any contract. These included Japanese and South Korean tuna companies working in cooperation with the Ghana government and Star-Kist International of Tema, a subsidiary company of Star-Kist Tuna of California. (See Table 8.3, for total catch and consumption.)

Serious management problems and lack of financial backing, especially in the areas of replacement equipment and other spare parts,

TABLE 8.3

Ghana Domestic Fish Catch and Consumption, 1968–72

	1968	1969	1970	1971	1972*
			Thousand Tons		
Total domestic marine catch	114.5	175.7	156.3	181.1	249.1
Lake Volta catch	50.2	60.5	39.9	39.2	32.4
Landings from foreign vessels on contract	11.0	7.8	16.6	20.5	0.5
Total catch	175.7	244.0	212.8	240.8	282.0
			Million Cedis		
Total domestic marine catch	26.2	36.8	28.8	27.5	45.9
Lake Volta catch	11.7	15.7	5.9	6.0	4.9
Landings from foreign vessels on contract	1.7	0.9	2.2	3.3	0.1
Total catch	39.6	53.4	36.9	36.8	50.9
Imports of fish and fish products	4.4	4.9	13.8	11.7	23.4
Total consumption	44.0	58.3	50.7	48.5	74.3

*Provisional.
Source: Department of Fisheries, Annual Report.

began to affect the fishing industry when the promotion of Ghanaian ownership began to take effect. The lack of import licenses to purchase equipment contributed to the decline of the number of fishing vessels. Furthermore, in 1973 the Ghanaian marine fish catch dropped because of the extension of territorial waters by neighboring countries, in whose waters the Ghanaian fishermen have constantly worked. Another factor that has affected the fishing industry is the virtual absence of factory ships. At the moment, because of the limited range of many of the boats and the absence of the factory ships, most of the vessels spend more time in transit, hauling small amounts of catch away from the fishing grounds and then returning, than they do in fishing.

Ghana has not exploited the tremendous fishing potential that exists in the Volta Lake, but some scientific work is being conducted under the Volta Lake Research Project. This organization is conducting investigations in fishery development and planning, as well as in fishery biology. Scientific studies are being conducted on practical methods of improving fishing equipment, fish processing, and the marketing sector.

A major technological innovation in the fishing industry has been the introduction of cold-storage facilities in Ghana. Ghana Cold Stores Ltd., which is owned jointly by the government and private companies, has the largest such facility in the country. Other facilities are owned by private firms. The State Fishing Corporation has a group of cold stores that serve as fishing depots for the corporation.

In 1972, with the development of the Operation Feed Yourself program, a scheme was adopted to save the several tons of fish that are normally allowed to rot during the peak fishing season. The storage of such excess fish was aimed at forestalling shortages during the low fishing seasons. Furthermore, the assurance of a constant supply of fish offered a sturdy market and stable prices throughout most of the year. To secure some form of such assurance, Ghana Cold Stores signed contracts with several boat owners to purchase and freeze fish for distribution throughout the major regions of the country, including inland areas. Ghana Cold Stores also felt that the need for high-quality fish depended on the efficient preservation of the fish at sea before they are unloaded. About two years ago a flake ice machine was installed in Accra to produce 50 tons of ice per day. In the late 1960s smoked fish accounted for 70 percent of all fish sold at retail and 90 percent of all frozen fish sold. The obvious advantage of freezing is that the fish can be smoked just before it is ready for consumption. Fish is rich in high-quality protein as well as fat, minerals, and vitamins. It is readily digestible and is not easily denatured by baking, smoking, or cooking. We can therefore conclude that the technology of refrigeration has made an immense contribution to the economy of Ghana through the fishing industry.

Rubber

One of the leading industrial raw materials produced in Ghana is rubber. Although the country started to export rubber as early as 1880, this commodity did not become important until the Nkrumah government developed a 20,000-acre rubber plantation in the Western Region. In 1967 the Ghana government entered into partnership with the Firestone Tire and Rubber Company to develop the existing plantation and to build a modern tire and tube plant at Bonsaso. With proper managerial expertise and some technological innovation in the production of the raw material, the Ghana Rubber Estates Ltd., which covers ten plantations, is operating some 25,000 acres, of which 4,000 are being tapped to produce rubber. The Ghana Rubber Estates (with Ghanaian estate managers and over 2,000 workers) are currently expanding their operations to cover 40,000 acres. To do this scientifically, nurseries have been established to supply the plantations. Most of the trees yield about ten pounds of latex a year. In addition the Estates operates its own extension service by giving tuition-free training to local farmers who wish to tap their own trees. The Estates also supply tapping equipment to farmers, which becomes their property after they have paid for it from the proceeds of the tapping. Whether this arrangement is totally satisfactory to the farmers is not well understood. Because of the lack of capital, however, the farmers have found this "barter trade" quite convenient.

It is important to acknowledge that since the tire plant was established in 1969, it has managed to produce about 90 percent of Ghana's tire and rubber tube requirements for cars, trucks, buses, and tractors. Furthermore, the tires and retreads manufactured in Ghana have been exported to Liberia, Sierra Leone, Gambia, and Togo for sorely needed foreign-exchange earnings. Firestone Ghana Ltd. and the Ghana Rubber Estates Ltd. are working together to expand and to develop the plantations and tire production in the country, so that the industry can grow to meet both the local and the export plans of the government.

Firestone Ghana has employed about 40 Ghanaians in staff positions. Local management development and training programs are offered by the company to ensure that the managerial problems that seem to beset all sectors of Ghana's economy are eliminated from the company's operations.

Cotton

The textile industry in Ghana has had a long history and constitutes an important sector of the economy. For several years it depended upon the importation of raw material from abroad (including the United States, under P. L. 48) to meet the needs of the factories, instead of developing cotton farms to meet local demand. When the financial position of the country became precarious, the government decided to encourage the development of a cotton industry. In June 1968 it established the Cotton Development Board with two main objectives in mind. First, it was recognized that the country had to become self-sufficient in cotton production to satisfy the needs of the many textile industries that were being established. Second, the government realized that, based on the natural environment of the country, Ghana could produce enough raw materials for home consumption and even have enough surplus of lint to enter the world cotton market. Furthermore, the possibility existed that edible oil could be produced from by-products such as cottonseed. Naturally the government's major concern was to conserve the considerable foreign exchange that the country has been investing in the importation of cotton from abroad.

The board established production extension services and research units to implement the objectives of the government. It has long been known that cotton grows rather well in the northern and upper regions of Ghana, as well as in the northern savanna areas of the Ashanti and Brong Ahafo regions and also in the grasslands of the Volta Region. By 1973, 8,600 farmers were growing cotton on more than 9,000 acres in about 650 villages. With the introduction of fertilizers and insecticides, 1.87 million kilos of cotton were produced in the 1972-73 cotton season. The board purchased it all from the farmers for 332,000 cedis. This quantity yielded about 720,000 kilos of lint, which was sold to the various textile factories in the country. A total of 820,000 kilos of seed were exported abroad after the board had met all its seed needs locally. Since the 1971-72 season the board has realized 1.5 million kilos of lint and 1.7 million kilos of seed for export.

Although the board has not achieved the targeted expansion, it has been able to encourage many farmers to participate in cotton production. As an added incentive it has been loaning large sums of money to farmers for the preparation of the land and the maintenance of their crops. Equipment to combat drought has been made available. In addition the farmers are able to purchase bullocks, tractors, and other farm equipment. The board is currently encouraging a few farmers to establish large-scale plantations so as to take advantage of the available machinery and equipment.

The board has not, however, lost sight of its original objective of producing edible oil from cottonseed; and serious consideration is being given to the production of edible oil as well as to the processing of the residue into oil cakes for animal feed.

The Interlocking Industrial Food Complex

One of the most exciting scientific and technological innovations in the economy of Ghana was established by the Nkrumah government in 1963, when it was decided to establish a food complex that would allow the country to begin in earnest the processing of locally produced and imported raw materials. The complex was designed to be profitable because cost and selling prices would be in accordance with world market prices, and the products would be of a high quality. But, most important, the capacity would be sufficient to cover the country's needs in each product handled by the complex. The initial focus of the food complex was the installation of a cereal flour mill having additional equipment for making cornmeal and extracting of corn oil. At the time Ghana consumed over 60,000 tons of flour a year, and imported virtually all the flour needed for bread making.

In 1970 the Busia government terminated the contracts under which the Drevici Group of Companies was to operate the food complex for the Ghana government. When the current government took office in 1972, however, the complex was incorporated into the Tema Food Complex Corporation. The new corporation quickly put various projects into operation; the flour mill, the animal feed plant, the conveyor system running from harbor to silos, the pneumatic ship-unloading plant, and quarries at Weija and in the Shai Hills. By 1973 the corporation had finished the previously uncompleted projects associated with the complex; the fish cannery, tin can plant, fish meal factory, oil mill, cold store, oil refinery, and margarine plant.

Besides the serious financial and, to some extent, managerial problems, it seems fair to conclude that it was a wise decision to salvage this excellent modern industrial complex, initiated by the Nkrumah government. The large-scale importation of manufactured food that has characterized the Ghanaian economy would certainly be a long-term detriment. Even now that the corporation is operating reasonably well, most of its earnings from the export trade are used to import food to supplement what is locally produced. This state of affairs, in which production of raw materials is not meeting demand, cannot continue indefinitely.

The complex is, however, making use of every available raw material. The by-products of one factory are used as the raw materials of another. In the final analysis, there is little waste at the completion of one cycle of production. The Ghana government realizes that the success of this corporation will undoubtedly encourage local and foreign investments in similar projects.

From the inception of the food complex, it was determined that the quality of products was to be high enough to compete with imported brands. The idea of manufacturing products with one quality for export and another, lower quality for the domestic market was rejected, in view of the existing mental block against made-in-Ghana products by the Ghanaian consumer. It was necessary to eliminate totally the notion of locally produced goods being inferior, by striving to produce and to sustain high-quality goods.

The establishment of the food complex is significant in four ways. First, it has demonstrated that developing countries can create an atmosphere for the systematic development of locally processed food items by importing appropriate technologies to suit local industrial production. Second, the complex has promoted the development of technical manpower for production and for managerial functions in the organization. Third, improved productivity and the quality of products from the food complex are rapidly giving the Ghanaian consumer assurance and pride in being self-reliant. Fourth, the food complex is economically sound, especially in terms of the jobs created, the savings of foreign exchange, and the fact that such an innovative establishment will encourage similar takeoff industries.

Infrastructure

Any attempt to improve the agricultural and industrial development of Ghana will depend on new investments in road, rail, water, and air transportation, as well as in an efficient communication system. During the colonial era little effort was made to develop a network of efficient transportation in the country. The few roads and rail systems established at that time were intended only to facilitate the shipment of raw materials for export. The Nkrumah era saw the beginning of a major attempt to improve transportation by the building of trunk and feeder roads. In fact, all the development plans initiated during the Nkrumah and the post-Nkrumah eras have included the massive construction of roads. In the mid-1960s the rapid influx of motor vehicles, accounting for four-fifths of all freight traffic, totally outstripped road development and maintenance, and led to the rapid deterioration of the existing roads.

Of the 20,000 miles of road in Ghana, only 2,500 miles are paved; the 4,000 miles of unpaved truck roads and the 13,500 miles of feeder roads are graveled. There have been several constraints in the development of good roads in Ghana, although the kinds of technology for building up-to-date roads in that environment do exist. The most unfortunate constraint has been the awarding of contracts to people whose technical competence in road building and maintenance is almost nonexistent. In recent years efforts have been made by the government to blacklist contractors who regularly default after handsome contracts have been awarded to them. The establishment of the Ghana Highway Authority by the current government will, it is hoped, introduce the kind of administrative discipline needed to bring significant improvement to the road systems.

There are certain financial and technical problems that the Ghana Highway Authority will have to overcome in order to perform its duties satisfactorily. First, there is a need to obtain both domestic and foreign funds to maintain the existing roads. There must be a planned maintenance program to rehabilitate the trunk and feeder roads. At the same time, it will be necessary to institute a systematic plan for building first-class roads that will reduce the maintenance cost of roads in the long run. The Accra-Tema Motorway, built during the Nkrumah era, is a good example to follow. Second, the Ghana Highway Authority should encourage the government to provide the kind of technical training needed by local contractors to improve their skills. In addition, local engineers should be encouraged to enter the road construction industries and thus enable the country to become self-reliant in this section of the economy. Third, the establishment and efficient operation of a bituminous plant will enable the country to reduce the cost of importing road-surfacing material that will prevent the erosion of unpaved roads during the rainy seasons. Fourth, the existing quarries in the country should be exploited to provide a major basic material needed in the road building industry. In addition, the research institute responsible for roads should show innovation by investigating the possibilities and practicalities of introducing new local materials in the road-building industry.

The railway system has been declining steadily since the Nkrumah era. The rail service is no longer able to provide cheap and efficient transportation. Furthermore, the standards established during the colonial era have fallen and the services in general have deteriorated. The equipment is old and poorly maintained. Calls have been made for streamlining its administration. In order for the rail system to assume an important role in the economic development of Ghana, it will be necessary for the government to rehabilitate and modernize the existing equipment. New trains must be purchased, and the system should be electrified. More important, a concerted effort should be made to extend the rail lines to northern Ghana, which is fast becoming the bread-basket of the country.

The creation of the Volta Dam has presented a good opportunity for the development of a cheap north-south water transportation system. The present government is encouraging the Public Works Department to provide ferry service to link isolated areas cut off by the creation of the Volta Dam.

Since the establishment of the Ghana Airways Corporation in 1958, the country has experienced a remarkable acceleration of internal and external travel. Air transport technology has had a profound effect on the mobility of Ghanaians. However, various management problems, including dependence on old aircraft and high operating and maintenance costs, have influenced the effectiveness of the air transport industry. Ghana has had great success in training its own pilots and engineers. What remains to be done is the acquisition of new planes and the establishment of an efficient and disciplined management.

The National Shipping Line is currently operating under the same constraints as the airways, because its holdings include old, slow, and small ships incapable of competing with faster and bigger vessels.

The telecommunications system in operation is equally old and inefficient. Recently the government has sought funds from the World Bank to improve and modernize the system. An efficient telephone system in a developing economy is essential. The importance of communicating with the outside world through the facilities provided by telex systems and radio, as well as inexpensive ground-based relay for satellite communication, should not be underestimated. These infrastructural facilities should not be considered as technologically unfeasible for Ghana. Developing countries have the advantage of a late start, of being able to avoid the large financial outlays that the developed countries had to make for R&D. What is needed now is the transfer and adaptation of these existing technologies to suit local conditions.

THE COUNCIL FOR SCIENTIFIC AND INDUSTRIAL RESEARCH

The history of scientific research in Ghana dates back more than half a century. The research objectives of the United Kingdom colonial government led to the establishment of regional research centers in the four West African colonies: the Gold Coast (Ghana), Nigeria, Sierra Leone, and Gambia. Two years after Ghana attained its independence, the West African Research Office was established in Accra to serve as the administrative head office of the regional institutes in those four countries; it dealt with cocoa, rice, and oil palm, as well as public health and building technology. In 1958, however, the Ghana government passed the Research Act, which established the National Research Council and gave it the responsibility of dealing with Ghana's scientific, economic, and industrial research. It fell to the council to establish an organizational structure as well as the physical facilities to carry out the mandate of the government. Before it could begin serious work, however, an executive instrument was issued in 1960 to nullify its activities. In October 1961, the West African Research Office was abolished by the mutual agreement of all countries concerned. In 1963 the Ghana Academy of Sciences was established and the National Research Council was merged into it, with a view to maximizing the utility of the available manpower.

After about six years of operation, the Academy of Sciences was reconstituted as a purely learned organization (now called Ghana Academy of Arts and Sciences) and was separated from the newly created Council for Scientific and Industrial Research (CSIR) upon the recommendation of a committee headed by the late Sir John Cockcroft of England. The release of the Cockcroft Committee report revealed the basic defects of the CSIR and suggested certain changes that, unfortunately, were not seriously implemented.

In 1972 the National Redemption Council (with its powers now invested in the Supreme Military Council) issued two directives aimed at improving the usefulness of the CSIR. The council requested the CSIR to make recommendations for reducing Ghana's dependence on imported food items by suggesting ways of increasing the production of locally grown crops for domestic consumption as well as for export, and increasing the production of raw materials to feed the cotton, bast fiber, and oil palm industries. These requests were in line with the government's overall objective of widening the basis of the Operation Feed Yourself and Operation Feed Your Industries programs.

Without going into great detail about the functions, stated aims, and actual contributions of the research institutes of the CSIR, I may offer the following observations. As research organizations go, CSIR is a young institution. It has, however, historical roots that go back to the colonial period. Many of the research objectives of the colonial era were transferred en masse to CSIR. The lines of research being pursued by a number of the institutes are not directly linked to the economic sectors of the Ghanaian economy.

Let us consider one of the major problems confronting the agricultural sector of the economy, the cause of postharvest deterioration in crops. Yams, for example, constitute a major staple crop in Ghana and in many tropical countries. The high percentage of waste resulting from normal storage practices has long been known to the research organizations. There is tremendous weight loss from dessication and respiration. The damage done by microorganisms and parasitic nematodes results in yam necrosis. Storing yams below 12.5° C. causes chilling injury and weight losses even higher than those caused by normal storage practice. Treatment of the tubers with fungicides has been attempted with some success elsewhere, but the cost of such an operation has not been analyzed very carefully.

One would expect the Food Research Institute to investigate the waxing of the tubers immediately after harvest, which reduces the level of fungal infection and the amount of weight loss, and at the same time makes the yams attractive in appearance. Waxing is used commercially in certain countries to reduce wastage of sweet potatoes, tomatoes, and fruits of all types. Until the research institutes take an investigative attitude towards their work, their usefulness will continue to be questioned by the public as well as the industrial community in Ghana.

In more general terms, one would have expected CSIR to have been actively involved in the establishment of small industries. Small-scale industries are often described as unproductive, so there is little incentive for organizations such as CSIR to promote and encourage their establishment. However, small industries can make many important contributions by using the most improved technology to produce components and supplies for large-scale industries. The usefulness of such small-scale industries has been demonstrated in many of the technologically advanced nations. Therefore, it would seem logical for certain institutes of CSIR to give scientific and technical guidance to "village-type" industries in their efforts to help the economy.

Nevertheless, some of the research institutes have demonstrated their understanding of the kinds of contributions required of them. The Soil Research Institute is a case in point. It has carried out a systematic survey of more than three-quarters of Ghana and has produced maps that show soil associations, vegetation, and present land use as well as areas considered suitable for crops and livestock development. Many special-project surveys have been undertaken by the institute in areas that have been designated for both small-scale and large-scale agricultural projects. In addition, the institute's Soil Fertility Section has been engaged in making fertilizer recommendations for a variety of crops in different ecological zones. The kinds of services rendered to the agrarian sector by the Soil Research Institute are aimed in the right direction. Unfortunately, other institutes within CSIR have not demonstrated similar devotion to their mission. One often reads in CSIR reports and in other government publications the clearly stated objectives of the institutes, but in reality their performance leaves much to be desired.

The problems facing CSIR and the institutes are many. First, there serious administrative problems have beset the operations of the secretariat, and therefore have affected the research perspectives.

Second, the ability of any research institute to perform admirably will depend on the kind of personnel that is available. The needs of the institutes, however, are not met by the number and kind of trained personnel available from the universities in Ghana. Most of the best students go to the medical school. The next best become university professors. The research institutes draw from what remains of this limited pool of skilled manpower.

Third, the knowledge and skills that the research staff should possess are not available in Ghana because of the limited range of the science programs offered not only in the elementary and secondary schools, but also at the university level. Furthermore, the senior personnel who are expected to give scientific direction to the staff are often too busy with administrative work, which is frequently preferred to the art of directing research programs. There are several bright young men and women currently working in the research institutes, but they have no guidance.

Fourth, the research facilities in the institutes are grossly inadequate. One would expect that if the government takes seriously the kinds of contributions the research institutes can make to the development of the country's economy, an attempt would be made to furnish them with enough funds to perform their functions effectively.

Science Policy

For a developing country like Ghana to make full use of its resources, it is necessary that it engage in the formulation of a science policy that takes into account the social and economic objectives of the country. The development of a science policy should not be viewed in isolation. In fact, it should be considered an integral part of the national development strategy. Since Ghana gained its independence, several attempts have been made to develop some sort of an integrated science policy. The performance of the science-policy bodies is a reflection of the major problems facing the nation. The need to organize science to play a major role in the economic and social development of Ghana has been recognized by both the government and the private sectors of the economy.

However, the makeup of the science policy bodies often lacks the broad representation of all the major sectors of the economy concerned with science and technology, including scientists, needed at the decision-making level. Most often, governments of developing countries inadvertently consider economists to be the only group of professionals whose input into national planning deserves attention. It is becoming increasingly clear that for a national planning body to be effective, it must depend upon the consensus of the collective knowledge of economists, scientists, administrators, and politicians. In addition, the body must have the backing of a permanent support staff capable of in-depth studies in all important areas of the economy.

It has been reported that Ghana needs such a support staff with sound background in systems analysis, operations research, and cost/benefit analysis, and one that is able to carry on an effective dialogue with both economic planners and scientists. Many visitors and consultants to the Ministry of Economic Planning have commented on the fact that while there are a number of highly trained professional economists and social scientists, there are practically no persons with advanced degrees and experience in science and technology in that all-important government ministry.

This realization led to the establishment of the Science and Technology Planning and Analysis Group within CSIR to provide the needed coordination and guidance in decision making. Its establishment was the result of the 1973 Workshop on the Role of CSIR in Determining Science Policy and Research Priorities for Ghana, organized by the U. S. National Academy of Sciences and the Ghana government. A project agreement between the Agency for International Development and CSIR made possible the necessary financial assistance to help the group perform its stated functions satisfactorily.

The functions of the Planning and Analysis Group are to provide analytical studies needed by CSIR to carry out its broad advisory function; to collaborate with the Economic Planning Division of the Ministry of Finance and Economic Planning, with a view to identifying priority areas of national development where scientific and technological advice is needed; to analyze and recommend assignment of identified projects to research organizations (institutes, agencies, universities, and private organizations) that can best deal with the problems; to maintain contact with the various research organizations and retrieve from them any findings (published or unpublished) likely to be of importance to national development, and to correlate and assess data from different sources for ultimate transmission to CSIR; to project the manpower and other resources needed for investigating problems; to recommend and organize ad hoc task forces for analytical studies authorized by CSIR; and to carry out other functions assigned by CSIR.

The above stated functions of the Planning and Analysis Group cannot be made a reality unless positive financial backing is obtained from the government to enable the group to exercise a certain amount of direct control in the implementation of its aims. Currently the nonscientific arm of the government controls the scientific community to the extent that there is an atmosphere of resentment and quiet noncooperation between these two bodies. While no one is suggesting that the scientific community be given absolute control over the budgetary procedures of the government, it is essential that their representatives be taken seriously. The success of the Five-Year Development Plan of Ghana (1975-80), for example, will depend largely on the effectiveness of the scientific and technical involvement in agriculture, industry, education, health, housing, and communications.

Without an effective national science policy, no amount of organizational reshuffling of the institutions will make an appreciable difference in the economic performance. Recognizing that the current government has a unique opportunity to present the best alternative to Ghana's economic policies, I would like to suggest, first, that the government must take the lead in explaining what the country wants to achieve, the kind of guidance it can offer in economic activity, and to what extent it intends to put the governmental machinery to work. Second, the government must show exactly how it intends to achieve its goals by mobilizing the full range of the resources of the private sector to take an active role in economic development. Third, the government must indicate the kinds of cooperation it intends to achieve with other governments and business concerns in capital investments, loans, technical assistance, and technological transfer.

Transfer of Technology

There is a general awareness in Ghana that the transfer of technology can help solve many problems associated with low productivity in agriculture and industry. The major question is to decide which form of technology transfer should be encouraged. The United Nations has set forth the following general guidelines for developing countries to follow in their attempt to acquire foreign technology: preinvestment studies, including preparation of a feasibility study and a detailed project report; basic and detailed engineering, including preparation of machinery specifications, plant design, and factory layout; selection of equipment, construction, erection, and installation of machinery and start-up of plant; acquisition of processing or manufacturing technology; and technical assistance during the postinstallation period, including training programs and various forms of management assistance.

Implicit in the above guidelines is the fact that any type of technology transfer must be adapted to the local environment and, to a certain extent, the local environment must be adapted to the technology. Since most technology transfer flows from the temperate developed world, it is essential that the tropical developing countries think in terms of the transfer of the principles of technology instead of transfer of technology per se. Unmodified transfer of technology has often ended in total disaster when applied in foreign environments. It is therefore essential that a genuine R&D program be instituted in Ghana before any substantial investment in technology is made. Quite often the acquisition of foreign technology assumes complicated dimensions because many developing countries have nontechnical personnel conduct their purchasing negotiations. The result is that often the wrong types of equipment are selected at unjustified cost. The reluctance of government procurement agencies to use the available trained scientific and technological manpower in the decision-making process has resulted in wasteful investments in equipment and services that are not appropriate for the country. It is important for Ghana to recognize that the bulk of technology available from suppliers in the developed countries has been constructed in a situation of high labor costs and relatively cheap capital. What is needed, therefore, is the development of technology suitable in an environment with relatively low labor costs and expensive capital.

A major way of bringing purposeful and rapid development into Ghana is for the government to encourage foreign industries to bring in their technology and technical assistance with the view of helping to raise the productivity of the country. An example of the kind of relationship I am advocating is the arrangement between the government and Firestone Ghana Ltd. , whereby the Firestone Tire and Rubber Company has provided the needed technology and the managerial expertise to enable Ghana to develop a rubber industry. The agreement also provides for the training of Ghanaians to participate at all levels of the industry.

THE CONCEPT OF AGRICULTURAL PRODUCTIVITY

The economic prosperity of Ghana and, for that matter, most of the developing countries will depend primarily on the nation's ability to produce food in abundance. For several years the Ghanaian economy has been heading toward more consumption and less production. The Ghana Commercial Bank's Monthly Economic Bulletin for 1976 indicated that the Central Bureau of Statistics figures showed the national price index at an all-time high. Using March 1963 as a base (100), the March 1976 price index was 533. 8. Although the report showed that all economic activities had contributed to the high price index, the greatest increase during the period was registered by prices of local food, including yams, rice, cassava, corn, and plantain, as well as other fruits and vegetables. The local food price index had risen by 550 percent since 1963.

These figures indicate that although Operation Feed Yourself and Operation Feed Your Industries have taken some steps forward, the programs have not advanced far enough to cure the nation's food problems. One would have expected that the agricultural technology available in the country would have had an impact on food production. Unfortunately, this has not happened.

The most important avenue for improving productivity in agriculture is genetic manipulation. It is therefore essential that agricultural research be geared to the development of local stocks into quick-maturing and high-yielding varieties suitable to the ecological conditions of the country, as well as high in protein. The genetic techniques available for the production of grain, as exemplified in agriculture and university experiment stations, often yield two or three times more grain per acre than the village farmers produce. The farmers have not been adopting the measures for increased efficiency of grain production that are available in the country. Part of the problem is the lack of well-organized extension programs to transfer and to translate the available knowledge.

Corn production is a case in point. Without a doubt all the agricultural experiment stations are aware of the spectacular increases in the yield of corn when improved seeds resulting from locally bred hybrids are planted. Most of the Ghanaian farmers who are familiar with the high-yield seeds do not obtain these hybrids because of the cost and the requirement that the seeds harvested from the hybrid not be used for planting. Although there is scientific knowledge that can result in the production of improved corn seeds without the necessity of buying seed for planting each year, agronomic difficulties preclude introduction of these seeds. Most improved seeds do well in fairly restricted areas, since they are bred to give high yields under specific climatic and edaphic conditions. Because of the uneven edaphic conditions that characterize the farmlands, the high-yield seeds often are less productive than improved local seeds. Such results have contributed to the farmers' lack of confidence in the scientific community. Other agronomic problems in corn production concern the ease of hybridization. In order to obtain successful crosses, an improved variety should be planted at the same time on several farms in a well circumscribed area.

Perhaps the most poorly solved breeding problem has been the inability of the crop research institutes to create high-yield varieties that will suit local consumer requirements. For example, preparation of kenkey (a very popular cornmeal dish) from high-lysine sweet corn commonly grown in the United States would be totally unacceptable to Ghanaians, although nutritionally it is a superior meal.

An improvement in the productivity of the principal cereals, vegetables, tubers, and roots (cassava, yams, and sweet potatoes) should be the major concern of the country's agricultural research organizations. Genetic improvement is a long-term process, and to obtain a required quality it is essential that a small number of the basic crops be selected for in-depth study. Research on local tubers and roots has so far been restricted in scope, although they constitute the staple food items of the population. The existing research centers, particularly those under the Ministry of Agriculture and CSIR, should be encouraged to study ways of improving productivity in these crops.

Very little headway in productivity will be recorded if research in crop improvement is not coupled with research, development, and utilization of improved agricultural practices for both traditional and new crops. The agricultural establishments should therefore focus on a program of study that would allow for the adoption and improvement of farming techniques that would increase crop yields and reduce production costs. The techniques generally used by most small farmers in the country are somewhat crude and often result in low productivity. A straightforward advance to full-scale mechanization by these farmers would certainly be unwise, in view of their limited technical skill and the heavy capital investments involved. There is, however, considerable potential in the introduction of intermediate technological devices that could prove beneficial to existing farms. The use of small power-traction equipment and even animal-drawn plows can achieve a remarkable improvement in agricultural productivity. The introduction of suitable mechanization would also permit the working of larger fields in minimum time and would reduce costs.

Currently the Ghanaian farmer's productivity is low for several reasons. One factor is that farms are often too small, averaging between one and five acres. It is therefore important that a certain amount of mechanization be used on as large an acreage as is practical. The northern savanna areas are well-suited to mechanized farming, and since the human population of the savannas is less than that of the forest zones, the question of unemployment would not arise.

Another factor that reduces productivity is the practice of shifting cultivation. Quite often a farmer does not farm the same area for more than two years. This means that the human and financial investment he makes in clearing the land has to be repeated every two or three years, thus adding to his expenditure.

Two methods that have been used to maintain soil fertility and reduce the practice of shifting cultivation are crop rotation and the use of fertilizer. The advantage of crop rotation is that it makes it possible for land to be rested by growing another type of crop on it instead of leaving it fallow. The perfect crop rotation system has not yet been attained, much less taught to the farmers. Nonetheless, the agricultural extension services should be able to explain the advantages and the efficacy of the idea to the satisfaction of the ordinary farmer.

The possibilities for increasing agricultural yields by using fertilizers of all types have been demonstrated in Ghana. It is only recently, however, that efforts have been made to introduce the best methods of application and to explain its economic effects to the farmers. The major problem has been the organization and promotion of efficient use of the available fertilizer. For its introduction to be successful, it is necessary to convince the farmer that the cost is not prohibitive and that his financial investment will result in reasonable profits.

The need for water management that will ensure a supply sufficient to support plant and animal life is an important consideration in any agricultural program. Problems of irrigation, drainage, flood control, and desalinization of brackish water need serious attention if Ghana's agriculture is to be improved. In the savanna areas of northern Ghana, for example, the single rainfall season from April to September deposits a considerable amount of water (40 inches), most of which runs off. This is succeeded by a severe dry season during which crops and livestock suffer severely. It is essential that research be undertaken to find better ways of using both surface water and groundwater for agricultural purposes. Maximum use can be made of the available water through irrigation by canals, ditches, and spraying.

The Dawhenya Irrigation Project, established some 25 miles east of Accra, was initiated in 1959 by the Nkrumah government as an integrated multipurpose livestock-improvement, irrigation, and settlement plan. The ultimate aim was to build a model rural village with a complete supply of pipe-borne water and electricity. The technological know-how needed to accomplish the objectives of the project was not locally available at the start. The government first appointed an Italian firm of engineers to serve as consultants and to start work on the project in 1962 by constructing the pump station, divider tank, and intake channel as well as canals and drains. Various difficulties prevented the realization of the original objectives. A team from the People's Republic of China was engaged to start the rice project on a few acres by establishing a threshing and polishing mill. The project was suspended during the 1966 change of government.

Not much was done until the current government incorporated the scheme in the Operation Feed Yourself program; the project was renewed with the help of students, government workers, and private individuals. The achievements of the Dawhenya Irrigation Project gave the country a renewed sense that much could be achieved in agriculture if the country was willing to work. Furthermore, it convinced the government that the people were ready to undertake the much larger Accra Plains Irrigation Project, estimated by the Kaiser Engineers in 1964 to cost U. S. $128 million.

Crop failures during the 1975-76 fiscal year, and especially the decline in rice production, made the government even more aware of the need for an integrated, well-organized, and well-managed irrigation program to supplement the available sources of water.

GENERAL OBSERVATIONS

Manpower

At the beginning of this paper I pointed out the need for a judicious exploitation of Ghana's manpower, both skilled and unskilled. Most "Ghana watchers" are often baffled by the seemingly ineffectual manner in which the country uses its manpower. It is common knowledge that Ghana can boast of having proportionately the best-educated elite and civil service in black Africa. The reasons for the poor use of manpower are various. Perhaps the most significant problem has been the lack of perception in the selection of qualified persons to fill strategic positions in the bureaucracy. Unfortunately the succession of governments has fallen prey to the selection of "fast talkers," lacking depth and experience, to fill key positions. For some inexplicable reason, when someone with experience comes to the attention of the government, that person is immediately overloaded with assignments, to such a point that his effectiveness is quickly diminished. Other qualified persons are often relegated to less fulfilling work. Those who become thoroughly disenchanted leave the country. It is common knowledge that Ghana has lost much of its best manpower to a number of world organizations and to other countries, simply because there seems to be no imaginative way of deploying most of such key personnel to the best advantage.

Although there is a strong appeal and support for scientists and others to emigrate, a major concern has been expressed in many quarters about the loss of technically trained people, particularly from the less developed to the developed nations and the international organizations. This free movement of individuals, commonly referred to as "brain drain," has presented a potential drain on the scientific and technological capabilities of the developing countries. Regardless of the patriotism that scientists and economists might feel toward their countries, they cannot contribute their best without adequate equipment, support, and some appreciation for their knowledge. The situation worsens if they are not allowed to work in an atmosphere conducive to imaginative thinking and productive work.

The selection of the wrong persons for important managerial positions has brought about serious mismanagement in many vital sectors of Ghana's economy. For example, the lack of proper deployment of many people who are unemployed points to a serious defect in the nation's economic planning. As E. P. Schumacher put it:

> A country's economic well being depends on how it uses
> the labour power of its ordinary people. If this labour is
> wasted in unemployment, the country will be desperately
> poor. If it is used up to the hilt, both quantitatively and
> qualitatively, the country will be well off and its economic
> development will be rapid. [8]

The exceedingly high rate of unemployment in Ghana provides a telling example of how unsuccessful its economic planning has been. So far the government has been unable to mobilize the labor force to form an important component in the development of the country. (See Table 8.4.)

TABLE 8.4

Recorded Unemployment in Ghana, 1968-73 (monthly averages)

	Registration	Placements (1,000)	Unemployment	Unemployment Index 1968 = 100
1968	35.3	5.2	17.6	100
1969	30.5	5.2	15.0	85
1970	33.9	6.3	16.5	94
1971	37.1	6.5	18.4	104
1972a	48.4	5.1	31.2	177
1971				
1st Qtr.	37.6	7.9	19.3	110
2nd Qtr.	38.1	7.4	18.5	105
3rd Qtr.	34.0	5.3	17.8	101
4th Qtr.	38.7	5.5	17.9	102
1972a				
1st Qtr.	52.1	6.2	41.4	235
2nd Qtr.	56.5	5.4	30.7	174
3rd Qtr.	42.9	4.1	26.6	151
4th Qtr.	42.2	4.6	26.2	149
1973				
1st Qtr.	42.9	5.3	27.1	154
2nd Qtr.	39.6	5.7	24.1	137
3rd Qtr.	41.7	4.5	27.2	154
4th Qtr.b	44.7	5.1	28.1	160

Note: Data collected on the basis of registrations at 50 employment centers throughout Ghana.
aBeginning in 1972, the data are affected by a change in the definition of recorded unemployment.
bAverage of two months.
Source: Ministry of Labor and Social Welfare, Employment Market Reports (monthly).

The migration of many people from the rural areas into the cities began soon after Ghana attained its independence and the government decided to begin its industrialization program in and around such metropolitan areas as Accra, Tema, Kumasi, and Takoradi. Reversal of the movement into the cities will require two positive commitments by the government. First, it must, as a matter of national policy, stress and encourage the establishment of labor-intensive industries, as far as this is practicable. Second, more industries should be established in the regions. This would help to avoid the unnecessary concentration of labor in the urban areas and encourage people to obtain gainful employment in the rural areas, thus reducing unemployment.

Another factor that aggravated the unemployment situation in the mid-1960s was that many state programs instituted soon after independence were closed down because of foreign-exchange difficulties. By 1970 more people had lost their jobs, adding to the nation's economic problems.

Short-Range and Long-Range Investments

Some of Ghana's major financial difficulties stem from the fact that the large initial capital investments made soon after independence were generally in community and social services that normally do not generate financial profit in the short run; this is especially true of education, health, transportation, and telecommunications (see Table 8.5). These investments have long-range benefits since, if well conceived, they facilitate the activities of industry and other income-generating concerns. Since independence there has been a tremendous expansion of education in Ghana, and the benefits of the initial investment have already manifested themselves in the number of educated persons available for deployment in the various sectors of the economy.

The educational system established by the British colonial government and continued by the subsequent Ghanaian governments did not, however, prove altogether beneficial because of the generally weak emphasis on technical education. For example, of all males attending school, only 1.4 percent are in commercial and technical schools. For females the percentage is 0.7. For some reason, when Ghana decided to embark on industrialization, a corresponding effort was not made to promote technical education. In fact, even the token technical education that existed at the elementary school level before independence, such as gardening, first aid, and woodwork, began to be dropped from the curricula. Only in recent years, when Operation Feed Yourself was initiated by the current government, has there been a revival of gardening in elementary schools, although it is not formally integrated into the school curricula.

TABLE 8.5

Functional Classification of Ghana Central Government Capital Expenditures, 1968/69–1972/73 (million cedis)

	1968/69	1969/70	1970/71	1971/72	1972/73 (est.)
General services	17.4	19.7	28.0	24.6	18.8
General Administration	7.6	10.5	19.2	13.6	9.9
Defense	8.7	8.3	6.7	6.8	7.3
Justice	0.6	0.6	0.3	1.5	} 1.6
Police	0.5	0.3	1.8	2.7	
Community services	16.7	28.0	35.8	32.2	32.3
Roads	12.3	20.7	29.8	24.1	23.8
Fire protection, water supply, sanitation	3.6	6.1	4.0	5.8	5.7
Other	0.7	1.2	2.0	2.4	2.8
Social services	13.2	16.1	19.1	21.5	29.8
Education	4.4	6.1	9.4	9.2	8.6
Health	2.2	4.8	5.7	8.4	6.5
Social security	—	—	—	—	—
Special welfare services	0.2	0.2	0.1	0.9	0.8
Other	6.4	5.0	3.9	3.0	13.9
Economic services	16.5	16.5	35.8	26.9	25.5
Agriculture, nonmineral resources	7.7	7.5	10.8	8.3	12.0

Fuel, power	0.3	1.6	4.2	2.2	1.8
Other mineral resources, manufacturing, construction	7.3	7.1	3.6	4.0	3.4
Transport, storage, communication[a]	0.5	0.8	2.3	2.7	1.9
Other[b]	0.7	-0.5	14.9	9.7	6.4
Unallocable (transfers to local government)	0.3	—	0.4	3.8	0.4
Total	64.1	80.3	119.1	189.0	106.7
Errors and omissions	—	-0.1	0.2	-0.5	—
Grand total	64.1	80.2	119.3	108.5	106.7

[a]Excluding Post and Telecommunications Administration.
[b]Including investments below the line (net).

Source: Central Bureau of Statistics; Report on Ghana's Government Finance Statistics, July 18, 1973.

There are even more fundamental problems with the educational system that Ghana inherited from the British. During the colonial era much emphasis was placed on the arts and humanities instead of on the natural and physical sciences. As I have pointed out, one could have expected that in a predominantly agrarian country such as Ghana, emphasis and high priority would be placed on the training of university graduates in the biological and technical sciences. [9] This is not the case. At present a substantial number of the competent students at the universities are encouraged to read law, political science, the classics, history, and the other humanities because the financial rewards in these fields exceed those in the sciences. The number of students who study the natural and physical sciences is relatively small, yet the need for a competent corps of biologists, medical doctors, and engineers far outweighs Ghana's need for lawmakers and political scientists. The educational system in the country is unconsciously designed to perpetuate dependence on foreign scientists and technologists.

The capital investment in health services has aided in the building of hospitals, health centers, health posts, and other medically related institutions. Although the health services in the major cities and towns have improved considerably since independence, much more needs to be done, especially in the rural areas. The 52 health centers currently being operated in the rural areas are not sufficient to care for the population. Nonetheless, the 10 million cedis or more that the government invests annually in the health sciences cannot be quantified on a profit basis. One expects, however, that sound planning will go into the allocation of government funds and will be designed to bring about a reasonable balance between long-range and short-range activities. The planning objectives of the government are the crucial factor in the overall development of the country. Maintaining a critical balance in allocating funds to short-range and long-range programs is a formidable task. Any serious mistakes in judgment will lessen the effectiveness of the development of the essential infrastructure and the productive sectors of the economy.

The Concept of Interdependence

Over the years many economists have stressed the need for interdependence between poor and rich nations. The tradition has been that the poor nations produce the basic raw materials by cultivating such items as cocoa, rubber, cotton, tea, and sugarcane on their best lands. The rich countries produce the manufactured goods to supply the poor nations. This arrangement seemed an amicable one throughout the colonial era, since production policies were exploitative rather than aimed at actual economic development. Since independence, developing countries have been questioning this arrangement, which is theoretically meant to encourage their economic progress. It now seems evident that this kind of developmental strategy is less economically fulfilling to the developing nations. Furthermore, such arrangements perpetuate the traditional dependence on the industrial countries, and do not allow the developing countries to adopt the strategies suited to their own economic and social conditions.

Let us examine the cocoa industry of Ghana in this light. Why should Ghana continue to export large quantities of cocoa beans to the developed countries while it is technically feasible for Ghana to process the beans to secondary, tertiary, and even final stages? If the developed countries really wish to help the economy of Ghana, they will be receptive to the idea of buying at least half of Ghana's cocoa production in processed form. Such an arrangement would bring about an industrialization program that would create employment and help Ghana to obtain better earnings for its investment of manpower and funds in the cocoa industry. The developed countries should be prepared to invest in such a venture if they are sincerely interested in rectifying the current imbalance in the concept of interdependence.

A truly integrated cocoa industry in Ghana should not necessarily be a predominantly capital-intensive venture. Properly conceived, it could be organized to achieve high rates of growth and inputs of both domestic and foreign capital, using the machinery and the technical skills available in the country. With careful use locally available production factors, scientific and technological applications could be made in small-scale industries, enabling them to make better use of the available financial resources as well as of skilled and unskilled manpower resources. The advantages of such small-scale industries (especially if they are situated in strategic rural areas) include the favorable ratio of generation of employment per unit of capital invested.

Although Ghana is a small country of only 92,000 square miles (about the size of the state of Oregon), the supply of its natural resources and its relatively small population of 9.9 million give it certain advantages that many developing countries do not have. If Ghana engages in a carefully planned economic strategy, a modest scientific and technological base to supply most of its basic needs can be developed. For example, in the agrarian sector, Ghana can manufacture many capital goods, such as simple machines, tools, and other farm equipment that are necessary for the proper exploitation of the land. The Tema Steel Works Division has the necessary capacity to produce agricultural implements that are directly suited to local conditions. There are several semiautomated activities in which the steel industry can engage by investing the human resources of the country to the maximum, instead of adopting the concept of economic growth based on highly automated machinery imported from the developed nations. Unfortunately, however, Ghana's development plans have not given a high priority to the steel industry.

While on a recent visit to the People's Republic of China, I had the opportunity to visit a commune steel factory in Shanghai. In comparing the physical development, and the quality and technical knowhow of the manpower of this commune establishment with the Tema Steel Division, the productivity of Tema is low. There is no doubt that the poor performance is directly attributable to the lack of commitment by the country to rely on its own manpower and natural resources to produce the basic equipment. So long as Ghana continues to import basic farm equipment and allows its economy to be subjected to the vagaries of the market forces in the developed world, the economic future of the country will become increasingly precarious.

Ghana's current economic problem is a variation on a theme that dramatizes the fundamental weakness of the economies of many developing nations. With proper management at all levels, Ghana can strengthen all its economic sectors by improving and modernizing the traditional low-capital-intensive technologies. If the proper atmosphere exists, it will be possible for local experts to serve as management consultants to help improve the economic and technological conditions of other developing countries, which often pay no more than lip service to the concept of self-reliance, when it comes to relying on the advice of their own qualified nationals. In his report Future Tasks for UNDP, Sir Austin Robinson, honorary president of the International Economic Association, observed:

> It is difficult also to understand why there should be so strong a reluctance to recruit as consultants the nationals of the country concerned who have left the country and have taken permanent jobs in other countries. Where they have relevant specialized knowledge, they possess also a local knowledge, an appreciation of local problems and an intimacy with local experts which will make them uniquely competent to advise. It is even more difficult to understand why, when there are competent consultancy firms in the developing country, there should be doubt about the propriety of inviting them to make feasibility or other studies. [10]

Finally, there is no doubt that many of the economic and social experiences of developed countries in northern temperate areas are mainly irrelevant to the needs of developing countries in the low latitudes. The sooner the developing nations recognize this fact, the better off they will be. It is equally important that a certain amount of

political stability be maintained in Ghana if there is to be any hope
for successful development. Political instability often encourages
entrepreneurial groups to invest in short-term financial activities
instead of in long-range ones. Furthermore, it tends to encourage
government involvement in business activities, in a way that often
stifles the economy and discourages innovations. As Ghana attempts
to break through the current economic difficulties, it is hoped that
the country will be guided by a quiet determination to follow the policy
of self-reliance while recognizing the potential for genuine cooperation and assistance from its friends abroad.

SUMMARY

A moderate yet potent introduction of science and technology
into the economic development of Ghana is essential; indeed, it is
vital. For this to become a reality will require judicious attempts by
the public and private sectors of the economy to make better use of
the available manpower and to exploit the country's natural resources
properly.

During the colonial era some scientific and technological methods
aimed at enhancing the extraction of raw materials and the production
of cash crops for the export trade were introduced. Little attempt was
made to develop Ghana through building on a scientific and technological basis to make the country economically independent.

The Nkrumah government gave a tremendous impetus to education and introduced science courses into the school systems as the
foundation for the development of technical personnel. Many national
development plans were initiated with opportunities for the application
of science and technology in specific sectors of the economy, but thus
far the use of science and technology in the overall development of
Ghana has been limited. A few areas, however, have benefited from
the introduction of intermediate types of technology, such as the packaging of cassava flour and the canning industry. But several chances
to make science and technology an effective medium for economic development have been missed. The establishment of the state enterprises was a good beginning for the introduction of science and technology, in relation to their impact on import substitution. Many of the
enterprises encountered difficulties, partly because of unsatisfactory
managerial and financial support.

The major introduction of science and technology during the
Nkrumah era was for the completion of the Volta River Project. Although many economists felt that Ghana was shortchanged in the financial negotiations, the fact remains that many construction companies were not willing at the time to make available what amounted

to high-risk capital investments without certain guarantees. Despite its critics the project has proved to be successful. Certainly its impact on the production of hydroelectricity has been most beneficial for the developing industries.

The post-Nkrumah era has seen several changes in government approach to the economic development of the country. The most significant was the introduction of Operation Feed Yourself and Operation Feed Your Industries. While these programs are in themselves important psychological acts, the initial enthusiasm seems to have waned because of ever-increasing prices of locally produced foods, a situation that the programs set out to correct. Even though inflation has affected production throughout the world, there is no justification for the current poor performance in the production of basic food items in Ghana. A close analysis of the situation reveals that, as was observed during the Nkrumah era, most of Ghana's problems are caused by poor management and gross lack of discipline at many levels of the economy, especially in the public sectors.

Among the most disappointing organizations in the country is CSIR. The pursuit of the kinds of research that could be linked to the economy of the country has never been realized. Since Ghana is an agrarian country, one would expect, for example, that CSIR would earnestly engage in research aimed at increasing food production and decreasing spoilage and waste in postharvest activities. CSIR should participate in the introduction of intermediate technology, especially in areas that could help the small farmer, and should encourage industries to engage in import substitution.

The problems of CSIR are due partly to the poor organizational structure and management of the institutes, the constant neglect of program orientation, and the fact that nobody seems to be accountable. Other factors involved are the poor support that the institutes get from the government, the fact that nobody has made it a point to involve the private sector in meaningful and dynamic support of the research and development activities of the institutes, and the need for obtaining high-caliber staff that can invent and innovate.

To make science and technology an effective component in the economic development of the country, it is essential that a realistic science policy be developed. Furthermore, for a development strategy to be effective, it is equally essential that a broad representation of qualified persons from all sectors of the economy, including scientists, be brought into the process of decision making. The establishment of the Science and Technology Planning and Analysis Group within CSIR is a step in the right direction, but the organization will be ineffective if the government pays only lip service to the support that it needs.

There is no doubt that Ghana can and should encourage the transfer of technology appropriate to its needs. More important, Ghana should think in terms of the transfer of the principles of technology instead of transfer of technology per se. It should be realized that the kinds of technology the nation needs should be, as far as practical, more labor-intensive, since capital-intensive technology will tend to aggravate the already bad unemployment situation.

Although several industries are beginning to introduce the kinds of technology needed to achieve rapid development, the general consumption rate in the country far exceeds its production. As a result there are undesirable cycles in the availability of such items as toiletries (specifically soap and toothpaste), milk, sugar, insecticides, and detergents.

Efficient and well-established infrastructure most often leads to economic prosperity. The transportation industry leaves much to be desired, for lack of transportation often creates artificial shortages in the country. It is not uncommon to observe simultaneous abundance of grain in the markets of northern Ghana and acute shortage in central and southern areas of the country.

The technological know-how needed to repair vehicles and roads exists in Ghana. What is needed is the will to institute well-managed establishments, with the appropriate financial backing, to keep the nation's transportation system in good repair. The railway system suffers from old equipment and poor maintenance, and needs a streamlined administration. In addition, the lake transport program should be greatly enhanced: the creation of the Volta Dam has presented an opportunity for such a development. Telecommunications technology needs to be properly established in Ghana to strengthen its contact with the rest of the world.

Although Ghana has rich natural resources, few serious efforts have been made to turn them to the betterment of the people. The human resources are equally underutilized. The current economic crises can be better solved by encouraging the public and private sectors of the economy to take science and technology seriously. The concept of self-reliance will become a reality only if Ghana makes a concerted effort to use efficiently all of its resources.

Whenever Ghanaians meet nowadays, there is an interminable discussion on the state of the economy. The reasons for Ghana's unenviable economic conditions are being questioned, since the country already possesses a relatively sophisticated civil service, a small but solid core of well-trained scientists, engineers, and medical doctors, and an affable population. One painful reality is that certain individuals in key public positions lack confidence in Ghanaian scientists and engineers because the country for years has relied on outside expertise. The successive civilian and military governments since independence have been supported principally by Ghanaian administrators who seem uneasy at entrusting responsibility to their own trained personnel. If Ghana is to develop through the utilization of science and technology, the lingering distrust of its trained scientists, technologists, and economists must end.

NOTES

1. F. Agbodeka, Ghana in the Twentieth Century (Accra: Ghana Universities Press, 1972).

2. E. Ashby, Universities: British, Indian, African; A Study in the Ecology of Higher Education (Cambridge, Mass.: Harvard University Press, 1966), p. 149.

3. K. Nkrumah, "Speech Delivered by Osagyefo the President at the Laying of the Foundation Stone of Ghana's Atomic Reactor at Kwabenya on 25th November 1964," Ghana Journal of Science 5, no. 1 (1965): 1-5.

4. Ibid.

5. K. Nkrumah, Speech delivered before the Ghana Parliament, March 4, 1959 (Accra: Government Printing Department, 1959).

6. K. Nkrumah, Seven Year Plan for National Reconstruction and Development Fiscal Year 1963/64-1969/70, p. v.

7. E. S. Ayensu and A. Bentum, Commercial Timbers of West Africa, Smithsonian Contributions to Botany no. 14 (Washington, D. C.: Smithsonian Institution Press, 1974).

8. E. P. Schumacher, "Intermediate Technology," Co-existence 5, no. 1 (1968).

9. E. S. Ayensu, "Science and Technology in Black Africa," in World Encyclopedia of Black Peoples, I (St. Clair Shores, Mich.: Scholarly Press, 1975), pp. 307-17.

10. Sir Austin Robinson, Future Tasks for UNDP (New York: U. N. Development Program, n. d.).

9

GAPS IN THE
SCIENCE-TECHNOLOGY-DEVELOPMENT
SEQUENCE: A COMMENT

Simon Kuznets

My intent is to present brief comments, so designed that the valuable chapters by the authorities on the experience of the seven countries, and the rich summary chapter by Professor Ranis, might be brought to bear upon questions that, to my mind, are still unanswered. Let me begin by dealing with the sequence suggested in the title of this symposium: science, technology and development (S–T–D).

The reason for beginning with this sequence is that, as Professor Ranis noted, the linkages among the three processes are not easily specified, and are as yet insufficiently illuminated by research on the history of science, technology, and economic growth. Professor Ranis mentioned that the linkage between science and technology is subject to a great deal of time variance and uncertainty. And it is true that in a number of cases in which one can date a major invention, the derivation of which lies in a scientific discovery made at another specific date, the time elapsed represents periods so long and so variable as to make it difficult to test analytical connections or use the data in considering even long–term policy decisions.

Yet one of the important questions to which we should seek an answer, as much for the benefit of the developed as for that of developing countries, is what are the major variables that affect the linkages between a scientific discovery (meaning an uncovering of a new aspect of the structure of the physical universe) and an invention (based on the discovery). This is particularly true of a major invention that, once converted into a successful technological innovation, sustains a high rate of economic growth and transforms the conditions of work and life for large proportions of the population.

Among these factors, three should be noted. First, a scientific discovery is never sufficiently specific in technological terms to permit the inventor-technologist to use it without a great deal of experimentation and trial and error, even assuming that the complementary discoveries and inventions have already been made. The scientist, in making the discovery, is concerned with the relation of what he has found to a theoretical framework summarizing the structure of the universe. He is not driven to examine some empirical parameters that often prove to be of crucial importance to the inventor and the technologist. We find that for many scientific discoveries, the discoverer-scientist, upon being asked what useful application can be made of it, says "none"; and we should not be surprised. It is difficult to visualize the eventual technological applications of a new scientific discovery, since they may be followed in a variety of directions; nor would the "basic" scientist be impelled in the applied direction by the analytical goals of his research. Thus, if my reading of the history of technology is correct, it took half a century from the time Heinrich Hertz demonstrated the existence and some properties of electromagnetic short waves, and the time when sufficient knowledge of their important empirical characteristics became available—partly as a result of invention, partly as a result of application of the invention—to be sure of the mechanism by which these rays were transmitted.

Second, if a great deal of input is required to exploit a scientific discovery for a major invention, such input will not be made unless there is the prospect of substantial demand for the product, unless the would-be inventor has a vision of the possible uses of the application of the discovery. In some cases these possible applications are obvious; in others they are not. But even when obvious, the applications may require a major effort not always assured of success. Hence, if there is a series of steps, involving time and resources, between a scientific discovery and an invention based on it (which, by the way, need not be true of much observational empirical evidence produced by sciences of a more empirical character, which evidence may be transmitted easily into practical application), there is likely to be a wide gap at this important link in the S-T-D sequence.

Third, one should note that a major invention, when first made, is only a skeleton. It is an outline that indicates, in a tested fashion, that the new process is feasible or that the new device is workable. It does not show the full potential, which will be uncovered only later; once the major invention is applied, a variety of subsidary inventions, innovations, and improvements are made and, in the course of mass application, the full potential of the major invention in efficient use is revealed. I am stressing this point because economic growth, or modern economic development in the sense in which Professor Ranis referred to it—a high rate of rise in product per worker and per capita, usually for a population that is also growing at a substantial rate—is sustainable only if there is a flow of technological innovations: not only inventions, but innovations sufficiently applied to spread and significantly affect the parameters of output and productivity. The link between invention and technological innovation is significant, and may be difficult. In some cases initial efforts to apply the invention may misfire, because of poor enterpreneurship, or insufficient attention to cost elements, or new technological obstacles discovered when the invention is applied. The resulting delays may widen the time span between the scientific discovery and its contribution to economic growth.

Among the conditions of success of a major technological innovation is the adjustment of social institutions, and of the patterns of work and life, to provide the proper channels. Thus, when the steam railroads, one of the major technological innovations based largely on inventions subsidiary to the basic invention of the steam engine, began to be established, it soon became apparent that the old form of organizing business units would not fit. No family or partnership could run a railroad company, with the enormous amount of fixed capital that it required and the continuity of management that had to be assured. It had to be organized as a modern corporation, with specifically formulated relations among the owners of the capital, the responsible managers, and (eventually) the government and the consuming public. But it took the United States almost half a century to evolve the modern business corporation as a new legal form. Likewise, a modern factory, similarly endowed with large fixed capital, could not be operated by a business unit typical of the handicraft shop of the Middle Ages or of the mercantilist period between the Middle Ages and the modern times. Nor could a modern factory be economically feasible in the countryside, for its (say) 5,000 workers, their families, and the people who would provide them with services would constitute a city; and to spread them over the country side would involve prohibitive costs of travel to and from work and of providing them with services. This means that, for efficient channeling, technological innovations required not only new forms of organization of economic units but also changes in conditions of work and life: a whole stream of changes in economic and social institutions and conditions of life and work of major groups in the population. The connection between technological innovations and such social and institutional changes was not fully rigid, in that some choice in the adjustment of society to the requirements of new technology remained. But the costs of many choices were prohibitive; and we find no developed country in which the sequence of modern technological innovations did not result in a major shift of the labor force to employee status (from self-employment, or small-scale entrepreneurship), and in which industrialization was not accompanied by significant urbanization, with a number of important corollaries of these major changes in conditions of work and of life.

Thus, in applying lessons from the past to the case of the developing countries today, we must recognize that the sequence from science to technology, within technology from invention to technological innovation, and within the latter from initial bare framework to efficient mass production, is also a sequence in which final or intermediate demand, and social and economic conditions, play a crucial role. They did so for the pioneer industrializing country, and for the early "followers." The early followers were successful in their attempts to borrow and adapt, partly because their social and economic conditions were already close to those of the successful pioneer.

Let us now briefly consider the S-T-D sequence in reverse (that is, D-T-S). The successful application of science-derived invention in the mass-production stage of technological innovations provides a major stimulus for further development of science (over and above the increase in resources available for it). This follows logically from what was said above. If in the course of making a scientific discovery, the scientist may not be interested in or concerned with certain empirical aspects of the phenomena in nature that he has uncovered, and then these aspects are explored in the course of application of invention, particularly in mass production, new data emerge, new tools are designed, and new puzzles provide further stimulus to science. Thus, the development of radio astronomy is the product of the development of the radio industry, because it was in the widening application of radio transmission that certain features of the rays used, in relation to the ionosphere, were discovered. And thus a tool was found by which new modes of observation of the skies were made possible in the implicit discoveries of industrial development. A major reason for concentration of science in the developed countries, particularly concentration of work on the scientific (and technological) frontiers, is not only that these countries are "rich" and have the resources, but even more because they are in a world in which their own production system continuously yields results in the search for greater efficiency in application of advanced technology, new data, tools, and scientific puzzles. It is hardly an accident that advanced industry and scientific tools are closely related, and that there has been a trend toward greater use of science in industrial research, with notable contributions to the flow of scientific discoveries.

In light of this brief discussion of the relations between science, technology, and economic development, as they have been observed in the developed countries, several implications bearing on the developing countries may be drawn. Such implications are necessarily tentative, but they should serve at least to indicate the directions of further exploration.

The first implication flows from the continuous interplay between science, technology, and development suggested above as characterizing the record of the developed countries since the late eighteenth century. As one considers the limited number of countries that became developed, one finds that they were continuously subjected to, or have continuously experienced, a flow of new scientific discoveries, new inventions, and technological innovations, and have had to adjust continuously to the promise of the innovations or to some of the problems that emerged with their extending mass application. When steam-engine technology, after almost a full century of development and extending application, began to reach its limits in the second half of the nineteenth century, electric power, the scientific base of which was already known in the early nineteenth century, began to come into its own; and the electric age, ushered in during the 1880s, then extended to the fractional motors and household appliances of the 1950s. Why this continuous shift from one major invention and innovation to another? The answer is that the exploitation of any major innovation reaches levels at which further economic gains are likely to be limited, while at the same time it is found subject to some limitations relative to newer inventions and newer potential innovations. Yet it is next to impossible to forsee in reliable fashion the key characteristics of a major innovation in its early stages and to forecast its eventually revealed limitations relative to a new and competing innovation, even if the latter can be identified. In general, technological innovation is by definition—the definition relating to the term "nov"—something new and, hence, partly unknown, before it is used extensively and thus becomes older and better known.

The important implication that follows is that the developed countries are under continuous pressure, and will continue to be under such pressure, to pursue science, inventions, and technological innovations oriented toward their problems—these problems being the obsolence of the older inventions, with whatever bottlenecks and negative externalities they may have created— and toward the promise of new technological innovations, continuously emerging from the feedback relation between past science, technology, and economic development. This means that the research agenda of the developed countries tends always to be relatively full. And there is a real question, which I raise with some hesitation, whether except in specialized cases mentioned below, a major shift of scientific and inventive resources of the developed countries to work on the problems of developing countries is possible or, in the longer run, advisable.

I am reluctant to argue that the developed countries are the carriers of advance in knowledge relevant, and potentially useful, to all mankind. Yet, once made, the products of discovery, invention, and innovation are accessible to the rest of the world. It is not only a question of direct transfer, of the type that occurred in control of death since World War II that sharply reduced death rates in almost all of the world's populations. It is also a question of borrowing through receiving advanced products in the normal course of international trade. One aspect that I noted about the country papers is the failure to emphasize that while the country in question may not have developed its own technology and its own science, it usually borrowed— and received the advanced products in trade. As I look around the world today, we really do not have fully traditional, technologically traditional, nations among most of the developing countries. They have all been affected, and mostly in a positive fashion, by living in a world in which mankind acquired the power and the possibility to raise productivity and utilize more power for mostly good human purposes (with the significant exception of those associated with national rivalries and frictions).

Second, if we assume that the possibilities of a large flow of scientific and inventive resources to work on the specific problems of the developing countries are limited, what is the implication for the development tasks of the developing countries? We may consider the question, while allowing that for specialized tasks, such as are suggested by the programs of the Wheat Institute or Rice Institute and similar research facilities, transfers of valuable human research resources from the developed countries can, and do, occur.

One may begin by referring to Professor Ranis' note that there is room for a much easier shift of parts of scientific and technological equipment toward the developing countries. The discussion above largely emphasized major scientific discoveries and their complement in invention and technological innovation. But science consists not only of this type of discovery, of basic paradigms. There is the powerful apparatus of data gathering, data checking, classification— with a number of observational and experimental sciences concerned with tests, measures, melting points, freezing points, maps, historical records of natural events, geological surveys, and the like. All of this is easily available; and its use does not require major transformations of economic and social institutions, except for education of a sufficient number of members of society to a point where they can utilize the vast stock of information relevant to problems of production. I would put a high valuation on the kind of activity that Professor Pastore referred to: geological surveys, surveys of coastlines and rivers; in short, the advance of tested information on the specific features of the country's land, soil, and waters. This is the kind of inquiry that, according to Professor Rosenberg's paper, the United States concentrated on through most of the first half, and much of the second half, of the nineteenth century, while the country was expanding toward the still not fully known West. The accumulation of this kind of empirical tested data, and inventory of the country's physical resources, is one prerequisite for the development of empirical science and of technology, the efficiency of which would be sustained by application of established knowledge of natural materials and processes to the specific natural resources of a given developing country.

Such beginning steps require the development within the developing countries of scientific and technological personnel capable of effectively contributing to the initial inventories, the latter demanding sufficient knowledge of, and expertise in, the major observational fields in the scientific curriculum. And this pressure for building up the country's scientific and technological human resources is all the more natural when one considers the urgent need for the proper choice among technologies available for borrowing, and for a variety of adaptive adjustments needed to modify such borrowed technology to better fit the country's resource endowment. There are, undoubtedly, various specific questions involved relative to the possible tradeoffs between developing the domestic human resources and continuing to purchase advanced technological goods. This is not the place, nor am I competent to discuss the general range of questions involved. But it does seem to me that the tested knowledge of the country's physical endowments and the capacity to modify technology available elsewhere to suit domestic needs are indispensable early in a country's economic development; and they ought to be largely undertaken by the country's scientists, engineers, and technologists, who would then provide the ground on which more advanced science and technology should flourish.

Finally, one should return to the comment made earlier that in the S–T–D sequence, the efficient development of all three requires transformation of economic and social institutions to provide proper channels. They are required not only for the more efficient organization of mass production embodying the contributions of major technological innovations (many derived from scientific discoveries). They are also required for the building up of the scarce scientific and inventive human resources that need for their flourishing a social and economic structure that encourages an objective view of the natural universe and of the role of man in it, as well as a concern with human material welfare involving the opportunity of relatively free choice for members of society whose welfare is the desideratum.

ABOUT THE EDITORS
AND CONTRIBUTORS

DR. WILLIAM BERANEK, JR. is Associate Director of Holcomb Research Institute, Butler University. He received his Ph. D. in chemistry from the California Institute of Technology. (As director of studies in the area of science and public policy, he has organized a nationwide symposium and published in the field.)

DR. GUSTAV RANIS is a professor of economics at Yale University and is affiliated with the Economic Growth Center at Yale University. He received his Ph. D. in economics in 1956 from Yale University. His published interests are in the fields of economic development and the economics of technological change.

DR. EDWARD S. AYENSU is presently director of the Endangered Species Program at the Smithsonian Institution. He earned his Ph. D. from the University of London. Dr. Ayensu has authored many scientific books and articles on systematic plant anatomy and tropical biology.

DR. DONALD STEPHEN LOWELL CARDWELL received his Ph. D. at the University of London in 1949. Currently he is reader in History of Science and Technology and head of the department, University of Manchester Institute of Science and Technology. His published interests are in the field of history of science and technology.

MRS. MÁRIA CSÖNDES is research associate of the Institute for Science Organization of the Hungarian Academy of Sciences, Budapest, Hungary. Research field of interest is on the interdependence of science and production.

DR. WOLFRAM FISCHER has been professor of economics and social history at Frei University of Berlin since 1964. He has published frequently in the area of regional and comparative studies of the industrialization process in Germany.

DR. SIMON KUZNETS obtained his Ph. D. in economics from Columbia University and has been a member of the Harvard faculty since 1960. A recipient of the Nobel Prize in economic growth, Dr. Kuz-his empirically founded interpretation of economic growth, Dr. Kuznets publishes extensively in this area.

DR. SHIGERU NAKAYAMA received his Ph. D. from Harvard University in 1959. He is presently lecturer in astronomy at the University of Tokyo, Tokyo, Japan. His writings have included works on the history of Japanese astronomy and Chinese scientific tradition.

DR. JOSÉ PASTORE is currently Professor of Sociology at the University of São Paulo. He received his Ph. D. in Sociology from The University of Wisconsin at Madison. His published interests are in the fields of agricultural development, higher education and employment as they affect social stability.

DR. NATHAN ROSENBERG is presently professor of economics at Stanford University. He received his Ph. D. from the University of Wisconsin in 1955. Currently his research interests include transfer of technology in the contemporary world, historical interrelationships between science and technology, and various topics in the history of economic thought.

DR. LAJOS SZÁNTÓ is presently Director of the Institute for Sciences, Budapest, Hungary, and associate professor of the Budapest University Eötvös Lóránd. His published interests are in the fields of science policy and science organization.

DR. PÉTER VAS-ZOLTÁN is Senior Research Associate of the Institute for Science Organization of the Hungarian Academy of Sciences, Budapest, Hungary. He has written numerous publications in the field of science policy and economics.

RELATED TITLES
Published by
Praeger Special Studies

*FOR A NEW POLICY OF INTERNATIONAL
DEVELOPMENT

Angelos M. Angelopoulos

A NEW INTERNATIONAL ECONOMIC ORDER:
Toward a Fair Redistribution of the World's
Resources

Jyoti Shankar Singh

POPULATION, PUBLIC POLICY, AND ECO-
NOMIC DEVELOPMENT

edited by
Michael C. Keeley

SCIENCE POLICIES OF INDUSTRIAL NATIONS:
Case Studies of the United States, Soviet Union,
United Kingdom, France, Japan, and Sweden

edited by
T. Dixon Long
Christopher Wright

*Also available in paperback.